Easy Pre-Calculus

STEP-BY-STEP

Easy Pre-Calculus
STEP-BY-STEP

Master High-Frequency Concepts and Skills for Precalc Proficiency—*FAST!*

Carolyn C. Wheater

New York Chicago San Francisco Athens London Madrid Mexico City
Milan New Delhi Singapore Sydney Toronto

1 2 3 4 5 6 7 8 9 LCR 23 22 21 20 19 18

ISBN 978-1-260-13511-4
MHID 1-260-13511-X

e-ISBN 978-1-260-13512-1
e-MHID 1-260-13512-8

Interior artwork by Newgen Imaging Systems

Other titles in the series:
Easy Mathematics Step-by-Step, William D. Clark, PhD, and Sandra Luna McCune, PhD
Easy Algebra Step-by-Step, Sandra Luna McCune, PhD, and William D. Clark, PhD

Contents

Preface

Easy Pre-Calculus Step-by-Step is an interactive approach to the mathematics necessary for pre-calculus. It contains completely detailed, step-by-step instructions for the skills and concepts that are the foundation for advanced mathematics. Moreover, it features guiding principles, cautions against common errors, and offers other helpful advice as "pop-ups" in the margins. The book leads you from basic algebra to a clear understanding of functions, explores fundamental trigonometry, and introduces you to the notion of limits. Concepts are broken into basic components to provide ample practice of fundamental skills.

Pre-Calculus, as its name implies, is meant to prepare you for calculus, but what does that mean? To succeed in calculus, or any advanced mathematics, you will need strong algebraic skills as well as a clear understanding of functions in general and the many different types of functions specifically: polynomial, rational, exponential, logarithmic, and more. You will want to build skill in graphing quickly by means of transformations, and you'll explore the polar coordinate system. From the construction of the trigonometric ratios in right triangles to the definition of the trigonometric functions in the unit circle, you will extend your knowledge of relationships and how to apply them. You may even investigate sequences and series and peek into the future by exploring limits.

Pre-Calculus may seem like too much to master, and the wide variety of topics may feel difficult to organize. With this step-by-step system, success will come. Learning pre-calculus, as with any mathematics, requires lots of practice. It requires courage to admit what you do not know, willingness to work on the problem, and a calm, orderly attempt to use what you do know. Most of all, it requires a true confidence in yourself and in the fact that with practice and persistence, you will be able to say, "I can do this!"

In addition to the step-by-step explanations and sample problems, this book presents a variety of exercises to let you assess your understanding. After working a set of exercises, use the solutions in the Answer Key to check your progress.

We sincerely hope *Easy Pre-Calculus Step-by-Step* will help you acquire the competence and the confidence you need to succeed in pre-calculus, calculus, and all your mathematical undertakings.

1

Graphs and the Graphing Calculator

A great deal of work in pre-calculus is centered on functions and their graphs. While the use of the graphing calculator has made it possible to explore many more, and many more complicated, functions than in the past, the ability to sketch the graph of a function by hand quickly is still an essential skill.

Step 1. Review Linear Equations

Your first introduction to graphing was linear functions, and they will always be important. Many of the skills you developed with linear graphs will carry over to other functions.

Recognize Horizontal and Vertical Lines

Recognizing the linear equations that don't behave in a typical fashion will save you time. Horizontal lines, because they have a zero slope, have an equation of the form $y = c$, where c is some constant. Vertical lines are the real oddity. They are not functions, and their equations don't fit the $y = mx + b$ standard. The equation of a vertical line is $x = c$, where c is a constant.

Graph Quickly

While constructing a table of values is always available as a graphing method, it is time consuming, and linear equations in particular allow for quick sketching methods.

- **Slope-intercept.** If the equation is in slope-intercept, or $y = mx + b$ form, or can easily be converted to that form, plot the y-intercept $(0,b)$, and then count the slope to locate other points on the line. Connect the points to create the line.
- **Intercept-intercept.** If the equation is in standard form, the quickest method is to determine the x- and y-intercepts, plot them, and connect. To find the x-intercept, let y equal 0 and simplify. To find the y-intercept, replace x with 0 and simplify.

Write Equations

Given a graph, or information about a graph, you may be asked to write the equation of the line. The simplest way to do that is to use the point-slope form: $y - y_1 = m(x - x_1)$

- **Point-slope form.** To write the equation of a line using point-slope form, you'll need to know a point on the line (x_1, y_1) and the slope m, or two points (x_1, y_1) and (x_2, y_2) from which you can calculate the slope, using the slope formula $m = \frac{y_2 - y_1}{x_2 - x_1}$. Then the equation can be written by replacing x_1, y_1, and m in the point-slope form $y - y_1 = m(x - x_1)$ with the known values. The point-slope form can be simplified and transformed to slope-intercept or standard form if desired.
- **Parallel and perpendicular lines.** To write the equation of a line parallel to a given line, determine the slope of the given line, usually by putting the given equation in slope-intercept form. Write the equation of the desired line by using point-slope form with the same slope as the given line and the point you want the desired line to pass through.

  To write the equation of a line perpendicular to a given line, you'll also want to determine the slope of the given line, but you'll use the negative reciprocal of the slope of the given line as the slope of the perpendicular line. Use point-slope form with that negative reciprocal slope and the point you want the line to pass through.

Exercise 1.1

Test your understanding by doing the following exercises.

1. Sketch the graph of $x = 6$.
2. Sketch the graph of $y = -3$.
3. Sketch the graph of $y = \frac{3}{4}x - 1$.
4. Sketch the graph of $2x - 4y = 12$.
5. Write the equation of a line with slope of –3 and a *y*-intercept of 5.
6. Write the equation of a line with slope of $\frac{5}{3}$ through the point (2,–1).
7. Write the equation of a line through the points (3,9) and (–4,2).
8. Write the equation of a line parallel to $5x - 7y = 35$ through the point (–7,4).
9. Write the equation of a line perpendicular to $4x - 2y = 11$ through the point (–2,5).
10. Find the equation of the perpendicular bisector of the segment that connects (3,–2) and (–5,6).

Step 2. Meet the Parents

If you are acquainted with the simplest function typical of a class, you'll find it easier to predict the behavior of functions you're trying to graph. These parent functions give you a place to start, and knowing a few key points on each parent graph will help you apply transformations.

Constant Function

The constant function $f(x) = c$, for some constant *c*, has a graph that is a horizontal line. Its domain is $(-\infty, \infty)$, and the range contains the single value, *c*.

Linear Function

Linear functions have the form $f(x) = mx + b$, where *m* is the slope and *b* is the *y*-intercept. The parent function is the function $f(x) = x$, with a slope of 1 and a *y*-intercept of 0. The domain is $(-\infty, \infty)$, and the range is $(-\infty, \infty)$. Key points on the parent graph are (–1,–1), (0,0), and (1,1).

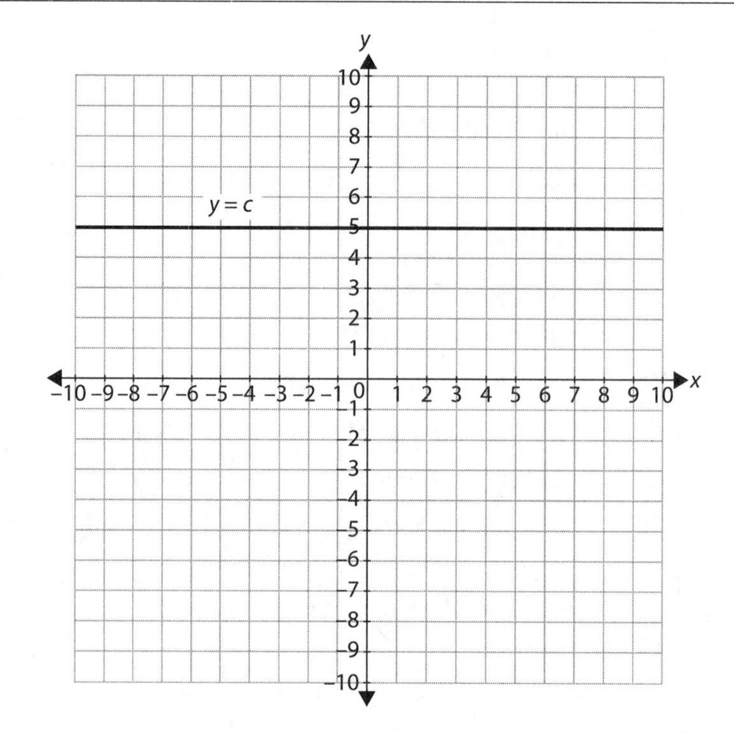

Constant function

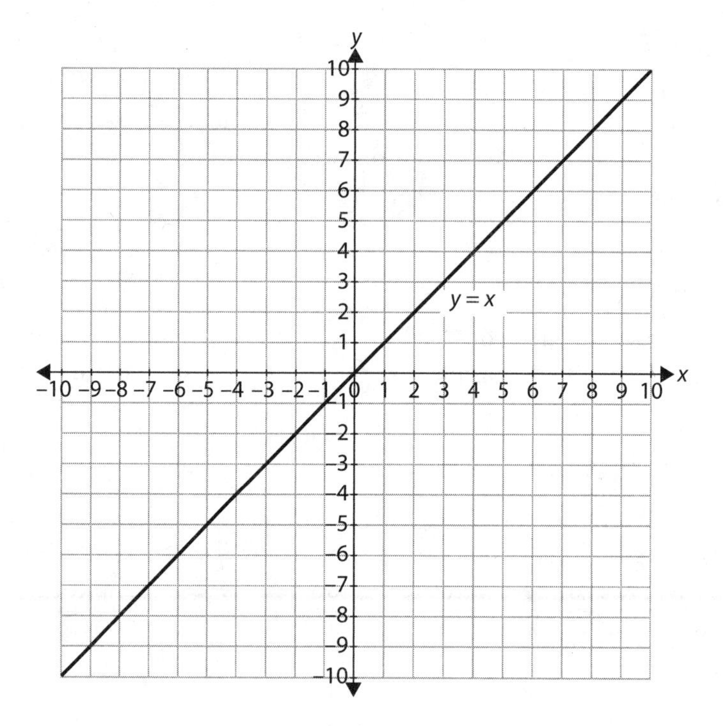

Linear function

Quadratic Function

The quadratic function has the form $f(x) = ax^2 + bx + c$, for real numbers a, b, and c, with $a \neq 0$, but is easier to graph when the equation is in vertex form, $y = a(x-h)^2 + k$. In this form, the vertex is (h, k). The parent function for the quadratic family is the function $f(x) = x^2$, with the vertex at the origin. The domain of the parent function is $(-\infty, \infty)$, and the range is $(0, \infty)$. Key points on the parent graph are (–1,1), (0,0), and (1,1).

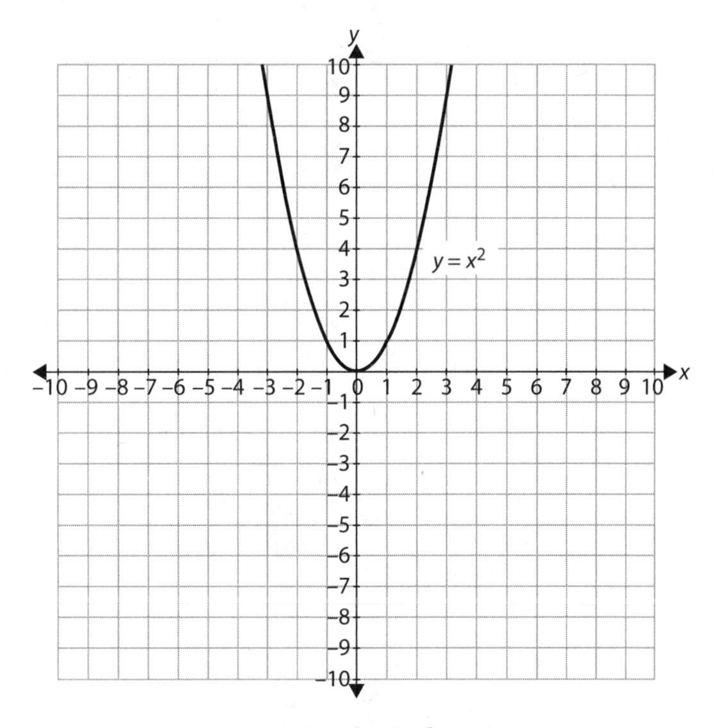

Quadratic function

Square Root Function

The function $f(x) = \sqrt{x}$ has a domain of $[0, \infty)$ and a range of $[0, \infty)$. Key points on the graph are (0,0), (1,1), and (4,2).

Cubic Function

Cubic functions have the form $f(x) = ax^3 + bx^2 + cx + d$, for real numbers a, b, c, and d, with $a \neq 0$, but here, too, the graph can be sketched more easily if expressed as $f(x) = a(x-h)^3 + k$. The parent function is $f(x) = x^3$ with a domain of $(-\infty, \infty)$ and a range of $(-\infty, \infty)$. The key points to remember are (–1,–1), (0,0), and (1,1).

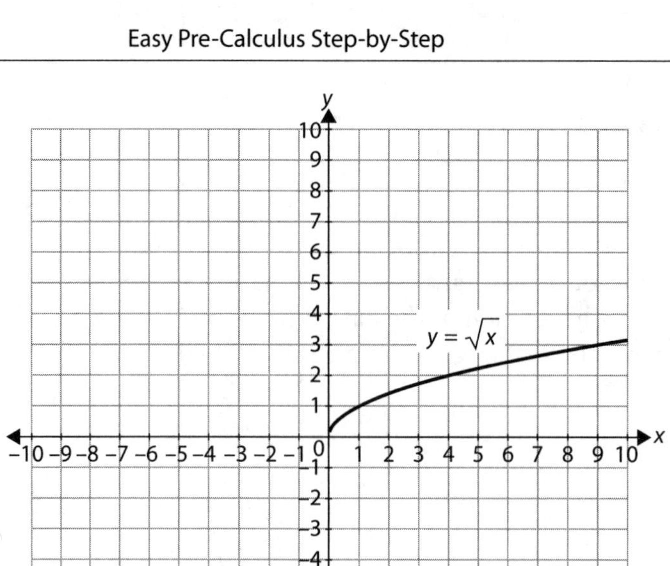

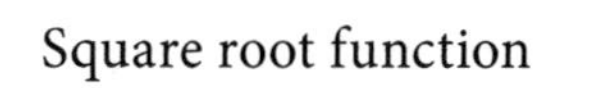

Square root function

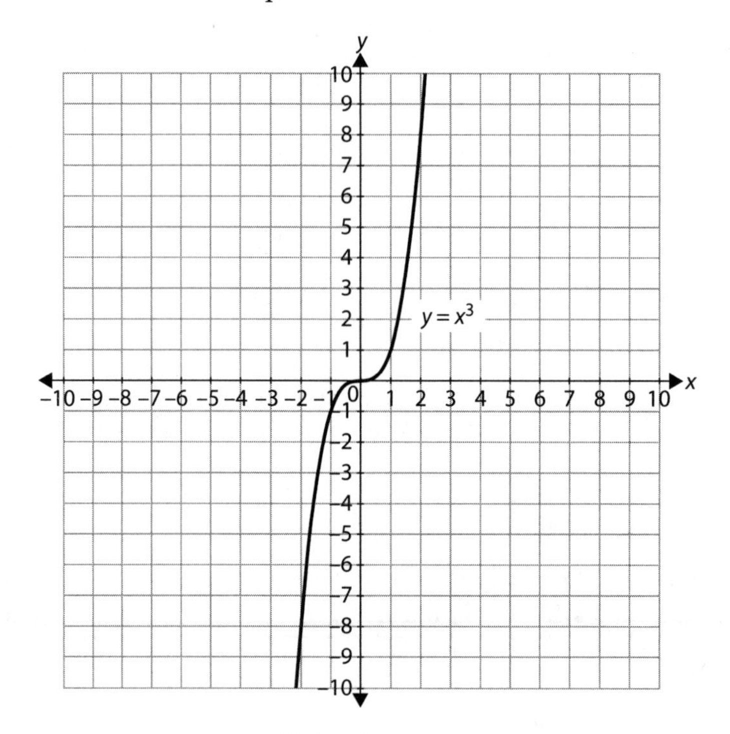

Cubic function

Cube Root Function

The function $f(x)=\sqrt[3]{x}$ has a domain of $(-\infty,\infty)$ and a range of $(-\infty,\infty)$. Key points on the graph are (–1,–1), (0,0), and (1,1).

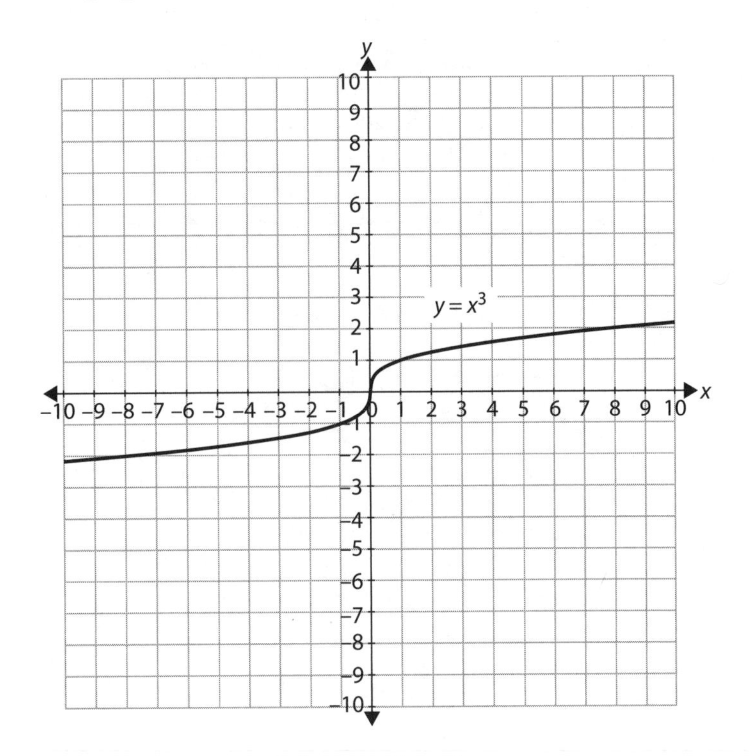

Cube root function

Exponential Function

The parent function for the exponential family, $f(x)=b^x$, for some constant base b, with $b>0$, has a domain of $(-\infty,\infty)$ and a range of $(0,\infty)$. The graph has a horizontal asymptote of $y=0$. Key points on the graph are $\left(-1,\frac{1}{b}\right)$, (0,1), and (1,$b$).

Logarithmic Function

The logarithmic function is the inverse of the exponential function. The parent graph is $f(x)=\log_b x$. The domain is $(0,\infty)$, and the range is $(-\infty,\infty)$. The graph has a vertical asymptote of $x=0$. Key points on the graph are $\left(\frac{1}{b},-1\right)$, (1,0), and ($b$,1).

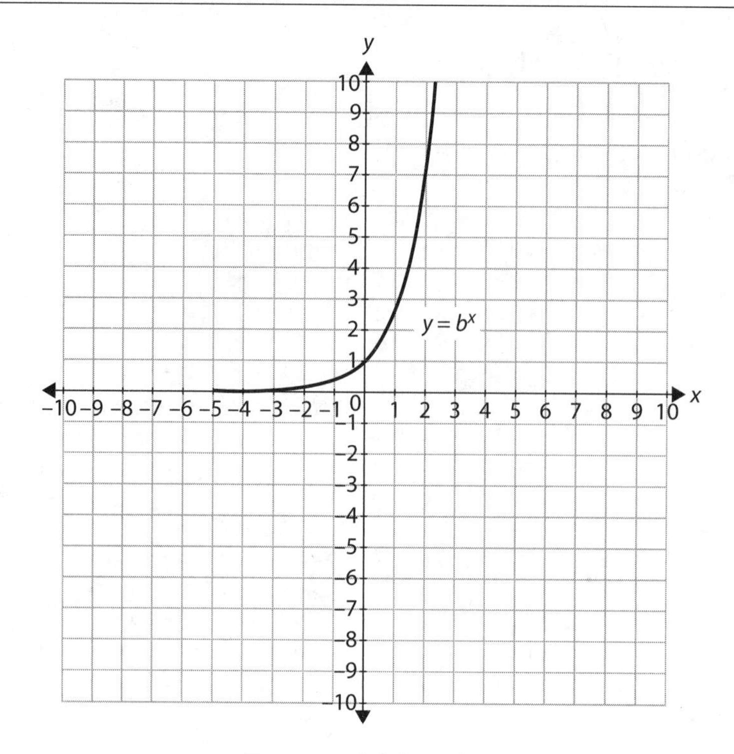

Exponential function

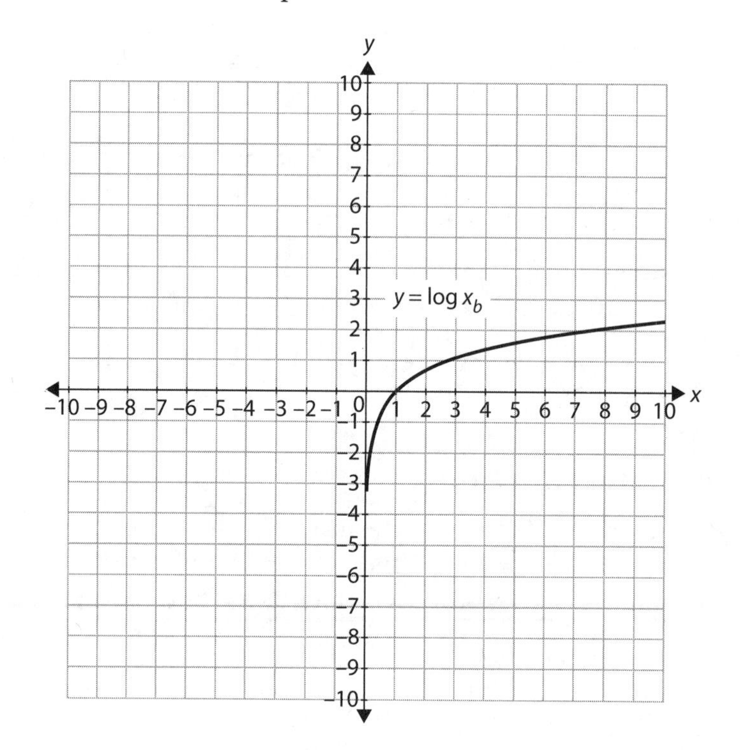

Logarithmic function

Rational Function

The parent function for the family of rational functions is the reciprocal function $f(x)=\frac{1}{x}$. The domain of the function is $(-\infty,0)\cup(0,\infty)$, and the range is $(-\infty,0)\cup(0,\infty)$. The graph has a vertical asymptote of $x=0$ and a horizontal asymptote of $y=0$. Key points on the graph are $\left(\frac{1}{2},2\right)$, $(1,1)$, $\left(2,\frac{1}{2}\right)$, $\left(-\frac{1}{2},-2\right)$, $(-1,-1)$, and $\left(-2,-\frac{1}{2}\right)$.

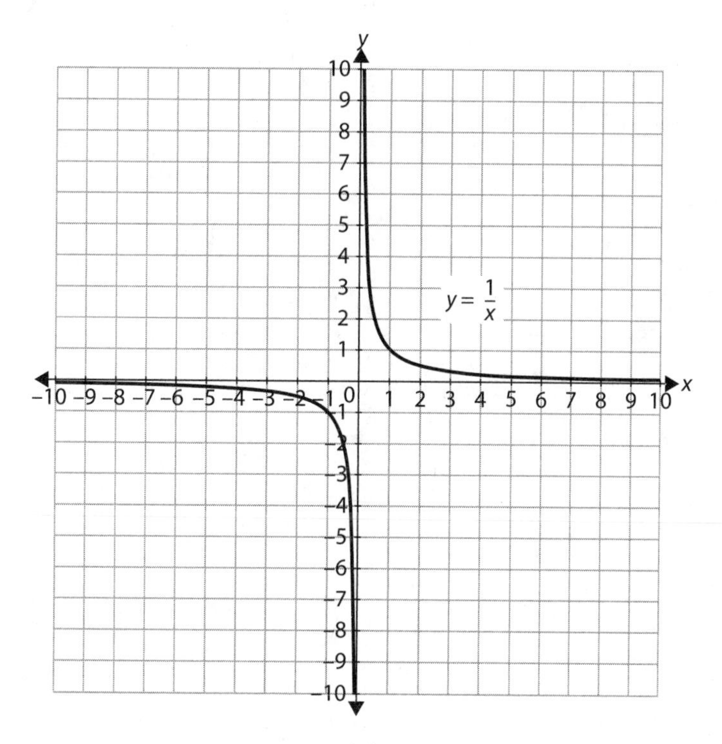

Rational function

Exercise 1.2

Identify the family from which each graph comes.

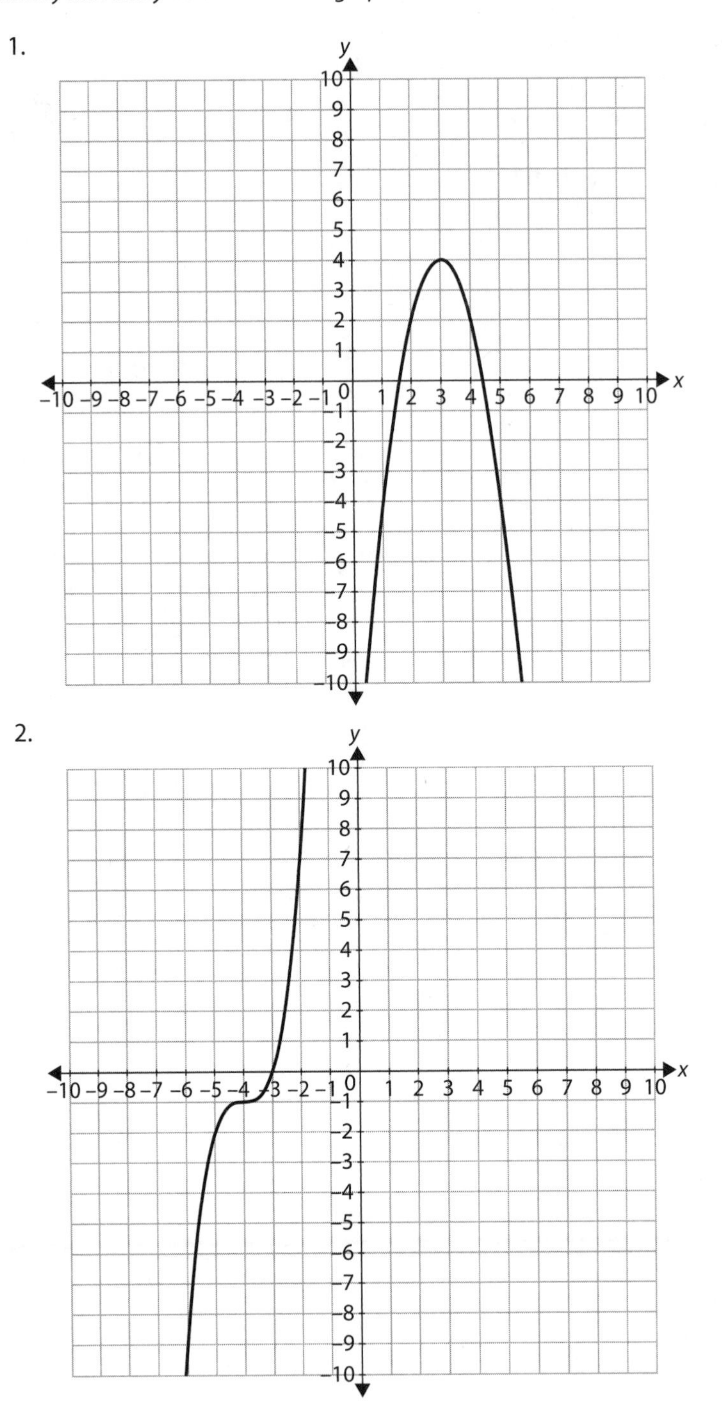

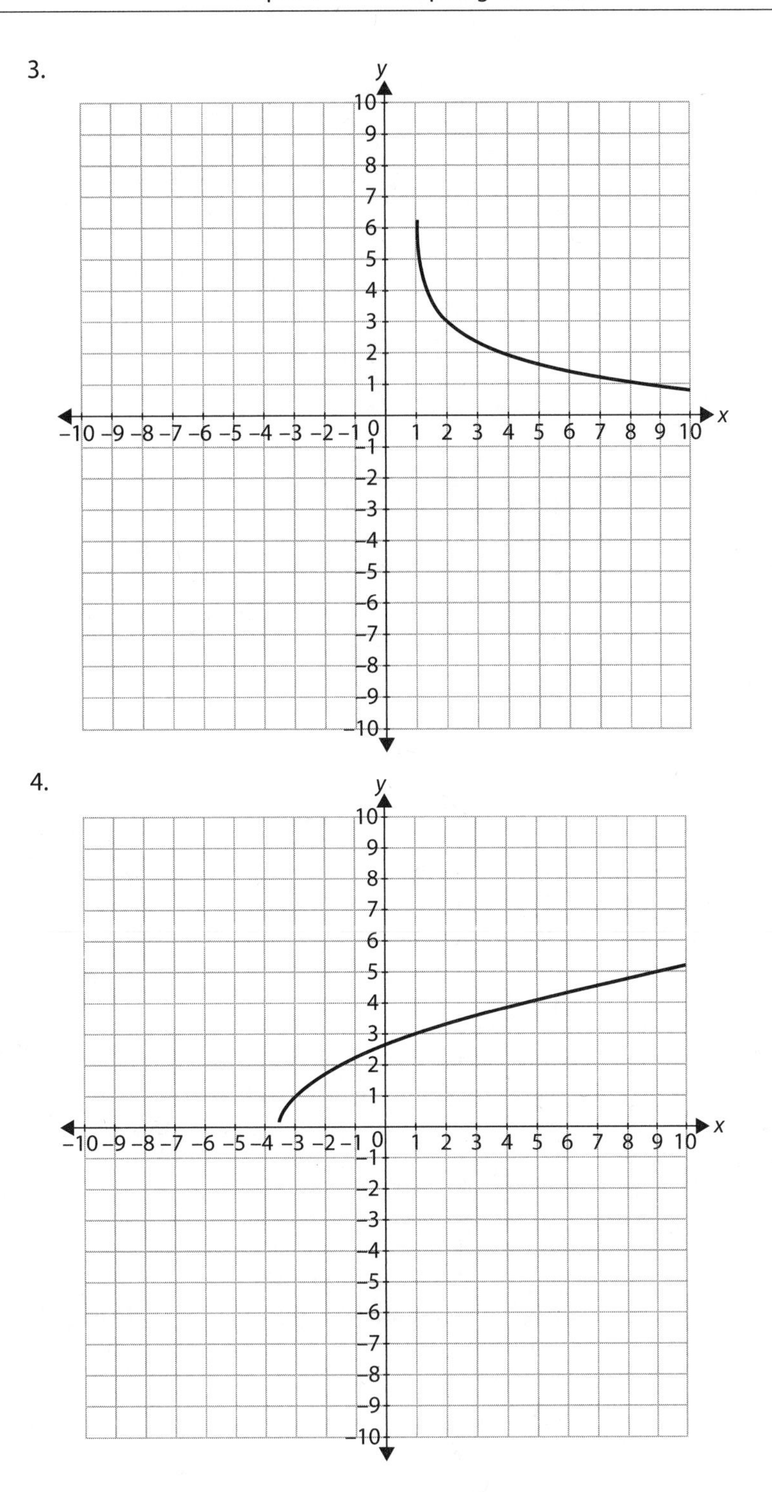
3.
y
x
4.
y
x

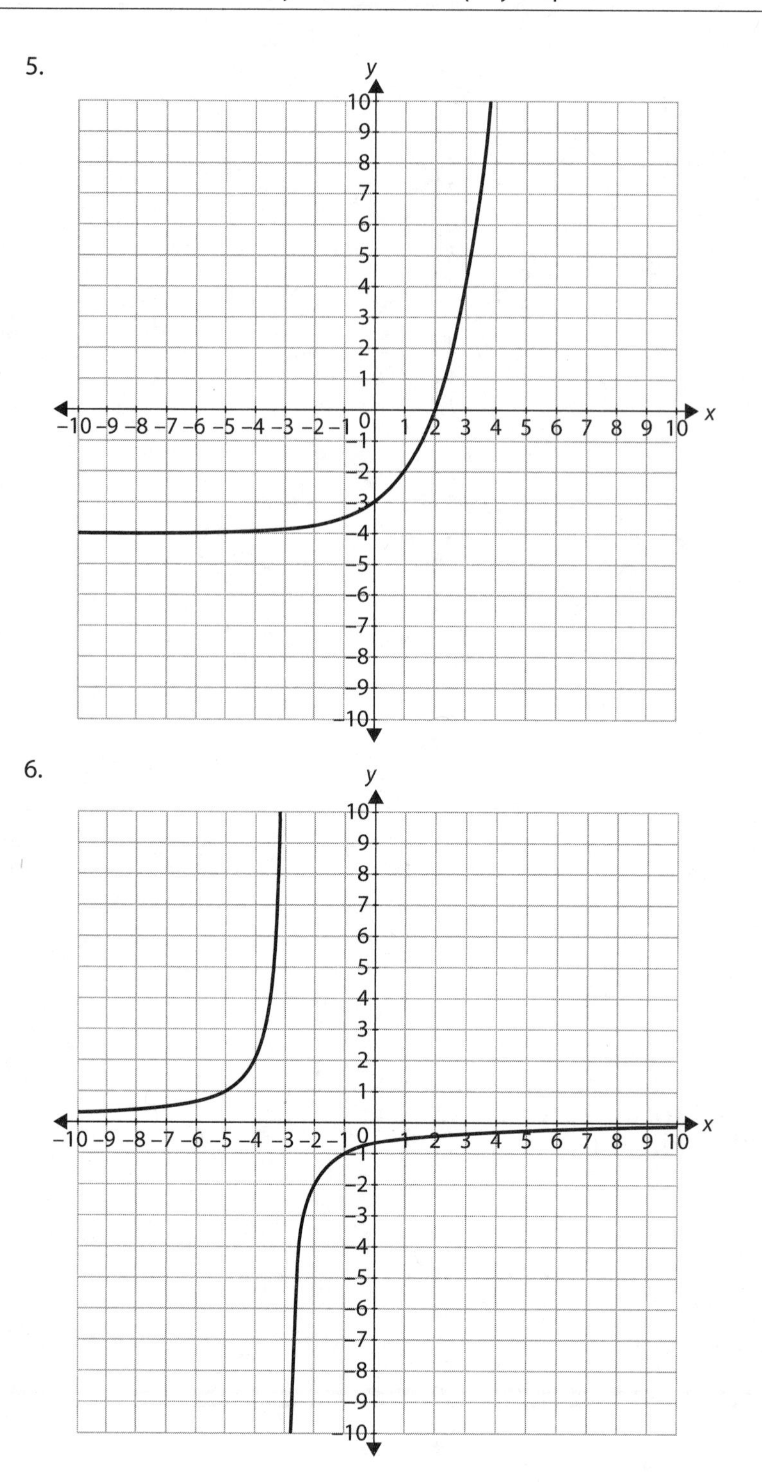
5.
y
x
6.
y
x

7.

8.

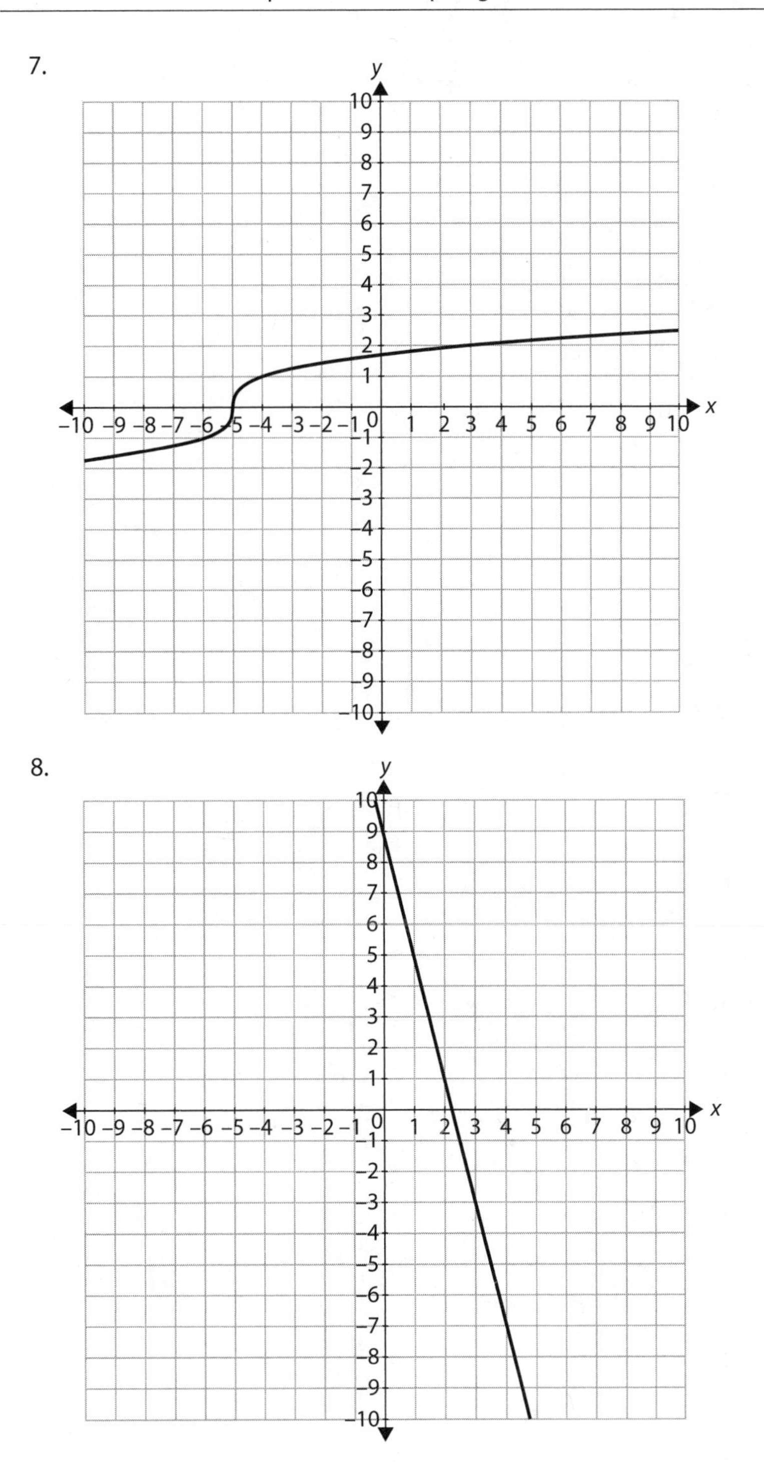

9.

10.

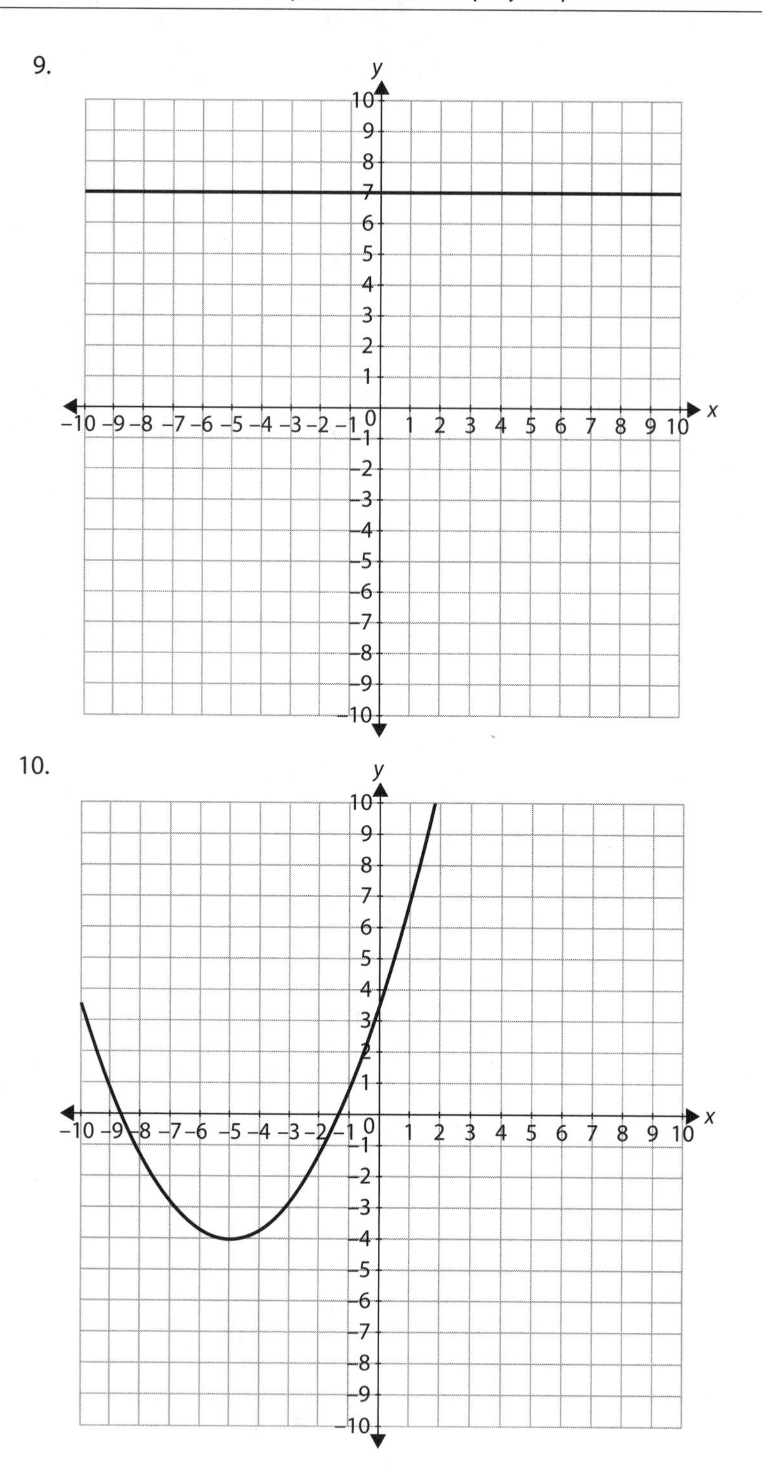

Step 3. Master the Fundamentals of Graphing

When you begin to sketch the graph of any function, it's wise to take a moment to think about essential information.

Identify the Domain

Begin with a domain of all real numbers, and then eliminate values, if necessary, according to the following checklist.

- Is there a denominator? If so, eliminate any values that make the denominator equal to 0.
- Is there a root with an even index? If so, eliminate any values that make the radicand negative.
- Is there a logarithm? If so, eliminate any values that make the argument of the log equal to 0 or a negative number.
- Is this function modeling a real situation? If so, consider what values make sense in that model. A function representing volume as a function of some dimension would not include negative numbers in the domain, because dimensions wouldn't be negative, even if it were algebraically feasible to evaluate the function for negative values.

Find Intercepts

While building a table of values is always available as a sketching technique, it is a time-consuming method. Finding x- and y-intercepts, however, is generally quick and easy and gives you a few key points to place in the graph.

To find the x-intercept, let y equal 0, and solve for x. You may find one or more x-intercepts, or you may find that there are no x-intercepts.

To find the y-intercept, substitute 0 for x, and simplify. Functions will have at most one y-intercept.

Analyze End Behavior

The behavior of the graph for very large positive values and very small negative values of x is referred to as the end behavior. For polynomial functions, a category that includes linear, quadratic, and cubic equations, as well as functions of higher degree, end behavior can be summarized by a few rules:

- If the polynomial function is of even degree, both ends of the graph will go to ∞ if the lead coefficient is positive and to $-\infty$ if the lead coefficient is negative.
- If the polynomial has an odd degree and the lead coefficient is positive, as $x \to \infty$, $f(x) \to \infty$ and as $x \to -\infty$, $f(x) \to -\infty$.
- For odd-degree polynomials with negative lead coefficients, as $x \to \infty$, $f(x) \to -\infty$ and as $x \to -\infty$, $f(x) \to \infty$.

The end behavior of simple rational functions is marked by horizontal asymptotes, but more complicated rational functions may behave in other ways. Exponential and logarithmic functions will behave like their parent functions, unless there is a reflection.

Exercise 1.3

Find the domain, range, intercepts, and end behavior of each function.

1. $f(x) = 9 - 2x$
2. $f(x) = 3x^2 - 4$
3. $f(x) = 4x^3 - 36x^2 + 4x - 36$
4. $f(x) = \sqrt{2x+5}$
5. $f(x) = 9 - 2^{x+5}$
6. $f(x) = 4\log_3(x-3)$
7. $f(x) = \dfrac{2}{3x-1}$
8. $f(x) = \sqrt[3]{8-x} + 4$
9. $f(x) = \dfrac{2x-3}{(x+4)(x-1)}$
10. $f(x) = 9x - x^2 - 14$

Step 4. Transform the Graphs

In many cases, you can sketch the graph of a function by recognizing that it represents transformations of the parent function. Changes made before the function is applied—for example, before squaring—will affect changes to the graph in the horizontal direction. Those made after the function have a vertical effect.

Translate

A translation is a rigid slide of the parent graph. The formula $f(x)=(x-h)^2$ slides the parabola h units to the right, and $f(x)=(x+h)^2$ moves it h units left; $f(x)=x^2+k$ translates the graph k units up, while $f(x)=x^2-k$ shifts it k units down.

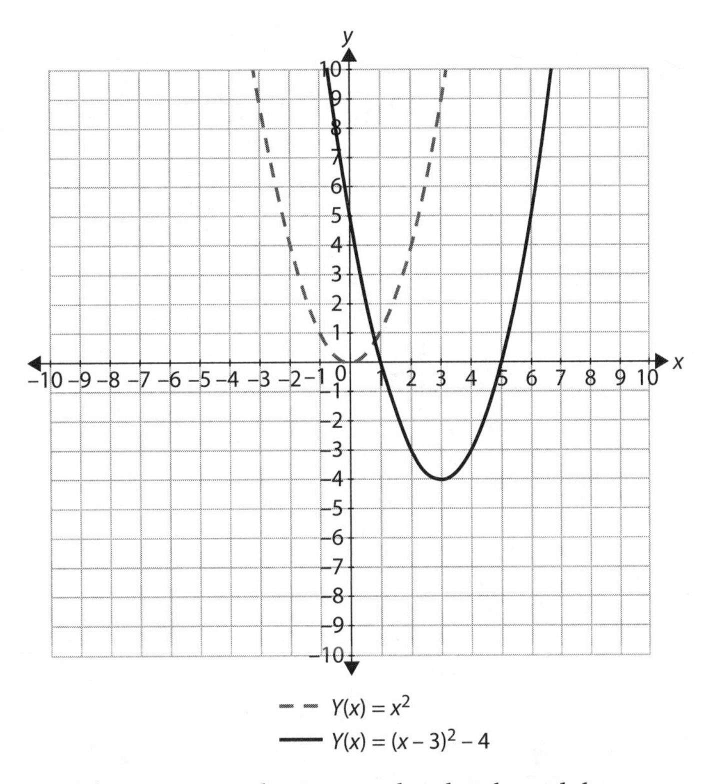

The parent quadratic, translated right and down

Reflect

Multiplying x by −1 before the function is applied, as in $f(x)=e^{-x}$, reflects the graph across the y-axis, but multiplying by −1 after the basic function works, as in $f(x)=-e^x$, reflects the graph over the x-axis. If the parent function is symmetric with respect to the y-axis, reflection about that axis will seem to have no effect, and for some functions, reflection across the x-axis and reflection across the y-axis will look identical. This is the case for the cubic function, for which $f(x)=(-x)$ and $f(x)=-x^3$ produce the same graph.

(a)

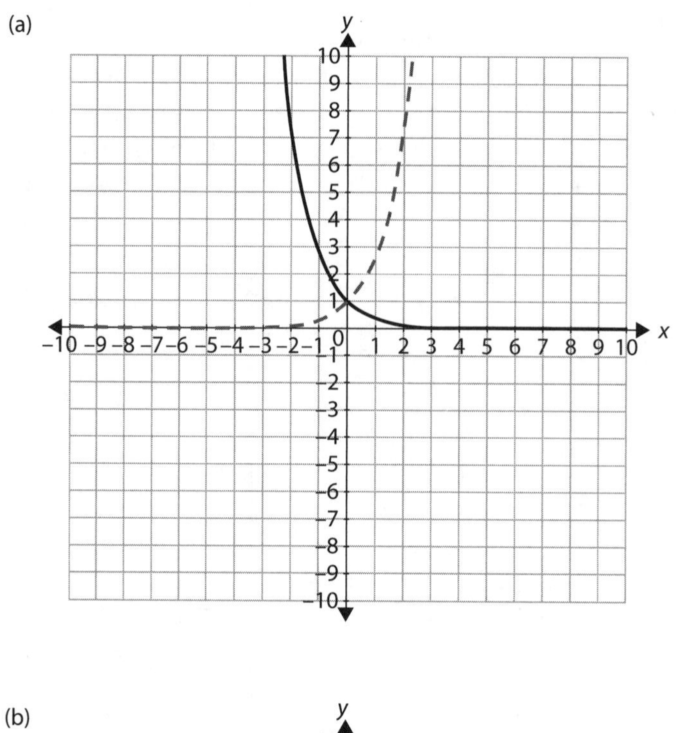

(b)

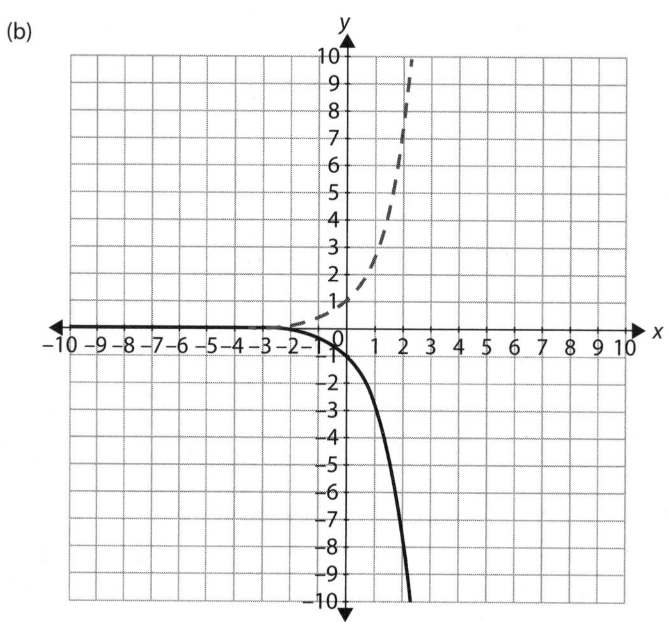

Reflection across the y-axis (a) and x-axis (b)

Stretch and Compress

Just as it is occasionally impossible to determine the direction of a reflection, it can sometimes be difficult to distinguish between a vertical stretch and a horizontal compression. Multiplying x by a constant greater than 1 before the function is applied causes a horizontal compression. Multiplying by a constant between 0 and 1 causes a horizontal stretch. Multiplying by a constant after the application of the function causes a vertical stretch if the constant is greater than 1 and a vertical compression if the multiplier is between 0 and 1. The formula $f(x) = (ax)^2$ squeezes the parabola in toward the y-axis, while $f(x) = ax^2$ stretches the parent graph upward, multiplying each y-coordinate by a. Simple algebra tells you that $f(x) = (ax)^2 = a^2x^2$, however, so the horizontal compression by a factor of a is identical to a vertical stretch by a factor of a^2. This confusion exists for many functions.

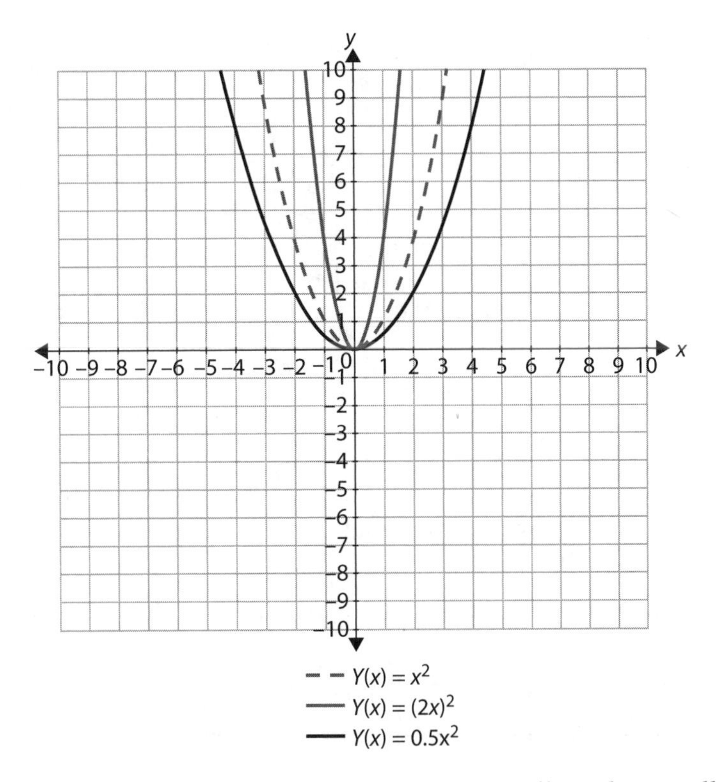

The quadratic function, compressed horizontally and vertically

Use Symmetry

Knowing when the graph is symmetric lets you sketch the function quickly.

- **About the x-axis.** Functions won't be symmetric about the x-axis; that would contradict the definition of function. You will sometimes want to graph curves that are symmetric about the x-axis, however. You can recognize them by replacing y with $-y$ and simplifying. If the result is identical to the equation you started with, the graph will be symmetric about the x-axis.

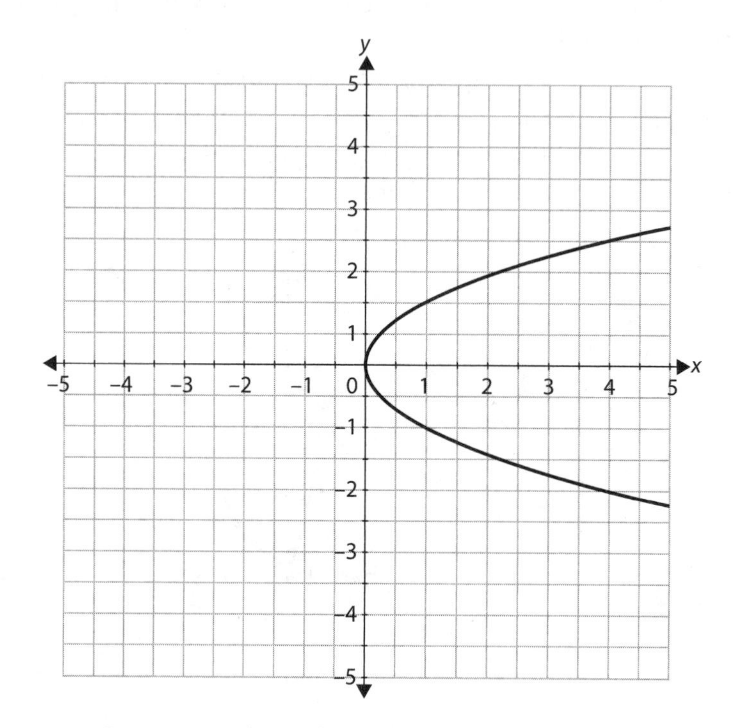

Symmetric about the x-axis but not a function

- **About the y-axis.** If $f(-x) = f(x)$, the graph of $f(x)$ will be symmetric across the y-axis.
- **About the origin.** If $f(-x) = -f(x)$, the graph of $f(x)$ will be symmetric about the origin.

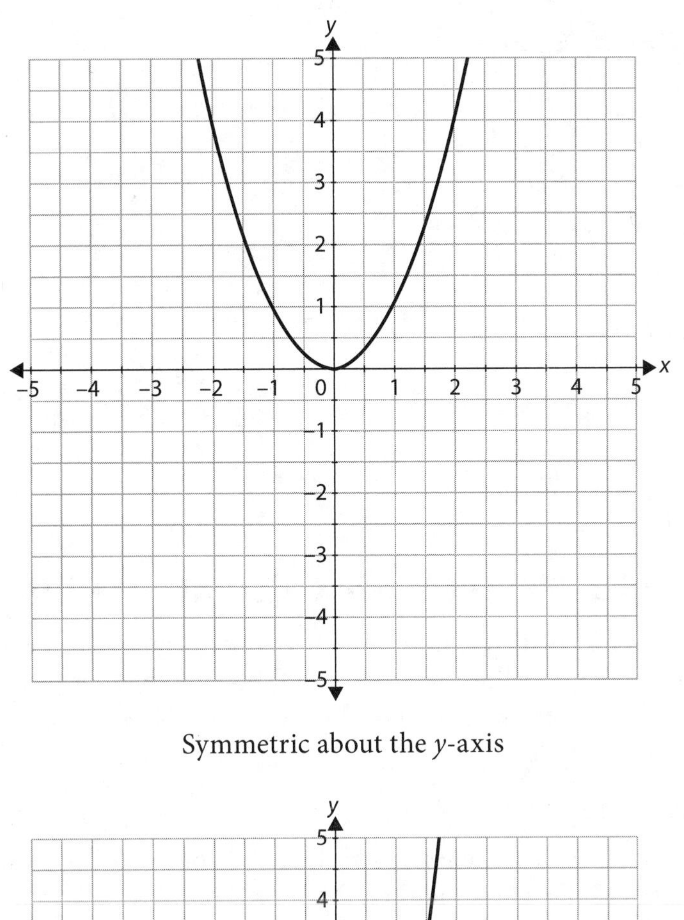

Symmetric about the y-axis

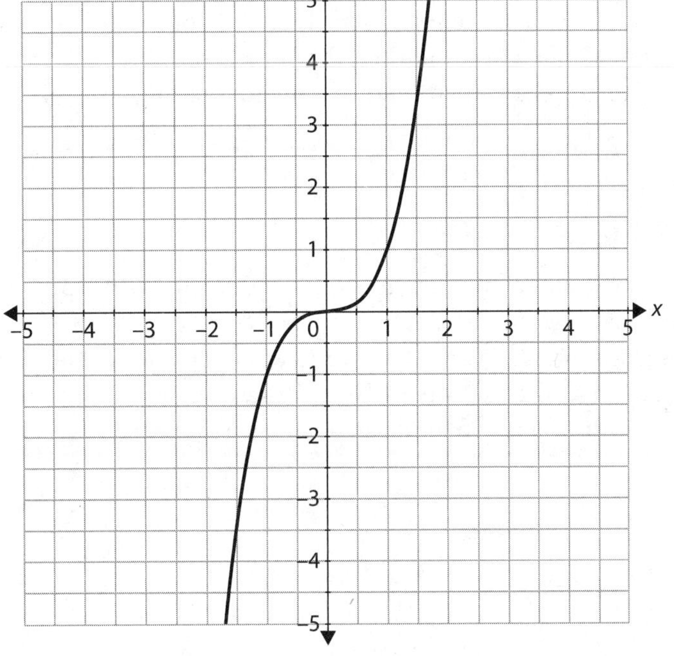

Symmetric about the origin

Exercise 1.4

Sketch each graph using transformations of the parent graph.

1. $f(x) = (x-5)^2 - 1$
2. $f(x) = 2x^3 + 4$
3. $f(x) = 1 - 2^x$
4. $f(x) = \ln(x+3) - 5$
5. $f(x) = \frac{1}{x-6}$
6. $f(x) = -3\sqrt{x-4}$
7. $f(x) = \frac{1}{2}\sqrt[3]{x-1} + 5$
8. $f(x) = -4 \cdot 3^{2x}$
9. $f(x) = 4\log_2(x-5) + 2$
10. $f(x) = 3 - \frac{5}{x+1}$

Step 5. Master Basic Calculator Skills

Graphing calculators have become standard equipment in pre-calculus, opening up the possibility of exploring functions that once were inaccessible. To use them, there are a few key skills you must have.

Graph Functions

You must solve any equation for y in terms of x in order to enter it into the calculator. Remember that the calculator is designed to graph functions. If you need to graph an equation that is not a function, such as the circle $x^2 + y^2 = 9$, you'll need to graph two equations: $y = \sqrt{9 - x^2}$ and $y = -\sqrt{9 - x^2}$.

Set Windows

Thinking about the domain of the function will help you set the minimum and maximum x-values to display. If you're having trouble estimating the range so that you can set the vertical dimensions of the window, hit the trace function and take note of the y-values displayed. You'll get an estimate of the minimum and maximum y-values to display.

Use Built-In Zooms

Your calculator should have an automatic setting for a standard viewing window, which is from –10 to 10 on both axes. In addition, look for a decimal zoom, which zooms in closer to the origin but is proportioned for easy tracing. When tracing in this window, each cursor movement is 0.1 units in the x-direction. If your graph of a circle looks like an oval, look for a square zoom, which adjusts the proportions of the display to equalize vertical and horizontal.

Find Intercepts

You can find a y-intercept on your calculator by hitting trace and entering $x = 0$. The y-value displayed is the y-intercept. The zero function should allow you to find x-intercepts quickly. Be sure your window is set so that you can see y-values both above and below the x-axis, because you'll need to set left and right bounds on the zero, and one should be positive and the other negative.

Find Intersections

The intersect feature will locate the point of intersection of two curves. You'll need to indicate which two curves as well as the approximate point of intersection.

Find Relative Extrema

Your calculator has functions for finding the maximum and the minimum values of a function in a region. You'll need to set a left and right bound and a guess at the value.

Exercise 1.5

Graph each function in an appropriate viewing window, and then find the x- and y-intercepts.

1. $f(x) = -7x + 18$
2. $f(x) = 8 - 6x - x^2$
3. $f(x) = x^4 - x^2 + 9$
4. $f(x) = x^3 - 7x + 3$

Find the point(s) of intersection for each system of equations.

5. $2x+9y=93$
   $5x-4y=43$

6. $7x+3y=12$
   $2x-5y=19$

7. $5x-2y=17$
   $3x+2y=16$

Find the relative maxima and minima for each function.

8. $f(x)=3x^2-7x+5$

9. $f(x)=x^3-4x^2+3x-2$

10. $f(x)=19-3x^2$

2

Functions

Pre-Calculus is a time to examine a variety of functions and become acquainted with the properties of functions in general and with the specific attributes of different types of functions. Begin with the information you need to deal with any kind of function.

Step 1. Analyze Relations and Functions

A relation is a pairing of elements from one set, called the domain, with elements of another set, called the range. A relation is a function if each element of the domain has a unique partner in the range. Each input is paired with just one output.

A common method for determining whether a particular relation is a function is to look at the graph and use the vertical line test. If any vertical line would intersect the graph more than once, the graph represents a relation that is not a function. The graph of a nonvertical linear equation is the graph of a function, but the graph of a circle represents a relation that is not a function.

The function notation, such as $f(x)$ or $s(t)$, tells you that you're dealing with a relation that is a function, gives the function a name (f or s), identifies the independent variable (x or t), and is often followed by a rule that defines the function, as in $f(x) = 3x - 7$ or $s(t) = -16t^2 + 40t + 5$. Replacing the independent variable with a constant in the function notation, as in $f(2)$ or $s(0.4)$, indicates that you should replace all occurrences of the independent variable with that value and evaluate. If $f(x) = 3x - 7$, then $f(2) = 3 \cdot 2 - 7 = -1$.

A function may be defined by a table of values or by a graph rather than by an equation. In that case, evaluating the function is a matter of looking up a value in the table or reading it from the graph. Functions defined in this way are generally defined only on a small interval and may be discrete functions, that is, functions consisting of separate, unconnected values.

When you encounter a new function, there are some routine observations you'll want to make.

Identify the Domain and Range

If the function is defined by a table or graph, the interval on which it is defined should be obvious. Take note of whether the function is discrete, in which case, you may be able to list the element in the domain.

If the function is defined by an equation, you'll need to determine the domain. Begin with a domain of all real numbers, and then eliminate values, if necessary, according to the following checklist.

- Eliminate any values that make denominators equal to 0.
- Eliminate any values that make radicands negative if there are even roots.
- Eliminate any values that make the argument of the log equal to 0 or to a negative number.
- If the function models a real situation, limit the domain to values that make sense in that model.

Anticipating the range of the function requires a little bit of experience with the function in question.

- Linear functions produce all reals, except for horizontal lines, which are constant functions with a single value in the range.
- Polynomial functions of odd degree have ranges of all reals, but those of even degree will have a limited range.
- Exponential functions can be expected to have a limited range, but logarithmic functions usually have a range of all reals.
- Simple rational functions generally have a single value missing from the range, but for more complicated rational functions, the range is less predictable.

Find Relative Extrema

Many of the functions you'll investigate will have turning points, as the quadratic function does. You probably learned to find the equation of the axis of symmetry by a simple formula and from that to find the vertex, or turning point, of the parabola. Other functions with turning points will not have convenient formulas to locate them, and the most effective method for finding them requires calculus. In pre-calculus, therefore, locating the turning points of a graph and identifying each as a relative maximum (the point where the graph stops increasing and starts decreasing) or a relative minimum (where it changes from decreasing to increasing) will be largely a calculator function.

> Axis of symmetry of the graph of $y = ax^2 + bx + c$ is $x = \frac{-b}{2a}$.

Of course, you'll want to make sure that you know how to find maxima and minima—relative extrema—on your calculator, but you'll also want to be sure that you set a good viewing window. If your window is too small, there may be relative extrema that you can't see. If it's too large, you may miss details. Thinking about the domain and the behavior of the parent graph for the family of your function will help you set an appropriate window.

When you specify a relative maximum or a relative minimum, you want to communicate both where it occurs, that is, at what x-value, and also how high or how low the point is, its y-value. You can give the extremum as a point, but you'll often hear it phrased as something like, "a maximum of 8 when $x = 2$."

Describe Increasing, Decreasing, and Constant Intervals

Once you know where the relative extrema, or turning points, of a graph are, you'll be able to describe where the function is increasing, where it's decreasing, and where it's constant. When you describe increasing, decreasing, or constant intervals, you want to give the x-values that define the intervals.

This information is commonly given in interval notation, although it can also be communicated by inequalities.

INTERVAL NOTATION	*INEQUALITY NOTATION*
$(-3, 4)$	$-3 < x < 4$
$[-3, 4]$	$-3 \leq x \leq 4$
$[-3, 4)$	$-3 \leq x < 4$
$(-3, 4]$	$-3 < x \leq 4$
$(-\infty, -3]$	$x \leq -3$
$(4, \infty)$	$x > 4$

> Parentheses (rather than square brackets) are always used around ∞ or $-\infty$. You can't include what goes on forever.

Functions that are decreasing over their entire domain or increasing over their entire domain are called monotonic functions. Linear functions and exponential functions are commonly monotonic, but polynomial functions will have one or more relative extrema and will not be monotonic.

Exercise 2.1

Complete the following exercises.

1. Find the domain and range of $f(x) = x^3 - x^2 + x + 1$.
2. Find the domain and range of $f(x) = \sqrt{4x - 9}$.
3. Find the domain and range of $f(x) = \dfrac{3}{(x-1)(x+5)}$.
4. Find all relative maxima and minima of $f(x) = \frac{1}{3}x^3 - 4x + 5$.
5. Find all relative maxima and minima of $f(x) = x^4 - 8x^2 - 9$.
6. Find all relative maxima and minima of $f(x) = 9 - x^2$.
7. Describe the intervals on which $f(x) = 9 - x^2$ is increasing, decreasing, or constant.
8. Describe the intervals on which $f(x) = \frac{1}{3}x^3 - 4x + 5$ is increasing, decreasing, or constant.
9. Describe the intervals on which $f(x) = \left(\frac{1}{2}\right)^{x-1}$ is increasing, decreasing, or constant.
10. Describe the intervals on which $f(x) = x^4 - 8x^2 - 9$ is increasing, decreasing, or constant.

Step 2. Perform Arithmetic of Functions

When functions are added, subtracted, multiplied, or divided, new functions are created. For the work you'll do in calculus, it's important to recognize when a function has been built in this way. Using what you know about the component functions will help you understand the new function that's been created.

The rules are simple: add, subtract, multiply, or divide; $f + g(x) = f(x) + g(x)$, $f - g(x) = f(x) - g(x)$, $f \cdot g(x) = f(x) \cdot g(x)$, and $\frac{f}{g}(x) = \frac{f(x)}{g(x)}$, provided that $g(x) \neq 0$. The functions may be presented in different ways, so you'll want to be familiar with all of them.

Use a Table

The functions may be presented to you by a table of values such as the example that follows. In this case, you'll be asked to evaluate a function such as $f + g(x)$ by finding the values of $f(x)$ and $g(x)$ and adding them.

x	0	1	2	3	4	5
f	5	3	3	0	4	2
g	−2	0	2	1	4	2

To evaluate $f + g(3)$, first find $f(3) = 0$ and $g(3) = 1$, and then $f + g(3) = 0 + 1 = 1$. In addition, $f - g(4) = f(4) - g(4) = 4 - 4 = 0$, and $f \cdot g(0) = f(0) \cdot g(0) = 5(-2) = -10$. You will not be able to evaluate any function for values of x other than 0, 1, 2, 3, 4, and 5, because the values of f and g at only those x-values are supplied for you.

Be careful when forming $\frac{f}{g}(x)$, because the division introduces a new complication: division by 0 is undefined. As a result, $\frac{f}{g}(1)$ does not exist, because it would equal $\frac{f(1)}{g(1)} = \frac{3}{0}$, which is undefined.

Use a Graph

If functions f and g are presented as graphs, the work of evaluating $f + g$, $f - g$, $f \cdot g$, and $\frac{f}{g}$ is done just as it was for the table of values. Inspect the graphs to find the value of $f(x)$ and of $g(x)$, and then add, subtract, multiply, or divide as necessary.

In addition, you may be able to sketch the graph of the new function by evaluating the new function for several values of x, plotting the new points, and sketching a graph through those points.

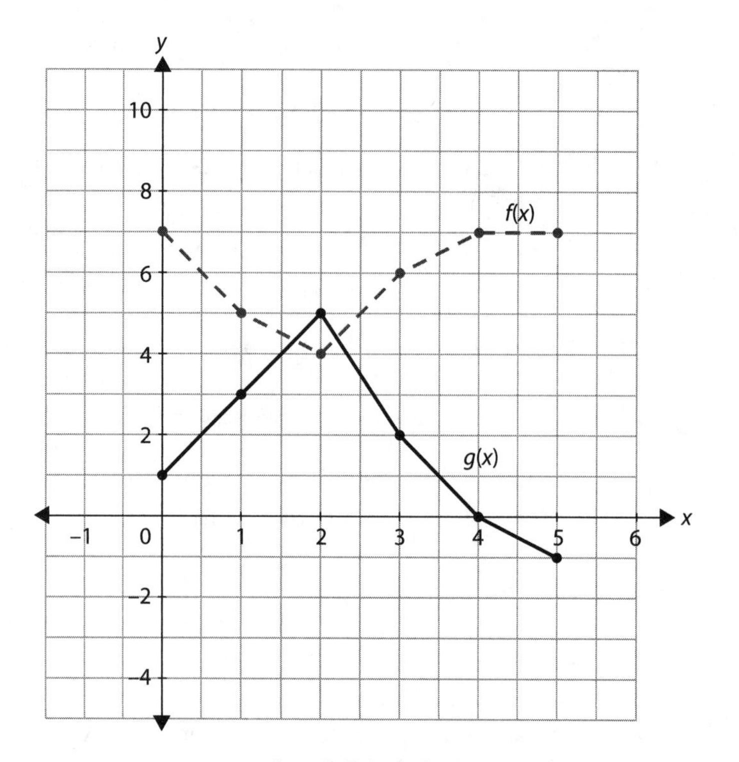

Graphs of $f(x)$ and of $g(x)$

The graphs of $f(x)$ and of $g(x)$ are shown in the preceding figure. By subtracting the values for all the integer values of x for which the function is defined, it's possible to sketch a graph of $f - g(x)$, as shown on page 31.

Form the Rule

If you're given the rules for $f(x)$ and $g(x)$, you can construct the rule for $f + g$, $f - g$, $f \cdot g$, or $\frac{f}{g}$ by adding, subtracting, multiplying, or dividing the expressions. If $f(x) = x^2 - 3$ and $g(x) = x - 3$,

$$(f + g)(x) = (x^3 - 3) + (x - 3) = x^2 + x - 6$$

$$(f - g)(x) = (x^3 - 3) - (x - 3) = x^2 - x$$

$$(f \cdot g)(x) = (x^2 - 3)(x - 3) = x^3 - 3x^2 - 3x + 9$$

$$\left(\frac{f}{g}\right)(x) = \frac{x^2 - 3}{x - 3}, \quad x \neq 3$$

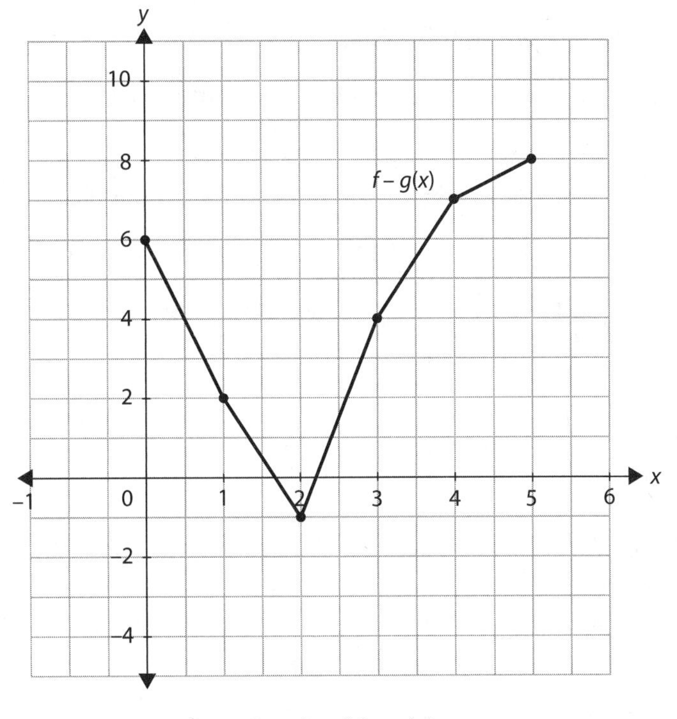

Graph of $f - g(x)$

You can simplify the expressions for the new functions wherever possible, and in the case of $\frac{f}{g}$, you can factor and cancel if there are factors common to the numerator and denominator.

Check the Domain

Whenever you think about a function, you should think about both the equation that defines the function and the domain of the function. When you create a new function from two existing functions by function arithmetic, the domain of the new function must respect the domains of the individual functions from which it's built.

If you start with $f(x) = \sqrt{x}$, which has a domain of nonnegative reals, or $[0, \infty)$, and $g(x) = x^2$, which has a domain of $(-\infty,\infty)$, and you form $(f+g)(x) = \sqrt{x} + x^2$, the new function will have the same domain restriction as f. The domain will be $[0,\infty)$. But even if the reason for the restriction disappears when you simplify, the restriction doesn't. If $f(x) = x^2$ and $g(x) = \frac{1}{x}$, $\frac{f}{g}(x) = \frac{x^2}{\frac{1}{x}} = x^3$, but the domain is $(-\infty,0)\ (0,\infty)$. Zero is not in the domain of $\frac{f}{g}$ because it is not in the domain of g. The domain of the new function is the intersection of the domains of the two functions from which it's formed. When you form $\frac{f}{g}$, there may be an additional domain restriction so $g(x) \neq 0$.

Exercise 2.2

Use the table to evaluate each function for the specified value of x.

x	−3	−2	−1	0	1	2	3
f	0	1	−1	2	−2	3	−3
g	2	0	−3	1	3	0	−2

1. $(f+g)(-2)$
2. $(f-g)(1)$
3. $\frac{f}{g}(2)$
4. $(f \cdot g)(0)$

Use the graphs of f and g on page 33 to draw the specified graph.

5. $f+g$
6. $f \cdot g$

Complete the following exercises.

7. If $f(x) = x^2 - 4$ and $g(x) = x + 2$, find the simplest expression for $\frac{f}{g}(x)$ and give its domain.
8. If $f(x) = 2x^2 - 3x + 1$ and $g(x) = x^2 - 2x - 7$, find the simplest expression for $(f-g)(x)$ and give its domain.

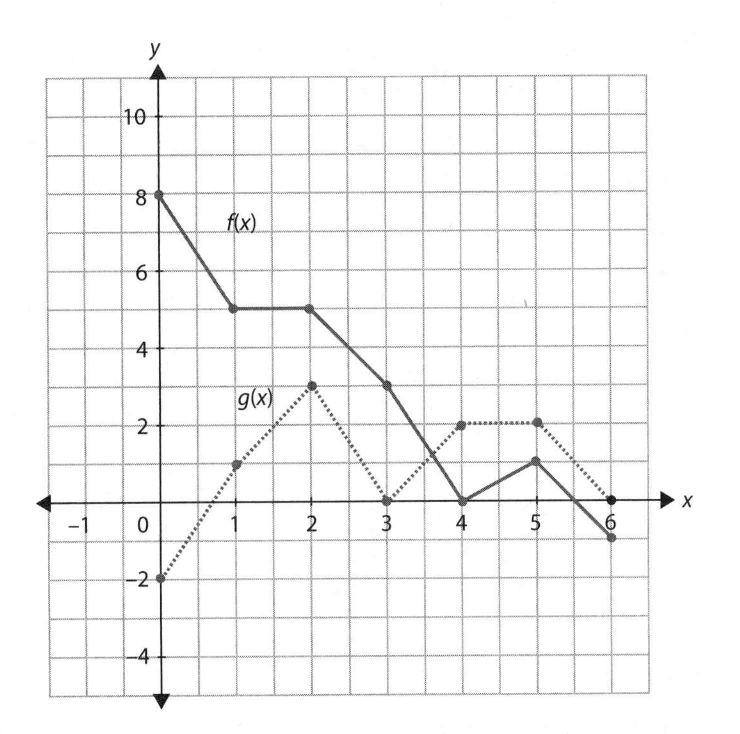

9. If $f(x) = \dfrac{1}{x-1}$ and $g(x) = \dfrac{1}{x+1}$, find the simplest expression for $(f+g)(x)$ and give its domain.
10. If $f(x) = \sqrt{x^2 - 3x}$ and $g(x) = \sqrt{x^2 + 3x}$, find the simplest expression for $(f \cdot g)(x)$ and give its domain.

Step 3. Compose Functions

The composition of two functions is a new function formed when two functions work in sequence. The function $f \circ g$ or $f(g(x))$ is the function that results when g works on an input, producing some output, and then f takes in that value and applies its transformation. If $f(x) = x^2 - 3$, and $g(x) = 2x + 5$,

$$f \circ g(4) = f(g(4)) = f(2 \cdot 4 + 5) = f(13) = 13^2 - 3 = 166.$$

The order in which the functions work is important; $f \circ g$ will not necessarily yield the same result as $g \circ f$.

$$g \circ f(4) = g(f(4)) = g(4^2 - 3) = g(13) = 2 \cdot 13 + 5 = 31.$$

Just as with arithmetic combinations of functions, questions about composition may be presented in a variety of ways.

Use a Table

If the functions are presented by a table of values, evaluating a composition means using the table to locate the output of the first function, then letting the output of the first become the input of the second, and locating the final output.

If f and g are functions defined by the following table, $f(g(1))$ can be found by first using the table to find that $g(1) = 0$ and then using the table to find that $f(0) = 5$.

x	0	1	2	3	4	5
f	5	3	3	0	4	2
g	−2	0	2	1	4	2

Use a Graph

If the functions are presented by graphs, you can evaluate a composite function by methods similar to those you used for functions presented in a table.

The graph on page 35 presents two functions, $f(x)$ and $g(x)$. To graph the composition, calculate the value of $f \circ g$ at a few key values.

x	$g(x)$	$f(g(x))$	
−1	3	0	(−1,0)
0	6	4	(0,4)
1	2	6	(1,6)
2	0	2	(2,2)
3	−1	3	(3,3)
4	4	1	(4,1)
5	6	4	(5,4)
6	5	−1	(6,−1)

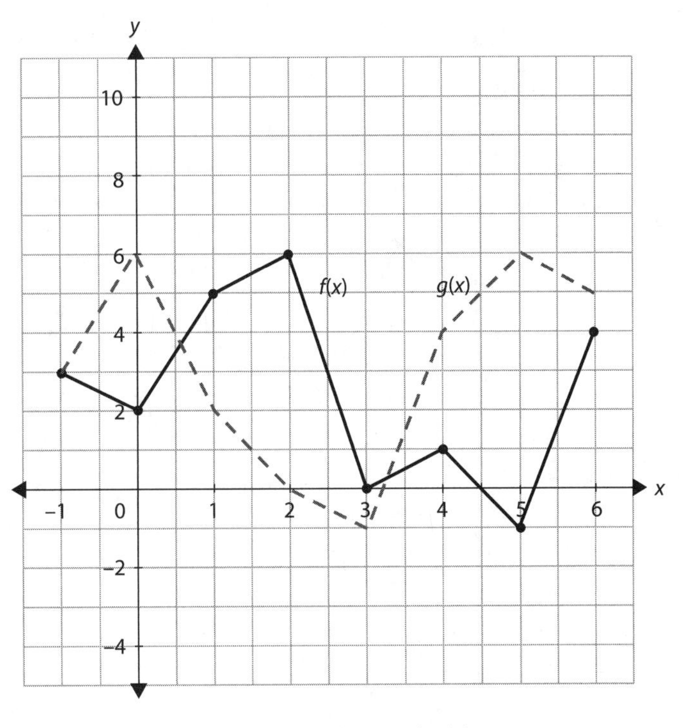

Graph of $f(x)$ and $g(x)$

Plot those points and connect, using lines if the graphs are linear, curves if one or both graphs curve. The resulting graph is shown on page 36.

Form the Rule

To form the rule for a composite function, replace the variable in the rule for the outer function with the expression for the inner function, and then simplify. The inner function takes in an x and operates on it and then hands the result to the outer function.

If $f(x) = x^2 - 3$ and $g(x) = 2x + 5$, and you want to form the rule for $f(g(x))$, the function g will work on x and produce $2x + 5$. It will turn that over to function f. Replace the x in $f(x) = x^2 - 3$ with $2x + 5$, and simplify.

$$\begin{aligned} f(g(x)) &= (g(x))^2 - 3 \\ &= (2x+5)^2 - 3 \\ &= 4x^2 + 20x + 22 \end{aligned}$$

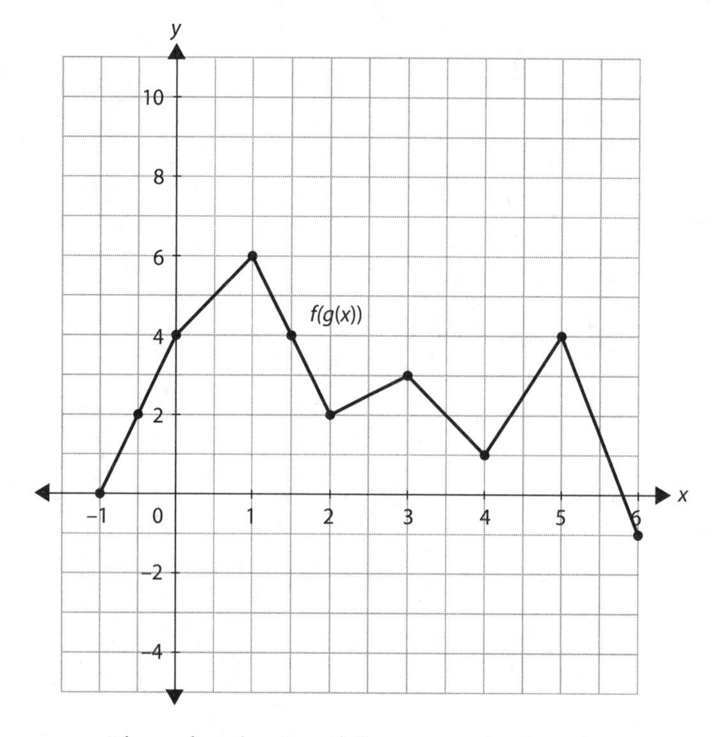

The ordered pairs of the composite function, connected with line segments

Don't assume that $f(g(x))$ and $g(f(x))$ have the same rule. They rarely do. To find $g(f(x))$, replace the x in g with $x^2 - 3$ and simplify.

$$
\begin{aligned}
g(f(x)) &= 2(f(x)) + 5 \\
&= 2(x^2 - 3) + 5 \\
&= 2x^2 - 1
\end{aligned}
$$

Check the Domain

When you define a new function, whether by arithmetic or by composition, you need to think about the domain of the function. To find the domain of a composite function, begin with the domain of the inner function. The domain of the composite cannot be any larger than that, but it may be smaller. If the inner function, operating on its whole domain, produces values not in the domain of the outer function, you'll need to restrict the domain further.

If $f(x)=\sqrt{x-5}$ and $g(x)=\frac{1}{x}$, and you form $f\circ g(x)$, you begin with the domain of g, which is all reals except 0. You'll never get that 0 back; $f\circ g(x)$ can't be defined where g is not defined. But then you need to look at the handoff from g to f, or in more mathematical terms, how the range of g compares to the domain of f. The range of g is all reals except 0, but the domain of f is $x\geq 5$. If you let g have its whole domain, it will produce values that f won't accept.

To deal with that problem, you need to restrict the domain of the composition. You want the result of g to be greater than or equal to 5.

$$\begin{aligned} g(x) &\geq 5 \\ \frac{1}{x} &\geq 5 \\ 1 &\geq 5x \\ \frac{1}{5} &\geq x \end{aligned}$$

You still have the $x\neq 0$ restriction, so the domain of the composition $f\circ g(x)$ is $(-\infty,0)\left(0,\frac{1}{5}\right]$. If you form the rule for $f\circ g(x)$, you get

$$\begin{aligned} f\circ g(x) &= f\left(\frac{1}{x}\right) \\ &= \sqrt{\frac{1}{x}-5} \\ &= \sqrt{\frac{1-5x}{x}} \\ &= \frac{\sqrt{1-5x}}{\sqrt{x}} \\ &= \frac{\sqrt{1-5x}\sqrt{x}}{x} \\ &= \frac{\sqrt{x-5x^2}}{x} \end{aligned}$$

That rule will still tell you that the domain can't include 0, because of the denominator of x, and must be restricted to values less than or equal to $\frac{1}{5}$

because of the radical. But don't depend on just the simplified version to tell you the domain of the composite.

If you compose $f(x)=\sqrt{x-5}$ and $h(x)=x^2$ to form $h\circ f$, the radical will disappear.

$$\begin{aligned} h\circ f(x) &= h\left(\sqrt{x-5}\right) \\ &= \left(\sqrt{x-5}\right)^2 \\ &= x-5 \end{aligned}$$

The restriction on the domain doesn't disappear, however. Even though the rule looks as though it could accept any real number, the domain is never larger than the domain of the inside function. The domain of $h\circ f$ is $[5,\infty)$.

Exercise 2.3

Use the table to evaluate the composite function for the given value of x.

x	−2	−1	0	1	2	3
f	3	1	0	−2	−1	2
g	−2	0	2	1	−1	3

1. $f(g(2))$
2. $f(g(-2))$
3. $g(f(1))$
4. $g(f(-1))$
5. $g(g(0))$

Find the rule for each composite function and give its domain.

6. Find $f(g(x))$ for $f(x)=x^2-2$ and $g(x)=\sqrt{x-4}$.
7. Find $g(f(x))$ for $f(x)=\frac{3}{x-1}$ and $g(x)=\frac{1}{x}$.
8. Find $f(g(x))$ for $f(x)=(x+1)^2$ and $g(x)=\sqrt{x}-1$.
9. Find $g(f(x))$ for $f(x)=x+\frac{1}{x}$ and $g(x)=x^2-\frac{2}{x}$.
10. Find $f(g(f(x)))$ for $f(x)=1-x$ and $g(x)=1-x^2$.

Step 4. Find Inverse Functions

Inverse operations undo one another's work. You use inverse operations to solve equations, undoing operations until the variable stands alone. Inverse functions are functions f and g such that $f(g(x)) = g(f(x)) = x$. The composition, in either direction, takes you back where you started from, to x. The inverse of a function, f, is a function, f^{-1}, that sends each element of the range of f back to the element of the domain from which it came.

If you're given two functions and asked if they're inverses, you can find both compositions. If both compositions, $f \circ g$ and $g \circ f$, equal x, the functions are inverses. If either or both are equal to anything other than x, the functions are not inverses. The functions $f(x) = 3x - 2$ and $g(x) = \frac{x+2}{3}$ are inverse functions because

$$f(g(x)) = f\left(\frac{x+2}{3}\right) = 3\left(\frac{x+2}{3}\right) - 2 = x + 2 - 2 = x \qquad \text{and}$$

$$g(f(x)) = g(3x-2) = \frac{3x-2+2}{3} = \frac{3x}{3} = x.$$

To find the inverse of a given function, you need to first verify that the function has an inverse function. Only functions that are one-to-one will have inverse functions. A one-to-one function is one in which each y-value comes from only one x-value as well as each x-value being sent to only one y-value. If the function isn't one-to-one, there will be values in the range that need to go back to two or more different values in the domain, and that will mean the inverse can't be a function.

Check Whether the Function Is One-to-One

The quickest way to determine whether the function is one-to-one is to use the horizontal line test on the graph of the function. The vertical line test tells you whether the graph represents a function, and the horizontal line test tells you whether the function is one-to-one. If any horizontal line intersects the graph in two or more distinct points, the function is not one-to-one.

If the graph is not easily available for the horizontal line test, inspect the equation of the function for signals that the function is not one-to-one. Those signs include even powers, which cause u-shaped graphs, and absolute value signs.

Don't give up if the function is not one-to-one, however. Try to restrict the domain to create a one-to-one function. The squaring function, $f(x) = x^2$, is not one-to-one, but the same rule restricted to $[0,\infty)$ or to $(-\infty,0]$ is one-to-one, and you can find an inverse for either of those. The inverse of $f(x)=x^2$ on $[0,\infty)$ is $f^{-1}(x)=\sqrt{x}$, and the inverse of $f(x)=x^2$ on $(-\infty,0]$ is $f^{-1}(x)=-\sqrt{x}$.

Find the Rule

To find the equation of the inverse function:

1. Replace $f(x)$ by y.
2. Swap the variables, that is, replace every y with x and every x with y.
3. Isolate y.
4. Replace y with $f^{-1}(x)$.

To find the inverse of the function $f(x)=\sqrt{3x-2}$, first note the domain and range of f. The domain of f, $\left[\frac{2}{3},\infty\right)$, will be the range of the inverse. The range of f, $[0,\infty)$, will be the domain of the inverse. Then replace $f(x)$ with y, swap the variables, and isolate y.

$$f(x)=\sqrt{3x-2}$$
$$y=\sqrt{3x-2}$$
$$x=\sqrt{3y-2}$$
$$x^2=3y-2$$
$$x^2+2=3y$$
$$\frac{x^2+2}{3}=y$$

The inverse of $f(x)=\sqrt{3x-2}$ is $f^{-1}(x)=\frac{x^2+2}{3}$, on the domain $[0,\infty)$.

Sketch the Graph

Because the inverse function sends each element of the range of f back to the element of the domain from which it came, every point on the graph of $f^{-1}(x)$ reverses the coordinates of a point on $f(x)$. If the point (2,1) is on the graph of $f(x)$, the point (1,2) will be on the graph of $f^{-1}(x)$.

This reversal means that the graphs of $f(x)$ and $f^{-1}(x)$ are reflections of one another across the line $y = x$.

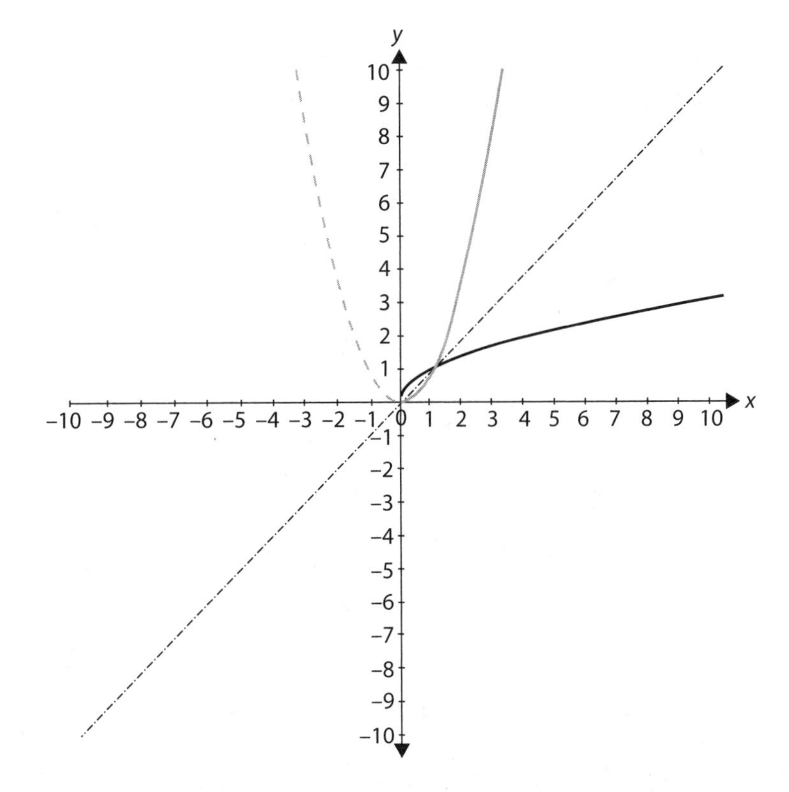

The graph of $f(x) = x^2$, limited to $[0, \infty)$ and $f^{-1}(x) = \sqrt{x}$, are reflections across $y = x$

To sketch the graph of $f^{-1}(x)$, choose key points on the graph of $f(x)$, reverse the coordinates, and plot the points. Connect the points, creating the reflection of $f(x)$ over the line $y = x$.

Exercise 2.4

Determine whether the given functions are inverses.

1. $f(x) = 2x - 5$ and $g(x) = \frac{1}{2}x + 5$
2. $f(x) = \sqrt{x - 3}$ and $g(x) = x^2 + 3,\ x \geq 0$
3. $f(x) = 2x^3 - 1$ and $g(x) = \dfrac{1}{2x^3 - 1}$

Determine whether the given function is 1–1.

4. $f(x) = 5x^2 - 4$

5. $f(x) = 2^x$

Find the inverse of each function, and specify the domain.

6. $f(x) = 3x - 1$

7. $f(x) = x^2 - 4,\ x \geq 0$

8. $f(x) = \dfrac{2x - 1}{3x}$

9. $f(x) = \sqrt{5x - 9}$

Sketch the graph of $f^{-1}(x)$.

10.

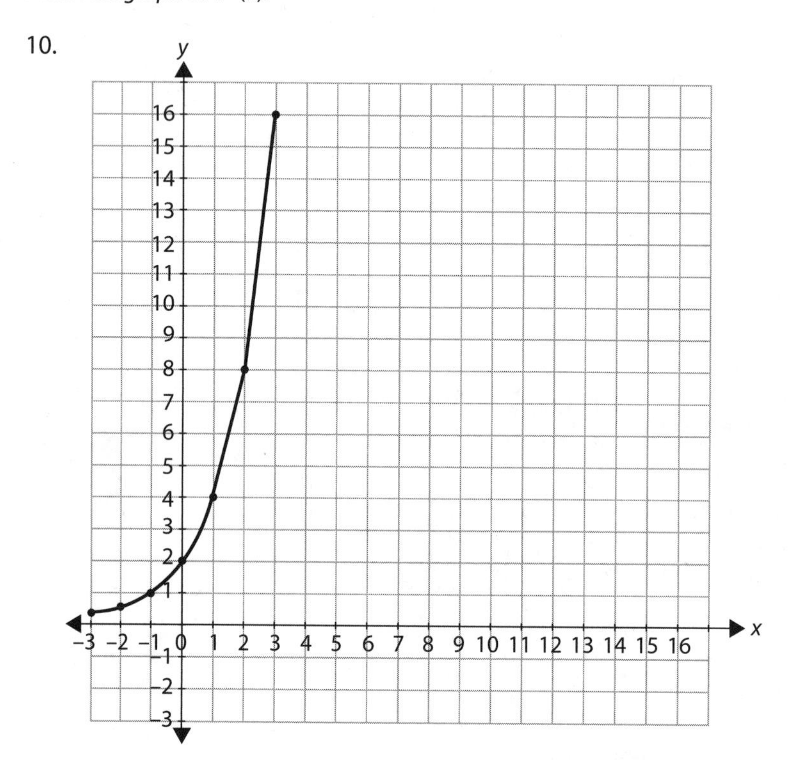

3

Quadratic Functions

Quadratic functions are functions of the form $f(x) = ax^2 + bx + c$, where a, b, and c are real numbers and a is nonzero. You'll find quadratic functions in vertical motion problems that discuss the height of an object dropped or thrown. They'll also show up when you investigate situations in which you're trying to maximize or minimize some quantity.

Step 1. Solve Quadratic Functions

You've probably already had some experience with solving equations of the form $ax^2 + bx + c = 0$. For all but the square root method, you want to be sure to have all nonzero terms on one side, equal to zero.

Take Square Roots

If the equation can be expressed as a perfect square equal to a constant, take the square root of both sides. The obvious use of this method is for equations of the form $x^2 = c$, for some constant c. So the solution of $x^2 = 81$ is $x = \pm 9$, the solution of $x^2 = 28$ is $x = \pm\sqrt{28} = \pm 2\sqrt{7}$, and the solution of $x^2 = -1$ is $x = \pm\sqrt{-1} = \pm i$. (We'll have more on complex numbers later.)

Don't forget: answers include both positive and negative square roots.

Simplest radical form:
$$\sqrt{28} = \sqrt{4 \cdot 7} = \sqrt{4}\sqrt{7} = 2\sqrt{7}$$

This method can also be used for equations that are in the form (or can easily be put in the form) of a quantity squared equal to a constant. So $(3x-1)^2 = 5$ can be solved by first taking the square root of both sides, to get $3x - 1 = \pm\sqrt{5}$, and then solving for x.

$$3x = 1 \pm \sqrt{5}$$

$$x = \frac{1 \pm \sqrt{5}}{3}$$

Complete the Square

Completing the square is a technique to force the equation into a form that can be solved by taking the square root. If you look at the equation $x^2 + 14x + 49 = 12$ and recognize that the left side is a perfect square trinomial, you can rewrite the equation as $(x+7)^2 = 12$, and then solve by the square root method.

$$(x+7)^2 = 12$$

$$x + 7 = \pm\sqrt{12} = \pm 2\sqrt{3}$$

$$x = -7 \pm 2\sqrt{3}$$

Completing the square turns one side of the equation into a perfect square trinomial. Follow these steps to complete the square:

1. Collect all terms on one side and simplify.
2. If the coefficient of x^2 is anything other than 1, divide through the equation by that coefficient.
3. Move the constant term to the other side.
4. Take half the coefficient of x, square it, and add that number to both sides. This creates a trinomial that is a perfect square.
5. Rewrite the perfect square trinomial as the square of a binomial, and simplify the constant.
6. Solve by the square root method.

To solve the equation $3x^2 + 15 = 24x$ by completing the square, move the terms to one side, divide by 3, and move the constant.

$$3x^2 - 24x + 15 = 0$$

$$x^2 - 8x + 5 = 0$$

$$x^2 - 8x = -5$$

Take half of –8, square it, and add 16 to both sides.

$$x^2 - 8x + 16 = -5 + 16$$

Rewrite as a square and solve.

$$(x-4)^2 = 11$$
$$x - 4 = \pm\sqrt{11}$$
$$x = 4 \pm \sqrt{11}$$

Use the Quadratic Formula

If you solve the equation $ax^2 + bx + c = 0$ by completing the square, the result is an equation called the quadratic formula: $x = \dfrac{-b \pm \sqrt{b^2 - 4ac}}{2a}$. Any equation of the form $ax^2 + bx + c = 0$ can be solved by substituting the values of a, b, and c into the formula and simplifying. Because completing the square often involves unfriendly fractions, many people prefer to use the formula.

Could you program your calculator to accept the coefficients and evaluate the quadratic formula? Here's the start of one version:

```
PROGRAM:QUADFORM
:INPUT "A =", A
:INPUT "B =", B
:INPUT "C =", C
:B²–4*A*C→D
:Disp D
:(-B+√(D))/(2*A)
→R
```

To solve a quadratic equation by the quadratic formula, collect all terms on one side equal to 0, and identify the values of a, b, and c. Substitute into the formula, and simplify carefully. To solve $5x^2 - 9x + 3 = 0$, use $a = 5$, $b = -9$, and $c = 3$.

$$x = \frac{-b \pm \sqrt{b^2 - 4ac}}{2a}$$

$$x = \frac{-(-9) \pm \sqrt{(-9)^2 - 4(5)(3)}}{2(5)}$$

$$x = \frac{9 \pm \sqrt{81 - 60}}{10}$$

$$x = \frac{9 \pm \sqrt{21}}{10}$$

This is the exact value. Use a calculator for the approximate decimals.

Use Factoring

Some quadratic equations can be solved by factoring ax^2+bx+c and applying the zero-product property, which says that if the product of two factors is 0, one or both of the factors is 0. It's critical that you first collect all nonzero terms on one side. Then factor, set each factor equal to 0, and solve.

$$6x^2+x-35=0$$
$$(3x-7)(2x+5)=0$$
$$3x-7=0 \quad 2x+5=0$$
$$3x-7=0 \qquad 2x=-5$$
$$x=\frac{7}{3} \qquad x=\frac{-5}{2}$$

Exercise 3.1

Complete the following exercises.

1. Solve by the square root method: $(2x-5)^2=16$
2. Solve by the square root method: $2(5x-7)^2-1=15$
3. Solve by completing the square: $x^2-12x-9=0$
4. Solve by completing the square: $2x^2+8x-3=0$
5. Solve by completing the square: $5x^2+15x+11=0$
6. Solve by the quadratic formula: $3x^2-7x+2=0$
7. Solve by the quadratic formula: $8x^2-9=4x$
8. Solve by the quadratic formula: $3-7x^2=12x$
9. Solve by factoring: $2x^2-5x-18=0$
10. Solve by factoring: $15x^2-13x-20=0$

Step 2. Explore Complex Numbers

When you first learned to solve quadratic equations, you were told the equation $x^2=-1$ had no solution, because it was impossible to take the square root of a negative number. It's true that the equation $x^2=-1$ has no *real* solutions, because it is impossible to take the square root of a negative number if you're operating in the real number system. But once you define the imaginary unit, $i=\sqrt{-1}$, you open the door to the complex number system, and equations such as $x^2=-1$ do have solutions.

Real, Imaginary, and Complex

Defining an imaginary unit $i=\sqrt{-1}$ not only provides a solution for that equation but also generates a whole system of imaginary numbers such as $3i$ and $-4.2i$. When you begin to combine these imaginary numbers with the real numbers, you form a new set of numbers, called complex numbers, that include the reals and the imaginaries and all the combinations. The complex numbers are all numbers of the form $a + bi$, where a is a real number and bi is an imaginary number. The real numbers have the form $a + 0i$, and the imaginaries have the form $0+bi$. Examples of complex numbers include $3-7i$ and $\frac{1}{2}+\sqrt{2}i$.

The Arithmetic of Complex Numbers

Doing arithmetic with complex numbers is similar to doing arithmetic with radicals. To add complex numbers, add the real parts and add the imaginary parts. To subtract, subtract the real parts and subtract the imaginary parts, but be vigilant about signs. Adding $3+5i$ to $6-9i$ gives you $9-4i$. Subtracting $(5+2i)-(7-4i)$ produces $-2+6i$.

When you multiply complex numbers, it's important to remember that $i=\sqrt{-1}$, so $i^2=-1$. So while $3 \cdot 5i$ is just $15i$, $-6i\cdot\frac{1}{2}i$ gives you $-3i^2$, and because $i^2=-1$, $-3i^2=-3(-1)=3$.

To multiply two complex numbers, use the FOIL rule (First, Outer, Inner, Last), as you would for binomials, but remember that $i^2=-1$.

$$
\begin{aligned}
(3+5i)(2-7i) &= 6-21i+10i-35i^2 \\
&= (6+35)+(-21+10)i \\
&= 41-11i
\end{aligned}
$$

For division by imaginary or complex numbers, the techniques of rationalizing denominators are used to simplify quotients. To divide by a pure imaginary, rationalize the denominator by multiplying the numerator and denominator by i and replacing i^2 with -1. The quotient $\frac{5+3i}{2i}\cdot\frac{i}{i}=\frac{5i-3}{-2}=1.5-2.5i$. To rationalize a denominator that is a complex, you'll need to multiply by the conjugate of the denominator. The numbers $a+bi$ and $a-bi$ are complex conjugates. If you multiply a complex number and its conjugate, for example, $(2+5i)$ and $(2-5i)$, the product will be a real number.

$$(2+5i)(2-5i) = 4-10i+10i-25i^2$$
$$= 4-25(-1)$$
$$= 4+25$$

Because the product of a complex number and its conjugate will always be a real number, multiplying the numerator and denominator by the conjugate of the denominator will rationalize the expression. The quotient

$$\frac{5-5i}{2-6i}\cdot\frac{2+6i}{2+6i} = \frac{10+30i-10i+30}{4+36} = \frac{40+20i}{40} = 1+\frac{1}{2}i.$$

Fundamental Theorem of Algebra

The fundamental theorem of algebra tells you that, over the complex numbers, a polynomial equation of degree n has n zeros, or solutions. For quadratics, this means that every quadratic equation has two solutions. Those solutions may be real or complex, but they will be either both real or both complex.

Some quadratics have a "double root"—the same solution twice, or "with a multiplicity of 2."

The Discriminant

The part of the quadratic formula under the radical, b^2-4ac, is called the discriminant, because it allows you to determine the nature of the solutions of the quadratic equation.

$b^2-4ac<0$	2 nonreal solutions
$b^2-4ac=0$	1 real double solution
$b^2-4ac>0$ and a perfect square	2 real, rational solutions
$b^2-4ac>0$ but not a perfect square	2 real, irrational solutions

Exercise 3.2

Simplify each expression.

1. $(3-7i)+(2+5i)$
2. $(11-2i)-(9+8i)$
3. $(4-7i)(6+2i)$
4. $\frac{4-9i}{2+3i}$
5. $2i\left(\frac{5+3i}{2-i}\right)-(4+3i)\left(\frac{6-5i}{2+i}\right)$

Determine the nature of the zeros of each function.

6. $3x^2 - 7x + 5 = 0$
7. $2x^2 + 5x - 4 = 0$
8. $4x^2 - 12x + 9 = 0$
9. $12 - 7x^2 = 10x$
10. $5x^2 - 8x + 3 = 0$

Step 3. Graph Parabolas

The graph of an equation of the form $f(x) = ax^2 + bx + c$ is a parabola, a cup-shaped graph defined as the set of all points equidistant from a fixed point, called the focus, and a fixed line, called the directrix. (More on that in Chapter 6.) For now, parabolas will open up or down. (Parabolas that open to the right or left will also come up in Chapter 6.) If you're familiar with the parent graph, $f(x) = x^2$, you can sketch the graph of any parabola with just a little extra information.

Find the Axis of Symmetry and Vertex

Every parabola is symmetric about some line. That line is called the axis of symmetry of the parabola, and it passes through the turning point, or vertex, of the parabola. In parabolas that open up or down, the axis of symmetry is a vertical line.

The equation of the axis of symmetry of the parabola $f(x) = ax^2 + bx + c$ is $x = \frac{-b}{2a}$. To find the vertex of the parabola, substitute $\frac{-b}{2a}$ into the original equation, to find the y-value. The parabola $f(x) = 3x^2 - 12x - 1$ is symmetric about the line $x = \frac{12}{2(3)} = 2$. The y-coordinate of the vertex is $y = 3(2)^2 - 12(2) - 1 = -13$, so the vertex is the point $(2,-13)$.

Convert Standard Form to Vertex Form

The $f(x) = ax^2 + bx + c$ form, or standard form, of the parabola can be converted to the vertex form $f(x) = a(x-h)^2 + k$. Vertex form takes its name from the fact that (h,k) is the vertex of the parabola, clearly visible in this form.

> These are alternate methods. No need to do both!

To change from standard to vertex form, do the following:

1. Move the constant to the $f(x)$ side.
2. Divide through by a.

3. Complete the square.
4. Isolate *y*.

If you wanted to put $f(x)=3x^2-12x-1$ in vertex form, rather than using the axis of symmetry, complete the square.

$$f(x)=3x^2-12x-1$$
$$f(x)+1=3x^2-12x$$
$$\frac{f(x)+1}{3}=x^2-4x$$
$$\frac{f(x)+1}{3}+4=x^2-4x+4$$
$$\frac{f(x)+13}{3}=(x-2)^2$$
$$f(x)+13=3(x-2)^2$$
$$f(x)=3(x-2)^2-13$$

Use Translation to Locate the Vertex

The parent graph for the family of parabolas has its vertex at (0,0). Once you know the vertex of your parabola, imagine sliding the parent graph to the new location. Translation, or sliding, is a rigid transformation. It will not change the shape of the graph but will only move it left or right, and up or down.

Use the Lead Coefficient to Set Direction and Shape

The change in the appearance of the graph comes from the value of a, the lead coefficient. The value of a modifies the graph in two ways: direction and stretch or compression.

If $a>0$, the parabola opens up. If $a<0$, it opens down. Once you know the vertex, a glance at the sign of the lead coefficient will tell you whether the parabola holds water or turns down, spilling water.

The value of a can also be used to set the width of the parabola. The effect of a is to stretch or compress the parabola vertically. The parent graph $f(x)=x^2$ has a pattern of change that you can modify to help sketch the graph.

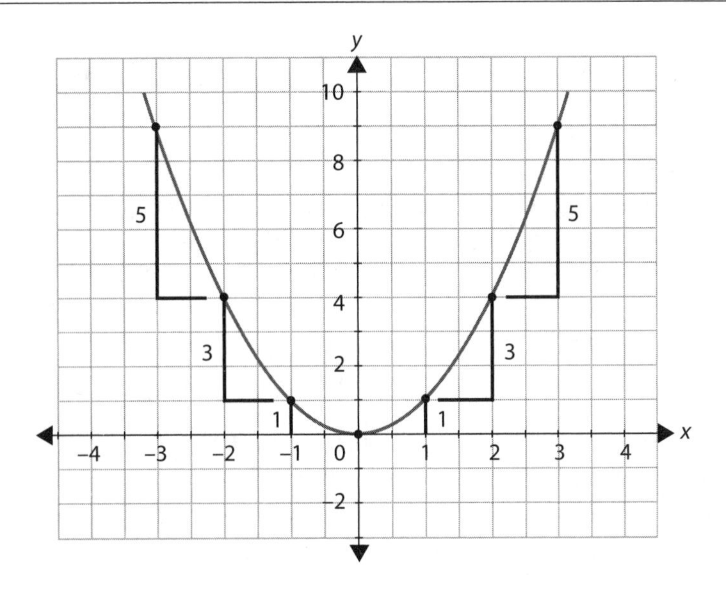

The rate of change of a parabola increases as you move away from the vertex

Each time you move one step to the left or right, the upward (or downward) motion is an odd number, and those odd numbers increase in sequence. This is true whenever $a = 1$. When a has a value other than 1, multiply each of the odds by a, and as you move left and right, move up or down $1a$, then $3a$, then $5a$, and so on.

Exercise 3.3

Graph each equation, and label the vertex and intercepts.

1. $f(x) = x^2 - 6x + 8$
2. $f(x) = -2x^2 - 9x + 5$
3. $f(x) = 3x^2 - 7x - 20$
4. $f(x) = 6x^2 + 7x + 2$
5. $f(x) = 35 + 2x - x^2$
6. $f(x) = \frac{1}{2}(x-2)^2 + 3$
7. $f(x) = -3(x+5)^2 - 1$
8. $f(x) = 2(x-1)^2 - 9$
9. $f(x) = -(x+4)^2 + 2$
10. $f(x) = -\frac{1}{3}(x-2)^2 + 4$

Step 4. Apply What You've Learned

Use the Vertical Motion Model

One of the most common applications of quadratic functions is the vertical motion model, which describes the height of an object dropped or thrown. For an object dropped from an initial height h_0, the height at time t is $h(t) = -\frac{1}{2}gt^2 + h_0$, where g is the acceleration due to gravity. For an object thrown with an initial velocity of v_0, from an initial height of h_0, the height is $h(t) = -\frac{1}{2}gt^2 + v_0t + h_0$.

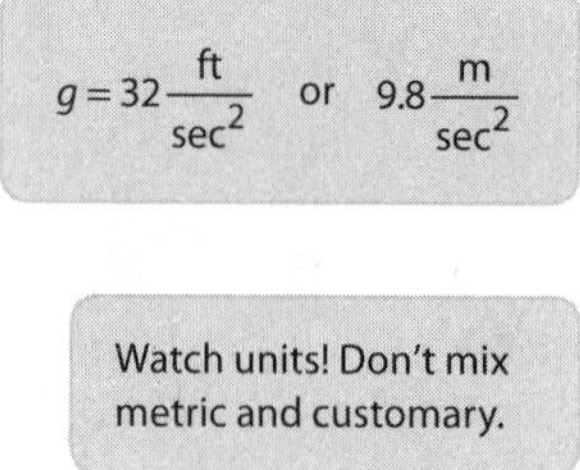

If you throw a ball up with an initial velocity of 40 feet per second while standing on a platform so that you release the ball at an initial height of 10 feet, the function that describes the height of the ball is $h(t) = -16t^2 + 40t + 10$.

Once you have the function, you can find the height of the object by substituting for t. The height of the ball 1 second after you release it is $h(1) = -16 + 40 + 10 = 34$ feet. You can find the time at which it reaches a specific height by substituting for $h(t)$ and solving the equation. To find out when the ball is 40 feet high, solve $40 = -16t^2 + 40t + 10$. You can graph to find the maximum height the ball will reach by finding the vertex of the parabola.

Find Maxima and Minima

You'll also see quadratic functions used in other problems, often problems that involve finding a maximum or minimum. To find the maximum rectangular area you can enclose with 300 feet of fencing, use the perimeter, which will be 300 feet, to express the dimensions of the rectangle as x and $150 - x$. The area will be $A(x) = x(150 - x)$, and finding the vertex of that parabola will tell you the maximum area and the dimensions that produce it. The vertex of $A(x) = x(150 - x) = -x^2 + 150x$ is (75,5625), so the maximum area is 5625 square feet, when the rectangle is 75 feet square.

Fit Equations to Data

If you're asked to find the quadratic function that fits data, your method will depend on what you know. If you know the vertex and one other point, start with vertex form. Substitute the coordinates of the vertex for h and k and the coordinates of the other point for x and y. Solve for a. Then go back to vertex form and write the equation using the known a, h, and k. For a vertex of (3,6) and a point (2,1) on the parabola,

$$y = a(x-h)^2 + k$$
$$1 = a(2-3)^2 + 6$$
$$1 = a + 6$$
$$a = -5$$

The equation of the parabola is $y = -5(x-3)^2 + 6$.

If you don't know the vertex, you'll need three points, and you'll have to solve a system of equations. You can use vertex form, but standard form may be easier. If the three points on the parabola are (2,5), (–5,12), and (0,–3), plug one set of coordinates into each copy of the equation $y = ax^2 + bx + c$. Solve the system:

$$5 = 4a + 2b + c$$
$$12 = 25a - 5b + c$$
$$-3 = 0a + 0b + c$$

You can also use the regression commands on your calculator to find an equation to fit the data.

You should find that $a = 1$, $b = 2$, and $c = -3$, so the equation is $y = x^2 + 2x - 3$.

Exercise 3.4

Complete the following exercises.

1. A ball is dropped from the top of a 120-foot tower. Write a quadratic function that represents the height of the ball at time t (in seconds), and use it to determine how long it takes for the ball to hit the ground.
2. A field goal kicker kicks a ball placed on the 35-yard line. His kick launches the ball with an initial velocity of 40 feet per second. Assuming the ball is on the ground when it is kicked, write a quadratic function that models the path of the football. What is the maximum height the ball will reach?
3. A ball thrown from the outfield leaves the outfielder's hand at a height of 7 feet, with an initial velocity of 33 feet per second. Write a quadratic function that models the height of the ball and use it to find when the ball hits the ground.

4. Eliana has 400 feet of fencing to enclose an area for her dog to play. The area is along the side of the house, so she needs to enclose only three sides. Write a quadratic function for the area of the dog run, and use it to find the dimensions that will give the maximum area.
5. Mr. Sanchez owns a company that manufactures packing cartons. For a particular contract, the company must produce a carton with a height of 6 inches and a rectangular base with a perimeter of 60 inches. Find the dimensions of the carton that will have the largest volume.
6. A rectangular pool 60 meters by 40 meters is to be bordered on all sides by a patio of constant width. The area covered by the patio is equal to the area covered by the pool. Find the width of the patio.
7. Find the equation of a parabola with vertex $(4,5)$ that passes through the point $(-3,-2)$.
8. Find the equation of a parabola with vertex $(-3,-2)$ that passes through the point $(4,5)$.
9. Find the equation of a parabola that passes through the points $(-1,10)$, $(2,4)$, and $(3,14)$.
10. Jena recorded the daily harvests of strawberries in her backyard garden. The data she collected are shown in the following table. Find a quadratic function that fits the data.

Day	1	3	5	7	9	11	13
Number of berries	5	16	25	28	33	26	21

4

Polynomial Functions

Polynomial functions are functions of the form

$$f(x) = a_n x^n + a_{n-1} x^{n-1} + \cdots + a_2 x^2 + a_1 x + a_0$$

The coefficients, a_i, are real numbers, and the exponents are integers. The degree of a polynomial is the degree of its highest degree term: $f(x) = x^3 - 5x^2 + 8x - 2$ is a third degree, or cubic, polynomial; $g(t) = t^8 - 1$ is a polynomial of degree 8. Linear functions are first-degree polynomials, and constant functions are polynomials of degree 0.

The domain of every polynomial function is all real numbers, $(-\infty, \infty)$. Any polynomial of odd degree has a range of all real numbers, but a polynomial of even degree will have a limited range. For a quadratic polynomial, the range extends from the y-value of the vertex upward if $a > 0$ or downward if $a < 0$. For polynomial functions of even degree greater than 2, the range is from the lowest turning point upward if $a > 0$ or from the highest turning point downward if $a < 0$.

Your primary tasks will be finding zeros of the function, sketching graphs, and finding relative maxima and minima.

Step 1. Factor Polynomials

You'll want to factor polynomials to find the zeros of the function, which are the x-intercepts of the graph. You'll find it useful to review factoring techniques from previous courses.

Greatest Common Factor

If you factor out the greatest common factor, you can often express a polynomial of higher degree as the product of a monomial and a polynomial that can be factored by simple techniques. The polynomial $12x^5+8x^4-10x^3$ has a greatest common factor of $2x^3$. Bring the common factor to the front, and in parentheses show the other factor.

$$12x^5+8x^4-10x^3=2x^3(6x^2+4x-5)$$

If you're trying to find the zeros of $f(x)=12x^5+8x^4-10x^3$, factoring to $f(x)=2x^3(6x^2+4x-5)$ will let you determine that there's a zero at $x=0$, and then use the quadratic formula on $6x^2+4x-5$ to find the two irrational zeros, $x=\frac{-3\pm\sqrt{34}}{6}$.

FOIL Factoring

The most common factoring technique in your repertoire is the reversal of the FOIL rule that rewrites a quadratic polynomial, or a polynomial of higher degree that is quadratic in form, as the product of two binomials. Polynomials that are quadratic in form are trinomials that have a constant term, a term involving x^n, and a term involving x^{2n}. The polynomial x^6-5x^3+6 is quadratic in form, because the degree of the lead term is twice the degree of the middle term.

To factor a polynomial that is a quadratic form, such as $f(x)=6x^{10}+19x^5-36$, you may find it helpful to make a variable substitution, replacing x^5 with a variable such as t, and x^{10} with t^2. The polynomial becomes $6t^2+19t-36$, which you can factor to $(2t+9)(3t-4)$. Then replace the original variable $f(x)=(2x^5+9)(3x^5-4)$ and you can find the real zeros by setting each factor equal to zero and solving.

Difference of Squares

The difference of squares, a^2-b^2 factors to $(a+b)(a-b)$, the sum and difference of the same two terms, and as with FOIL factoring, this pattern can also be extended to polynomials of higher degree: $9t^8-16=(3t^4+4)(3t^4-4)$ and $x^6-64=(x^3+8)(x^3-8)$. Sometimes, as in the previous example, it will be possible to factor one or both of the factors further.

Sum and Difference of Cubes

The sum of cubes $a^3 + b^3$ and the difference of cubes $a^3 - b^3$ factor to the product of a binomial and a trinomial. The factors of the sum differ from the factors of the difference only in the placement of signs.

$$a^3 + b^3 = (a+b)(a^2 - ab + b^2)$$
$$a^3 - b^3 = (a-b)(a^2 + ab + b^2)$$

The sum of cubes $x^3 + 64$ factors to $(x+4)(x^2 - 4x + 16)$ and the difference of cubes $27x^3 - 125 = (3x-5)(9x^2 + 15x + 25)$. If you apply these factoring patterns to the example from the previous section,

$$\begin{aligned} x^6 - 64 &= (x^3+8)(x^3-8) \\ &= [(x+2)(x^2-2x+4)][(x-2)(x^2+2x+4)] \\ &= (x+2)(x-2)(x^2-2x+4)(x^2+2x+4) \end{aligned}$$

Set each factor equal to zero and you'll find real zeros at $x = \pm 2$, and each of the trinomials will give two complex zeros if you use the quadratic formula.

Factoring by Grouping

You can sometimes factor cubic polynomials with four terms by breaking the polynomial into two binomials and removing a common factor from each. That factoring may reveal a factoring for the cubic polynomial.

The cubic polynomial $2x^3 + 6x^2 - 9x - 27$ can be factored by first examining the binomials $2x^3 + 6x^2$ and $-9x - 27$.

$$2x^3 + 6x^2 - 9x - 27$$
$$2x^2(x+3) - 9(x+3)$$

Because both factorings reveal the same $(x+3)$, you can factor that out as a common factor.

$$2x^3 + 6x^2 - 9x - 27 = (x+3)(2x^2 - 9)$$

Exercise 4.1

Factor each polynomial.

1. $x^3 - 5x^2 + 9x - 45$
2. $8x^3 - 27$
3. $3x^3 + 21x^2 - 54x$
4. $x^4 - 5x^2 - 14$
5. $10x^2 - 13x - 3$
6. $125 + 64x^3$
7. $x^3 + 5x^2 - 4x - 20$
8. $3x^7 + 21x^4 - 24x$
9. $9x^2 - 25$
10. $16x^2 - 8x + 1$

Step 2. Use Synthetic Division to Find Zeros

> Synthetic division is sometimes called synthetic substitution because it can be used to find the value of the function at a particular value of *x*.

> Because synthetic division can be used only with a linear divisor, you'll still sometimes need to use long division instead, but most people prefer synthetic division for its speed and simplicity and because addition leads to fewer errors than the subtraction involved in long division.

The factoring techniques you learned in earlier courses may not always be adequate to find the zeros of polynomial functions. The process of looking for zeros can become something of a trial-and-error process, and the quickest method of testing whether a particular value is a zero of a polynomial is by synthetic division. The method of synthetic division simulates division of a polynomial by a linear factor, using only the coefficients and constants, by a simple routine of multiplication and addition.

To test if $x = 4$ is a zero of $f(x) = 4x^4 + 8x^3 - 97x^2 - 62x + 264$, organize the terms in order from highest degree down to the constant, inserting zeros if any terms are missing. Place the coefficients on a line, with the possible zero to the side, usually the left.

$$\underline{4|}\quad 4 \quad 8 \quad -97 \quad -62 \quad 264$$

Bring down the first coefficient, multiply by the possible zero, and place the result under the second coefficient. Add.

$$\begin{array}{r|rrrrr} 4 & 4 & 8 & -97 & -62 & 264 \\ & \downarrow & 16 & & & \\ \hline & 4 & 24 & & & \end{array}$$

Repeat the multiply and add routine, working across the line.

$$\begin{array}{r|rrrrr} 4 & 4 & 8 & -97 & -62 & 264 \\ & \downarrow & 16 & 96 & -4 & -264 \\ \hline & 4 & 24 & -1 & -66 & \underline{|0} \end{array}$$

The final number is the remainder. The zero remainder in this example indicates that $x = 4$ is a zero. The other numbers on the bottom row are the coefficients of the quotient, or reduced polynomial. Because $x = 4$ is a zero of $f(x) = 4x^4 + 8x^3 - 97x^2 - 62x + 264$, $x - 4$ is a factor, and $f(x) = 4x^4 + 8x^3 - 97x^2 - 62x + 264 = (x-4)(4x^3 + 24x^2 - x - 66)$. You can continue testing possible zeros using the reduced polynomial. If a number is a zero of the reduced polynomial $4x^3 + 24x^2 - x - 66$, then it is a zero of

$$f(x) = 4x^4 + 8x^3 - 97x^2 - 62x + 264.$$

> The remainder is also the value of *f*(4). The remainder theorem says that the remainder when *f*(*x*) is divided by $x - c$ will be *f*(*c*). That's why it's also called synthetic substitution.

> The reduced polynomial is always one degree lower because you divided by a linear, or first-degree, factor.

Synthetic division can be used to test irrational numbers or complex numbers as well as rational numbers. Just spread out the coefficients to give yourself room and combine only like terms. To test whether $2 - 3i$ is a zero of $f(x) = x^3 - 5x^2 + 4x - 1$, use synthetic division.

$$\begin{array}{r|rrrr} 2-3i & 1 & -5 & 4 & -1 \\ & \downarrow & 2-3i & -15+3i & -13+39i \\ \hline & 1 & -3-3i & -11+3i & \underline{|-14+39i} \end{array}$$

Because the remainder is nonzero, you know that $2 - 3i$ is not a zero.

The process involves a lot of multiplying of complex numbers (or irrational numbers involving radicals), and that can be tedious. Many people prefer to count on the fact that irrational or complex zeros will occur in conjugate pairs and use long division. If, for example, you suspect that $2 + \sqrt{3}$ is a zero, you anticipate that $2 - \sqrt{3}$ will be a zero as well. The factors $x - 2 - \sqrt{3}$ and $x - 2 + \sqrt{3}$ multiply to $(x - 2 - \sqrt{3})(x - 2 + \sqrt{3}) = x^2 - 4x + 4 - 3 = x^2 - 4x + 1$. You may find it simpler to use long division to determine if $x^2 - 4x + 1$ is a factor.

Exercise 4.2

Find the quotient and remainder.

1. $(x^3 - 4x^2 + 5x - 2) \div (x - 2)$
2. $(x^5 + 4x^3 - 3x^2 + 5) \div (x + 1)$
3. $(x^4 - 25) \div (x - 5)$

Determine whether the binomial f(x) *is a factor of the polynomial* p(x).

4. $f(x) = x - 3,\ \ p(x) = x^3 - 2x^2 + 3x - 3$
5. $f(x) = x + 5,\ \ p(x) = x^3 - x^2 - 28x + 10$
6. $f(x) = x - 1,\ \ p(x) = x^3 - 2x^2 - 11x + 12$
7. $f(x) = x - 3,\ \ p(x) = 2x^4 - x^3 + x^2 - 3x - 6$
8. $f(x) = x + 4,\ \ p(x) = x^5 - 32$
9. If $p(x) = x^5 - 2x^4 + 3x^3 - x^2 + 2x + 1$ find $p(-2)$.
10. If $p(x) = x^4 - 2x^3 + x^2 - 5$ find $p(4)$.

Step 3. Use Long Division to Factor Polynomials

When zeros are irrational or complex, factoring and synthetic division may not be enough, but you can turn to long division, especially if you want to test for quadratic factors. The linear factors that correspond to irrational or complex zeros may be cumbersome for synthetic division. You may find that multiplying a pair of conjugate factors together and using long division with the resulting quadratic proves to be easier.

Long division of polynomials is modeled on the algorithm for long division that you learned in arithmetic. As with synthetic division, it can be used to test if one polynomial is a factor of another, and if the remainder of the division is 0, the divisor is a factor of the dividend. Unlike synthetic division, it's not limited to linear factors.

You may want to use this when you're searching for irrational zeros, such as $\pm\sqrt{3}$, or complex zeros, such as $2 \pm 3i$. In both cases, you can anticipate the pair of zeros, and therefore the pair of factors, such as $x + \sqrt{3}$ and $x - \sqrt{3}$, or $x - 2 + 3i$ and $x - 2 - 3i$, which multiply to $x^2 - 3$ or $x^2 - 4x + 13$.

To divide $x^4 + 8x + 10$ by $x^2 - 3$, arrange the dividend and the divisor in standard form, highest power to lowest, and insert zeros for any missing powers to make it easier to line up like terms. Divide the first term of the dividend by the first term of the divisor, and place the result as the first term of the quotient. Multiply the entire divisor by the term you just placed in the quotient, aligning like terms under the dividend. Subtract, and bring down any remaining terms in the dividend.

$$\begin{array}{r} x^2 \\ x^2+0x-3 \overline{)\, x^4+0x^3+0x^2+8x+10} \\ \underline{x^4+0x^3-3x^2 } \\ -3x^2+8x+10 \end{array}$$

Repeat those steps, but use this new expression formed by subtracting and bringing down as your dividend.

$$\begin{array}{r} x^2-3 \\ x^2+0x-3 \overline{)\, x^4+0x^3+0x^2+8x+10} \\ \underline{x^4+0x^3-3x^2 } \\ -3x^2+8x+10 \\ \underline{-3x^2+0x+9} \\ 8x+1 \end{array}$$

You can express the remainder as a fraction by putting the remainder as the numerator of the fraction and the divisor as the denominator.

$$(x^4+8x+10) \div (x^2-3) = x^2-3+\frac{8x+1}{x^2-3}$$

Because the remainder is not 0, you know that x^2-3 is not a factor of $x^4+8x+10$.

Exercise 4.3

Use long division to divide p(x) ÷ f(x). *Find the quotient* q(x) *and the remainder* r(x). *Express* p(x) *as* q(x) · f(x) + r(x).

1. $p(x)=2x^3-3x^2+4x-3,\quad f(x)=x^2-5$
2. $p(x)=x^3-9x+1,\quad f(x)=x^2+2$

3. $p(x) = 4x^5 - 6x^4 + 2x^3 - 5x^2 + 2x - 1, \quad f(x) = 2x^2 - 1$
4. $p(x) = 9x^4 + 7x^2 + 8, \quad f(x) = 3x^2 + 2$
5. $p(x) = x^6 - x^4 + x^2 + 1, \quad f(x) = x^3 + 1$

Determine whether f(x) *is a factor of* p(x).

6. $f(x) = x^2 + 2, \quad p(x) = x^4 - 3x^3 + 3x^2 - 6x + 2$
7. $f(x) = 2x^2 - 1, \quad p(x) = 2x^4 + 8x^3 - 11x^2 - 5x + 2$
8. $f(x) = 3x^2 + 4, \quad p(x) = 6x^4 + 23x^2 + 11x + 12$
9. $f(x) = x^2 + x + 1, \quad p(x) = x^4 + x^3 - 3x^2 - 4x - 4$
10. $f(x) = x^2 - 2x + 3, \quad p(x) = 2x^4 - x^3 - x^2 + 11x - 3$

Step 4. Find the Zeros of Polynomial Functions

When you're asked to find the zeros of a polynomial function, you'll want an organized plan of attack, and that plan will vary depending on whether or not you have access to a graphing utility. Your calculator will likely have a function to find real zeros, but remember that it will determine those zeros by an approximation technique. If the zero is rational, it should find the exact value, but if the zero is irrational, the calculator will give you a decimal approximation. Finding the exact zero is up to you, and so is finding nonreal zeros.

The approximation is done by the bisection method, once commonly done by hand before the use of calculators. When you are asked for left and right, or lower and upper bounds, you're supplying a value of x at which the function is negative and one at which it's positive. The intermediate value theorem tells you that somewhere in between, there's a zero. The bisection method finds the midpoint of that interval and tests to see if the function is positive or negative at that midpoint value of x. This allows you to find an interval half as large in which the zero must fall. Repeating the process cuts the interval in half, and in half again, and again, and again, until you've got a good approximation of the zero. It's tedious to do by hand, but the calculator can do it for you quickly.

Apply the Fundamental Theorem of Algebra

The fundamental theorem of algebra says that over the complex numbers, a polynomial function of degree n has n zeros. Linear equations have one solution, quadratics have two, cubics have three, and so on.

You've seen quadratics that have what are sometimes called double roots: the same zero occurring twice. When that happens, the zero has a multiplicity of two. When the fundamental theorem of algebra counts zeros, it counts a zero with a multiplicity of two as two zeros. For functions of higher degree, you may find zeros with even larger multiplicities.

The fundamental theorem of algebra talks about zeros over the complex numbers. If you restrict your search for zeros to the real numbers, you may not find all n zeros for an nth-degree polynomial, because some of its zeros may not be real. You know, however, that nonreal zeros will occur in conjugate pairs, so if you're looking at a fifth-degree polynomial, you may find five real zeros, or three real zeros and a pair of complex zeros, or one real zero and two pairs of complex.

List Possible Rational Zeros

Once you have a sense of what you're looking for, you need some possibilities. The possible rational zeros of any polynomial function will be ratios of factors of the constant term to factors of the lead coefficient. You should consider both positive and negative ratios. These are possible zeros, so you'll need to test them to see if they are actual zeros, and they're only rational numbers, so there may be other irrational zeros and nonreal zeros.

Use Descartes' Rule of Signs

If you have access to a graphing utility, looking at the x-intercepts of the polynomial will help you narrow the list of possibilities before you start testing. If you don't have that option, there are some other ways to narrow the list. You can look at $f(x)$ and count the number of times the sign of the coefficients change; $f(x) = x^3 - 4x^2 + 3x - 1$ has three sign changes, but $g(x) = x^3 + x^2 + x + 1$ has none. The number of sign changes in $f(x)$ is the maximum number of positive real zeros. There may be fewer zeros than sign changes, but there won't be more.

To find the maximum number of negative real zeros, look at the sign changes of $f(-x)$. For $f(x) = x^3 - 4x^2 + 3x - 1$, $f(-x) = -x^3 - 4x^2 - 3x - 1$ and has no change in signs. For $g(x) = x^3 + x^2 + x + 1$, $g(-x) = -x^3 + x^2 - x + 1$ has three sign changes.

If you're asked to find the zeros of $f(x) = x^3 - 4x^2 + 3x - 1$, you'll want to test the positive numbers on your list of possibilities and not waste time on the negative ones, but to find the zeros of $g(x) = x^3 + x^2 + x + 1$, you'll want to do just the opposite, because $g(x)$ has no positive zeros.

Reduce the Polynomial

Narrow the possibilities as best you can, and then use synthetic division to test possible zeros. Once a zero is found, write the polynomial in factored form and continue searching. Don't go back to the original polynomial, but use the reduced polynomial, because any zeros of that reduced polynomial will be zeros of the original. Continue searching for zeros by synthetic division until the reduced polynomial is a quadratic, if possible, and then use factoring or the quadratic formula.

To find the zeros of the polynomial function $f(x)=x^4-x^3-7x^2+x+6$, start with a list of possible rational zeros. The factors of the constant term 6 are 1, 2, 3, and 6, and the lead coefficient of 1 has only 1 as a factor, so the possible rational zeros are ±1, ±2, ±3, and ±6. The function $f(x)=x^4-x^3-7x^2+x+6$ has two sign changes, and $f(-x)=x^4+x^3-7x^2-x+6$ has two sign changes, so you'll want to try both positive and negative zeros. Try 1 using synthetic division.

$$\begin{array}{r|rrrrr} 1 & 1 & -1 & -7 & 1 & 6 \\ & \downarrow & 1 & 0 & -7 & -6 \\ \hline & 1 & 0 & -7 & -6 & \underline{|0} \end{array}$$

The remainder of zero signals that 1 is a zero of the function, and you can write the function in factored form as $f(x)=(x-1)(x^3-7x-6)$. Continue trying possible zeros, using the reduced polynomial.

$$\begin{array}{r|rrrrr} 1 & 1 & -1 & -7 & 1 & 6 \\ & \downarrow & 1 & 0 & -7 & -6 \\ \hline -1 & 1 & 0 & -7 & -6 & \underline{|0} \\ & \downarrow & -1 & 1 & 6 & \\ \hline & 1 & -1 & -6 & \underline{|0} & \end{array}$$

You know that −1 is also a zero and can write the polynomial as $f(x)=(x-1)(x+1)(x^2-x-6)$. At this point, you can factor x^2-x-6, or use the quadratic formula, or continue trying possibilities with synthetic division. By whatever method, you should find $f(x)=(x-1)(x+1)(x-3)(x+2)$, so the zeros of the function are −2, −1, 1, and 3.

Exercise 4.4

For each function, state the number of zeros over the complex numbers, and list the possible rational zeros.

1. $f(x) = 3x^3 - 17x^2 + 15x - 25$
2. $f(x) = x^5 - 13x^4 - 120x + 80$
3. $f(x) = 6x^3 - 7x + 3x^4 - 10x^2 + 2$

For each function, use Descartes' rule of signs to list the maximum number of positive real zeros and the maximum number of negative real zeros.

4. $f(x) = 2x^3 - 15x^2 + 27x - 10$
5. $f(x) = x^4 - 7x^2 + 12$
6. $f(x) = 6x^4 - 11x^3 - 51x^2 + 99x - 27$

Reduce each polynomial to a product of linear factors.

7. $f(x) = x^3 - 5x^2 - 15x + 27$
8. $f(x) = 2x^3 - 5x^2 + 12x - 5$
9. $f(x) = x^4 + 10x^2 + 9$
10. $f(x) = x^4 - 4x^3 - 2x^2 + 12x - 16$

Step 5. Graph Polynomial Functions

One of the traditional applications you'll learn in differential calculus is curve sketching. You'll use what you learn about the derivative to discover information about the graph of a function. Without that calculus to call upon, your ability to sketch graphs now will have some limitations, but there's still a lot you can determine with your current skills.

Find Intercepts

The y-intercept of the graph of a polynomial function is simply the constant term. The x-intercepts are the real zeros of the function. Finding those intercepts will give you a good start on the graph of the function. In an earlier example, you found that the zeros of the function $f(x) = x^4 - x^3 - 7x^2 + x + 6$ are –2, –1, 1, and 3. Those are the x-intercepts of the graph of the function. Its y-intercept is the constant term, 6.

Plot the Basic Shape

You can anticipate some key features of the shape of the graph of a polynomial function. The graph of a polynomial function of degree n will have a maximum of $n-1$ turning points. That means that a quadratic has one turning point and a cubic may have two, while a quartic or fourth-degree polynomial may turn three times. Multiplicities may make those turns difficult to see, as in the graph of $f(x)=x^3$, and for distinct zeros, the closer together zeros fall, less rise or fall will occur between intercepts. The function $f(x)=x^4-x^3-7x^2+x+6$ has four distinct real zeros, so the graph will cut through the x-axis four times.

If a zero has an odd multiplicity, the graph will cut through the x-axis at that intercept, but at zeros that have even multiplicities, the graph will just touch the axis and turn away without passing through. As a result, you'll see a turning point at that zero. The function $f(x)=x^4+x^3-3x^2-5x-2$ has a zero of -1 with a multiplicity of 3, and a zero of 2 with a multiplicity of 1. Its graph cuts through the x-axis at both zeros, but you'll see a flattening at -1 because of the multiplicity.

The function $f(x)=x^3-3x-2$ also has zeros of -1 and 2, but the zero of -1 has a multiplicity of 2. The graph just touches the x-axis at -1.

When the function has nonreal zeros, you'll see a turning point that seems to float above or below the x-axis. A minimum above the x-axis or a maximum below the axis are signs of complex zeros. When you're trying to sketch the graph of a function with complex zeros, you might want to plot a few points to try to find that turning point. The graph of the function $f(x)=x^3-2x^2+x-2$ on page 68 has a real zero of 2, and two nonreal zeros.

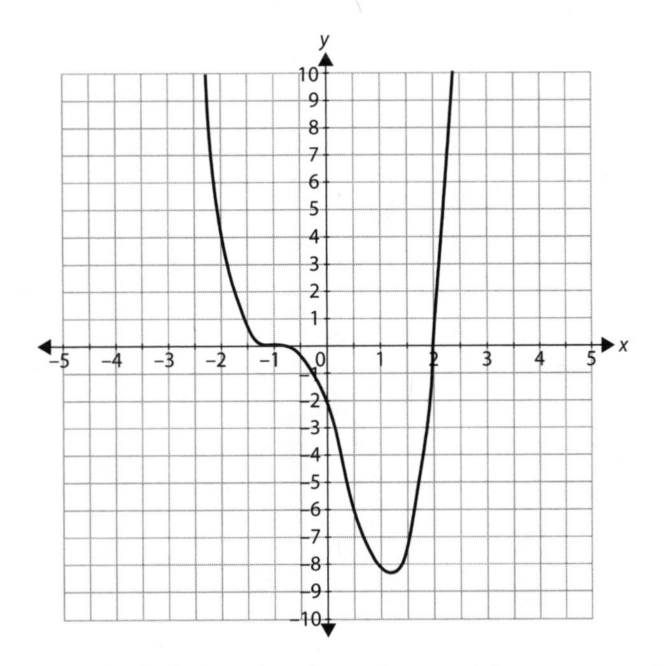

The graph of $f(x) = x^4 + x^3 - 3x^2 - 5x - 2$ has zeros at −1, with a multiplicity of 3, and at 2

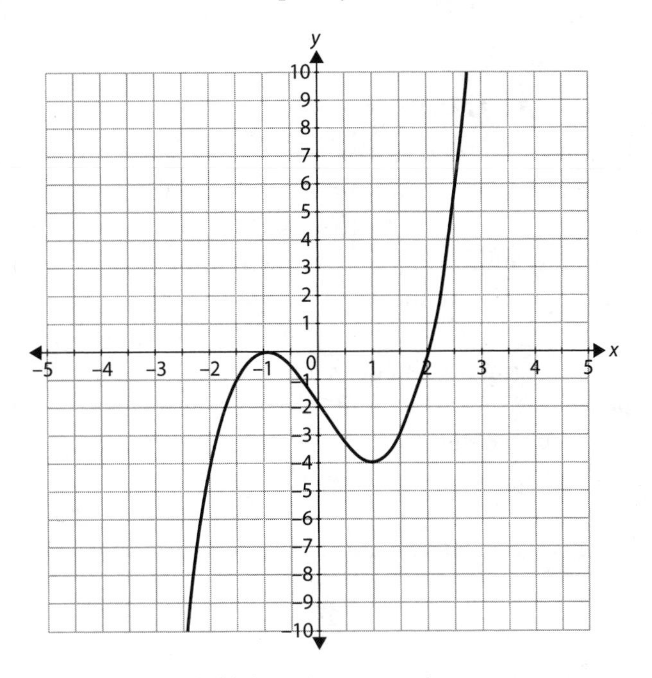

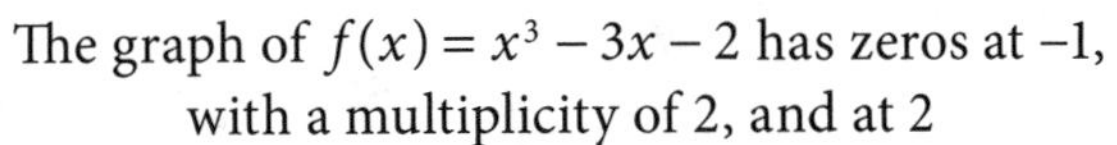
The graph of $f(x) = x^3 - 3x - 2$ has zeros at −1, with a multiplicity of 2, and at 2

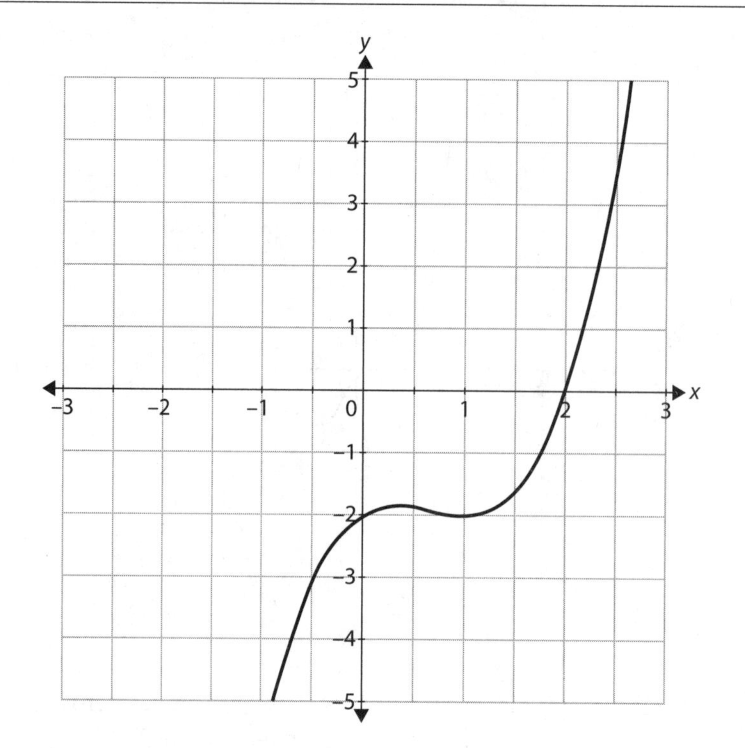

The graph of $f(x) = x^3 - 2x^2 + x - 2$ has only one real zero

Find Relative Extrema

Without calculus, you can't say exactly where the turning points, or relative maxima and relative minima, occur. You won't be able to say exactly how high or how low they are either, without the help of a graphing calculator. If you work at connecting your known points with a smooth curve, however, you can approximate them.

The function $f(x) = x^4 - x^3 - 7x^2 + x + 6$ on page 69 has four distinct real zeros, so the graph will cut through the x-axis four times. It has three turning points, two minima, and one maximum, each of which occurs roughly midway between zeros. The rise and fall of the graph above and below the x-axis varies.

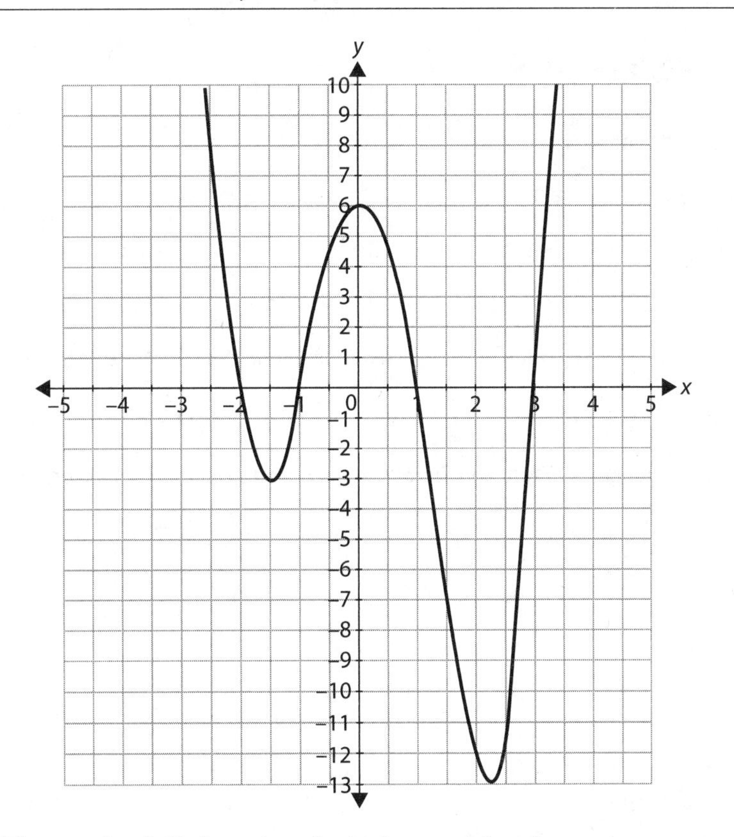

The graph of $f(x) = x^4 - x^3 - 7x^2 + x + 6$ has four x-intercepts, one relative maximum, and two relative minima

Use the Lead Coefficient to Set End Behavior

The graphs of polynomial functions of odd degree will increase without bound on one end and decrease without bound on the other end. If the lead coefficient, the coefficient of the highest degree term, is positive, the graph will increase on the right and decrease on the left, as a line with a positive slope. If the lead coefficient is negative, the graph falls to the right and rises to the left, as a line with a negative slope.

Polynomial functions of even degree behave as parabolas: either both ends go up or both ends go down. If the lead coefficient is positive, both ends go up, but if the lead coefficient is negative, both ends go down.

Exercise 4.5

Sketch a graph of each function.

1. $f(x) = 2x^3 - 3x^2 - 12x + 8$
2. $f(x) = 3x^4 + 4x^3$
3. $f(x) = x^5 - 5x^2 + 4x$
4. $f(x) = (x-2)^3 + 1$
5. $f(x) = -x^3 + 3x - 2$

Find the x- and y-intercepts of each function, and describe the end behavior.

6. $f(x) = 1 - x^4$
7. $f(x) = x^5 + x^3 - 6x$

Use a graphing utility to find the relative maxima and relative minima of each function.

8. $f(x) = x^3 - 3x^2$
9. $f(x) = 4x - x^3$
10. $f(x) = x^4 - 2x^3 + x^2$

5

Rational Functions

Rational functions are quotients of two polynomials. The name rational in this context comes from ratio. If $f(x)$ is a rational function, it can be written as $f(x)=\frac{p(x)}{q(x)}$, where $p(x)$ and $q(x)$ are polynomial functions and $q(x)\neq 0$. The function $f(x)=\frac{x-3}{x+5}$ is a rational function, as is $g(x)=\frac{1}{x}$, but $h(x)=\frac{4x-3}{\sqrt{x^2-x+1}}$ is not, because the denominator is not a polynomial. Functions such as $h(x)=\frac{4x-3}{\sqrt{x^2-x+1}}$ are usually described by the more general term *algebraic fraction*.

Because rational functions have denominators there are generally values that must be excluded from their domains, although it is possible to have a rational function, such as $f(x)=\frac{x+1}{x^2+1}$, that has a domain of all reals. Rational functions may have ranges that include all real numbers or ranges that exclude some values. Each function will need to be examined for domain and range.

Step 1. Operate with Rational Expressions

Before you begin graphing rational functions or looking for the zeros of rational functions, it's wise to review operations with rational expressions.

Simplify

To simplify a rational expression, factor both the numerator and denominator and cancel any factors they have in common.

When you're working with rational expressions, you should always be aware of the domain of the expressions. Even if the domain is not explicitly stated, values of the variable that would make the denominator equal to zero must be eliminated. In the example, the domain is all real numbers except 4 and −4.

$$\frac{x^2-4x}{x^2-16}=\frac{x\cancel{(x-4)}}{(x+4)\cancel{(x-4)}}=\frac{x}{x+4}$$

Multiply and Divide

The basic rule for multiplication of rational expressions mimics the rule for multiplying fractions: multiply the numerators and multiply the denominators, and then simplify. As with fractions, however, you can often get some of the simplifying done in advance, so factor all numerators and all denominators before multiplying. Cancel any factor that appears in both a numerator and a denominator.

$$\frac{x^2-x+12}{x^2+5x+4}\cdot\frac{x^2+2x+1}{x^2-9}=\frac{(x-4)\cancel{(x+3)}}{(x+4)\cancel{(x+1)}}\cdot\frac{\cancel{(x+1)}(x+1)}{\cancel{(x+3)}(x-3)}$$

$$=\frac{x^2-3x-4}{x^2+x-12}$$

To divide rational expressions, first invert the divisor, and then multiply.

$$\frac{x^2+2x+1}{x^2-16}\div\frac{x^2+5x+4}{x^2-7x+12}=\frac{x^2+2x+1}{x^2-16}\cdot\frac{x^2-7x+12}{x^2+5x+4}$$

$$=\frac{(x+1)\cancel{(x+1)}}{(x+4)\cancel{(x-4)}}\cdot\frac{\cancel{(x-4)}(x-3)}{\cancel{(x+1)}(x+4)}$$

$$=\frac{x^2-2x-3}{x^2+8x+16}$$

Add and Subtract

In order to add or subtract rational expressions, you must have a common denominator. To convert rational expressions to a common denominator, first factor each denominator. Your common denominator should be the lowest-degree polynomial that includes all the factors in the denominators. The lowest common denominator for $\frac{x+1}{x^2-16}$ and $\frac{x+4}{x^2-7x+12}$ is much easier to see when the denominators are factored. For $\frac{x+1}{(x+4)(x-4)}$ and $\frac{x+4}{(x-4)(x-3)}$, the common denominator is $(x+4)(x-4)(x-3)$. Don't bother multiplying that out.

Convert each rational expression to that common denominator by multiplying numerator and denominator by any factors of the common denominator that are missing from the denominator.

$$\frac{x+1}{(x+4)(x-4)} = \frac{x+1}{(x+4)(x-4)} \cdot \frac{x-3}{x-3} = \frac{x^2-2x-3}{(x+4)(x-4)(x-3)}$$

$$\frac{x+4}{(x-4)(x-3)} = \frac{x+4}{(x-4)(x-3)} \cdot \frac{x+4}{x+4} = \frac{x^2+8x+16}{(x+4)(x-4)(x-3)}$$

Add or subtract the numerators, paying special attention to signs when subtracting. If possible, factor and simplify. To add $\frac{x+1}{x^2-16}$ and $\frac{x+4}{x^2-7x+12}$, change to a common denominator and combine like terms in the numerators.

$$\frac{x+1}{x^2-16} + \frac{x+4}{x^2-7x+12} = \frac{x^2-2x-3}{(x+4)(x-4)(x-3)} + \frac{x^2+8x+16}{(x+4)(x-4)(x-3)}$$

$$= \frac{2x^2+6x+13}{(x+4)(x-4)(x-3)}$$

To subtract $\frac{x+1}{x^2-16} - \frac{x+4}{x^2-7x+12}$, change to a common denominator, then remember that the fraction bar acts as a grouping symbol, and distribute the minus before combining like terms.

$$\frac{x+1}{x^2-16}-\frac{x+4}{x^2-7x+12}=\frac{x^2-2x-3}{(x+4)(x-4)(x-3)}-\frac{x^2+8x+16}{(x+4)(x-4)(x-3)}$$
$$=\frac{(x^2-2x-3)-(x^2+8x+16)}{(x+4)(x-4)(x-3)}$$
$$=\frac{x^2-2x-3-x^2-8x-16}{(x+4)(x-4)(x-3)}$$
$$=\frac{-10x-19}{(x+4)(x-4)(x-3)}$$

The fraction bar is a "vinculum" and acts as a grouping symbol. It has the effect of putting parentheses around the numerator and around the denominator.

Solve Rational Equations

Solving equations involving rational expressions becomes a question of solving a linear, quadratic, or polynomial equation once you've done some simplifying. To get to that point, there are two tactics you can use.

Cross Multiply. If your rational equation is two equal rational expressions, or can be easily put in that form, you can treat it as if it were a proportion and solve by cross multiplying. To solve $\frac{4x-3}{x+5}+\frac{5x}{x+1}=0$, move one expression to the other side of the equal sign, and then cross multiply.

$$\frac{4x-3}{x+5}+\frac{5x}{x+1}=0$$
$$\frac{4x-3}{x+5}=\frac{-5x}{x+1}$$
$$(4x-3)(x+1)=-5x(x+5)$$
$$4x^2+x-3=-5x^2-25x$$

The result is a quadratic equation, which can be solved by factoring.

$$4x^2+x-3=-5x^2-25x$$
$$9x^2+26x-3=0$$
$$(9x-1)(x+3)=0$$

$$\begin{array}{ll} 9x-1=0 & x+3=0 \\ 9x=1 & x=-3 \\ x=\frac{1}{9} & \end{array}$$

Clear Denominators. Often the simpler method for solving rational equations is to multiply the entire equation by the common denominator of all the expressions in the equation. This multiplication will clear the denominators. To solve $\frac{3x-1}{x+2}-\frac{2x+3}{x-1}=\frac{x-25}{x^2+x-2}$ or $\frac{3x-1}{x+2}-\frac{2x+3}{x-1}=\frac{x-25}{(x+2)(x-1)}$, multiply through the equation by the common denominator of $(x+2)(x-1)$, and solve the resulting quadratic equation.

$$(x+2)(x-1)\left[\frac{3x-1}{x+2}-\frac{2x+3}{x-1}\right]=\left[\frac{x-25}{x^2+x-2}\right](x+2)(x-1)$$
$$(3x-1)(x-1)-(2x+3)(x+2)=x-25$$
$$3x^2-4x+1-(2x^2+7x+6)=x-25$$
$$3x^2-4x+1-2x^2-7x-6=x-25$$
$$x^2-11x-5=x-25$$
$$x^2-12x+20=0$$
$$(x-10)(x-2)=0$$
$$x=10 \qquad x=2$$

You could clear one denominator at a time by multiplying through by each one, but using the least common denominator gets the job done faster.

Exercise 5.1

Complete the following exercises.

1. Simplify: $\frac{x-2}{x^2-2x-3}\cdot\frac{x^2-6x+9}{x^2-4}$

2. Simplify: $\frac{x^2-2x-3}{x+2}\div\frac{x^2-9}{x^2+5x+6}$

3. Simplify: $\frac{3x}{x+5}+\frac{1}{x-2}$

4. Simplify: $\frac{x+3}{x}-\frac{2}{x+3}$

5. Simplify: $\frac{1+\frac{1}{x+3}}{3-\frac{1}{x-1}}$

6. Solve: $\frac{3x}{x-2}+\frac{5}{x+5}=\frac{7x}{x^2+3x-10}$

7. Solve: $\frac{3}{x+2}-\frac{39}{x^2+2x}=\frac{-2}{x}$

8. Solve: $\frac{4x}{x+4}+\frac{3}{x-1}=\frac{15}{x^2+3x-4}$

9. Solve: $\frac{2x}{4-x}=\frac{x^2}{x-4}$

10. Solve: $\frac{2}{x-10}-\frac{3}{x-2}=\frac{6}{x^2-12x+20}$

Step 2. Deal with Discontinuities

The essential feature of a rational function, the denominator, means that the function will often have discontinuities, or values that must be excluded from the domain because the function is undefined for that value. Your first consideration when examining a rational function is to identify the domain. It will be easier if you factor the denominator, but if that's not possible, set the denominator equal to 0 and solve.

> While rational functions that are defined for all real numbers do exist—the function $f(x)=\frac{x-3}{x^2+1}$ is an example—most rational functions will have restrictions on the domain, specifically, the values that make the denominator equal to 0.

The discontinuities you encounter in rational functions fall into two categories. Essential discontinuities are breaks at which vertical asymptotes occur. Removable discontinuities are holes. To identify discontinuities and distinguish between the two types, express the numerator and denominator of the rational expression in factored form. Each value that makes a factor of the denominator equal to 0, and therefore makes the denominator equal to 0, is a point of discontinuity, a value at which the function is undefined.

Removable Discontinuities

If a factor of the denominator can be canceled with a factor of the numerator, the discontinuity represented by that factor is a removable discontinuity or hole. The function $f(x)=\frac{x^2-x}{x^2-1}=\frac{x(x-1)}{(x+1)(x-1)}$ has two discontinuities, $x=1$ and $x=-1$. The factor $x-1$ occurs in both the numerator and denominator, so the discontinuity $x=1$ is a removable discontinuity. The graph of the function $f(x)=\frac{x^2-x}{x^2-1}$ is identical to the graph of $g(x)=\frac{x}{x+1}$ except that f is missing the point $\left(1,\frac{1}{2}\right)$ that occurs on the graph of g.

Essential Discontinuities

Factors of the denominator that don't cancel out cause essential discontinuities. For values of the independent variable near an essential discontinuity, the function either increases without bound or decreases without bound. The common expression is to say that the function "goes to infinity" or negative infinity.

Vertical asymptotes are vertical lines that the graph of the function approaches but does not touch or cross. They occur at essential discontinuities. The graph of the rational function $g(x)=\dfrac{x}{x+1}$ has a vertical asymptote because it has an essential discontinuity at $x=-1$. The equation of the vertical asymptote is $x=-1$. As x approaches -1 from below, the function increases without bound, and the graph turns upward, drawing closer and closer to the line $x=-1$. As x approaches -1 from above, however, the function decreases without bound, with the graph nearing the other side of the line $x=-1$.

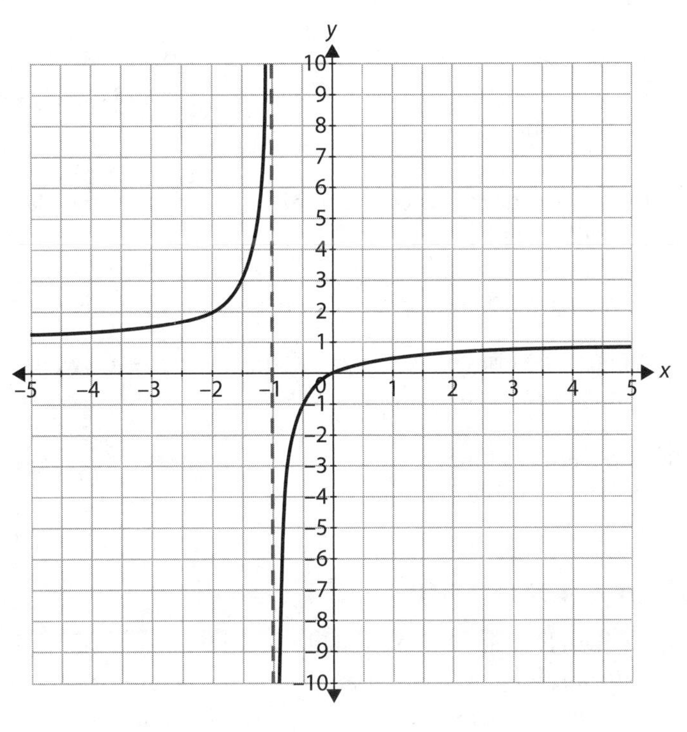

The vertical line $x=-1$ is the vertical asymptote for the graph of $g(x)=\dfrac{x}{x+1}$

Exercise 5.2

Complete the following exercises.

1. Find the domain of $f(x) = \dfrac{5}{x-3}$.

2. Find the domain of $f(x) = \dfrac{1}{x^2}$.

3. Find the domain of $f(x) = \dfrac{x+5}{x^2+9}$.

4. Find the vertical asymptote(s) of $f(x) = \dfrac{x-1}{x^2-x-30}$.

5. Find the vertical asymptote(s) of $f(x) = \dfrac{3x}{12x^2-27x}$.

6. Identify the discontinuities of $f(x) = \dfrac{2x^2+3x-2}{x^2-1}$, and classify each as essential or removable.

7. Identify the discontinuities of $f(x) = \dfrac{6}{x^2-9}$, and classify each as essential or removable.

8. Identify the discontinuities of $f(x) = \dfrac{x^2-4x+3}{x^2+x-12}$, and classify each as essential or removable.

9. The function $f(x) = \dfrac{4x-2}{6x^2+x-2}$ has a removable discontinuity. Find a function that is identical to $f(x) = \dfrac{4x-2}{6x^2+x-2}$ everywhere except at the removable discontinuity.

10. Find a function that is identical to $f(x) = \dfrac{x^2-16}{x^2+8x+16}$ everywhere except at the removable discontinuity.

Step 3. Examine End Behavior

For extremely large and extremely small values of the independent variable, rational functions will either approach a constant or will increase or decrease without bound. In many cases, the function approaches an asymptote, either horizontal or oblique.

To determine the end behavior of a rational function, consider the degree of the numerator and the degree of the denominator.

- If the degree of the numerator is less than or equal to the degree of the denominator, the graph will have a horizontal asymptote.
- If the degree of the numerator is greater than the degree of the denominator, the graph will sometimes have a slant (oblique) asymptote.

Horizontal Asymptotes

If the degree of the numerator is less than the degree of the denominator, the graph will have the x-axis as its horizontal asymptote. The equation of the horizontal asymptote is $y = 0$. Because the degree of the numerator is lower than the degree of the denominator, as x becomes extremely large, the denominator grows so much more quickly than the numerator that the quotient approaches 0. The function $f(x) = \frac{4x}{x^2 - 1}$ has a horizontal asymptote of $y = 0$ because the numerator is first degree, but the denominator is second degree.

If the degree of the numerator equals the degree of the denominator, highest-power terms in the numerator and the denominator quickly overwhelm any other terms as x becomes very large or decreases to very negative values. The horizontal asymptote can be found by looking at the ratio of the coefficients of the lead terms. The function $f(x) = \frac{4x - 3}{2x + 1}$ has a horizontal asymptote of $y = \frac{4}{2} = 2$, and the function $g(x) = \frac{5 - 8x + 3x^2}{2 - 9x - 5x^2}$ has a horizontal asymptote of $y = -\frac{3}{5}$.

Slant Asymptotes

If the degree of the numerator is one more than the degree of the denominator, the function will have a slant asymptote, also called an oblique asymptote. To find the slant asymptote, divide the numerator by the denominator, using long division if necessary. The equation of the slant asymptote is $y =$ the quotient of that division. Ignore the remainder.

To find the oblique asymptote for the function $f(x) = \frac{6x^3 + 9x^2 - 4x + 1}{3x^2 - 1}$, use long division.

$$
\begin{array}{r}
2x+3 \\
3x^2-1\overline{\big)\,6x^3+9x^2-4x+1} \\
\underline{6x^3 \qquad\quad -2x} \\
9x^2-2x+1 \\
\underline{+9x^2 \qquad\quad -3} \\
-2x+4
\end{array}
$$

The slant asymptote is the line $y = 2x + 3$.

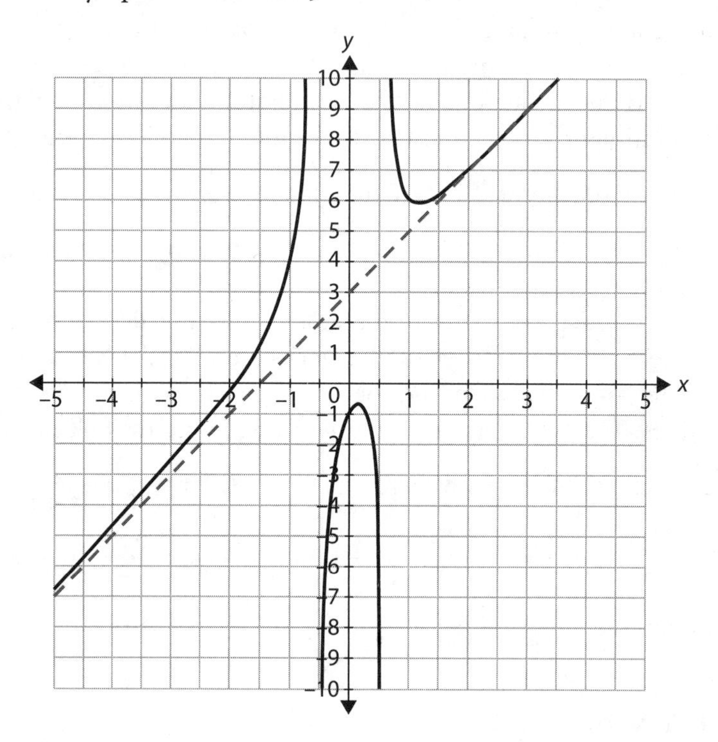

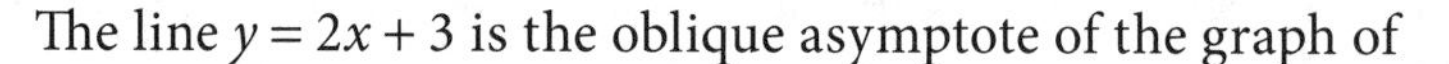

The line $y = 2x + 3$ is the oblique asymptote of the graph of

$$f(x) = \frac{6x^3 + 9x^2 - 4x + 1}{3x^2 - 1}$$

If the degree of the numerator is more than one degree larger than the degree of the denominator, there is neither a horizontal asymptote nor a slant asymptote. You can make a statement about the end behavior, however, based on the difference in the degrees of the numerator and denominator. If the difference is even, the end behavior will be similar to that of a polynomial function of even degree, with both ends going to infinity or to negative infinity. If the difference is odd, the ends behave as a polynomial of odd degree, with one end going to infinity and the other to negative infinity.

If the end behavior resembles that of a polynomial of even degree, look at the ratio of the lead coefficients. If the ratio is positive, both ends go to ∞. If it's negative, both ends go to $-\infty$.

If the end behavior of the function resembles that of a polynomial of odd degree, and the ratio of the lead coefficients is positive, the function goes up to the right and down to the left. If the ratio of the lead coefficient is negative, the function goes up to the left and down to the right.

Horizontal and slant asymptotes talk about the end behavior of the function. They are asymptotes for the ends. Simple rational functions do not cross their horizontal or slant asymptotes, but more complex functions may cross the horizontal or slant asymptotes for smaller values of x.

Exercise 5.3

Complete the following exercises.

1. Identify the horizontal asymptote of $f(x)=\frac{-4}{x+3}$.
2. Identify the horizontal asymptote of $f(x)=\frac{3}{2x-1}$.
3. Identify the horizontal asymptote of $f(x)=\frac{4x+1}{5-2x}$.
4. Find the slant asymptote of $f(x)=\frac{x^2-8x+1}{x-2}$.
5. Find the slant asymptote of $f(x)=\frac{x^2-9}{x}$.
6. Find the slant asymptote of $f(x)=\frac{2x^3-3x^2-x-1}{x^2+2}$.
7. Describe the end behavior of $f(x)=\frac{x^3}{x-1}$.
8. Describe the end behavior of $f(x)=\frac{8-x^3}{2x+1}$.

9. Describe the end behavior of $f(x)=\dfrac{x^5-1}{2-3x^2}$.

10. Describe the end behavior of $f(x)=\dfrac{5x^3}{3x^5}$.

Step 4. Find Intercepts

Finding the y-intercept of a rational function is simply a matter of substituting 0 for x and simplifying, but finding any x-intercepts may require a bit of algebra. Technically, you're looking for the values of x that make the rational expression equal to 0, but you can break that down to the values of x that make the numerator 0 but don't make the denominator 0. You already identified the values that make the denominator 0 when you found the domain and the discontinuities, so you know what values to avoid. You simply need to set the numerator equal to 0 and solve.

Don't be surprised if you find that you have no x-intercepts. Remember that you may have a horizontal asymptote of $y=0$.

For the function $f(x)=\dfrac{x^2-7x-30}{x-3}$, the y-intercept is $\dfrac{-30}{-3}=10$, or (0,10), and there is a vertical asymptote at $x=3$. The x-intercepts are the values that make $x^2-7x-30=0$, so solve

$$\begin{aligned} x^2-7x-30&=0\\ (x-10)(x+3)&=0\\ x=10\quad x&=-3 \end{aligned}$$

The x-intercepts are (10,0) and (−3,0).

Exercise 5.4

Complete the following exercises.

1. Find the x- and y-intercepts of $f(x)=\dfrac{2}{4-x}$.

2. Find the x- and y-intercepts of $f(x)=\dfrac{3x-5}{x+2}$.

3. Find the x- and y-intercepts of $f(x)=\dfrac{x^2-4x-21}{x+7}$.

4. Find the x- and y-intercepts of $f(x)=\dfrac{2x-1}{x}$.

5. Find the x- and y-intercepts of $f(x)=\dfrac{x^2-x}{x+1}$.

Step 5. Graph Rational Functions

If you put together the results of your analysis of the rational function, sketching the graph shouldn't be difficult.

Draw Asymptotes

Begin by sketching the vertical asymptotes and horizontal or slant asymptotes. The parent graph for the family of rational function, the graph of $y=\dfrac{1}{x}$, has a vertical asymptote of $x=0$ and a horizontal asymptote of $y=0$.

Graphs with horizontal and vertical asymptotes in other places are translations of the parent graph. If you divide the numerator by the denominator and express the remainder as a fraction of the divisor, you can see these translations in the equation. Once you have placed the asymptotes you can expect the pieces of the graph to fit into sections of the plane created by the asymptotes.

The rational function $f(x)=\dfrac{2x-1}{x-3}=2+\dfrac{5}{x-3}$ has a vertical asymptote at $x=3$ and a horizontal asymptote of $y=2$.

Plot Intercepts

By plotting the x- and y-intercepts, you locate the graph in the plane and position it in the sections created by the asymptotes. If the function does not have x-intercepts or a y-intercept, you may choose to evaluate the function for several values of x.

The function $f(x)=\dfrac{2x-1}{x-3}=2+\dfrac{5}{x-3}$ has an x-intercept of $\left(\dfrac{1}{2},0\right)$ and a y-intercept of $\left(0,\dfrac{1}{3}\right)$.

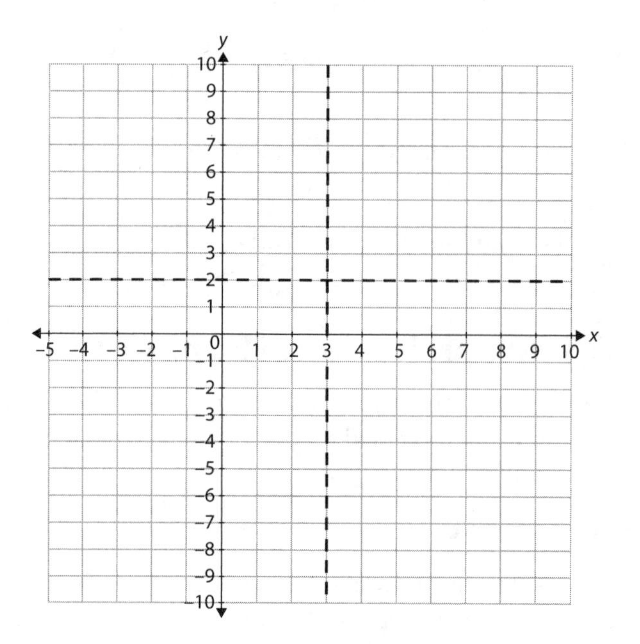

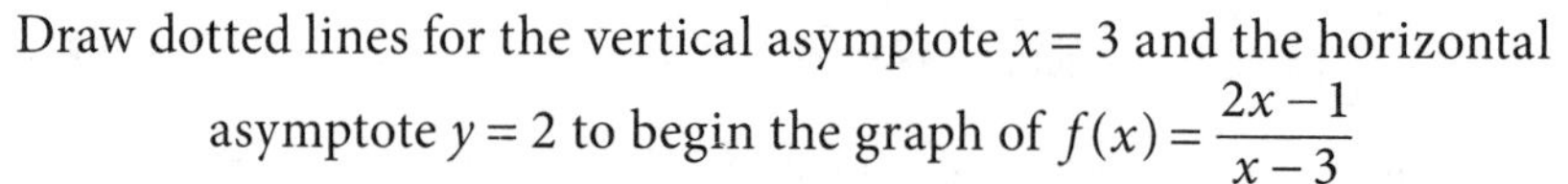

Draw dotted lines for the vertical asymptote $x = 3$ and the horizontal asymptote $y = 2$ to begin the graph of $f(x) = \dfrac{2x-1}{x-3}$

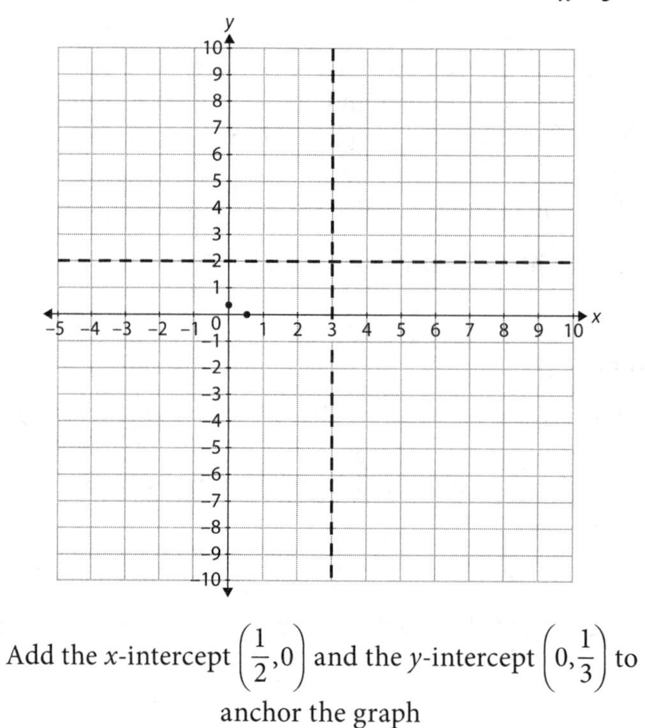

Add the x-intercept $\left(\dfrac{1}{2}, 0\right)$ and the y-intercept $\left(0, \dfrac{1}{3}\right)$ to anchor the graph

Know the Shape

Simple rational functions are hyperbolas. The parent function $y = \frac{1}{x}$ is a hyperbola with one wing in the first quadrant and one in the third quadrant, and the graph of $y = \frac{1}{x^2}$ has wings in the first and second quadrants. A simple rational function will have a graph that is a hyperbola with its wings in two of the "quadrants" created by its asymptotes. If the denominator is an odd degree, the two pieces will be placed diagonally from one another. If the degree of the denominator is even, they'll be side by side.

Check for Reflections and Stretches

Multiplying the numerator by −1 causes a reflection over the x-axis. Multiplying it by a constant greater than 1 stretches the graph vertically, while multiplying by a constant between 0 and 1 compresses it.

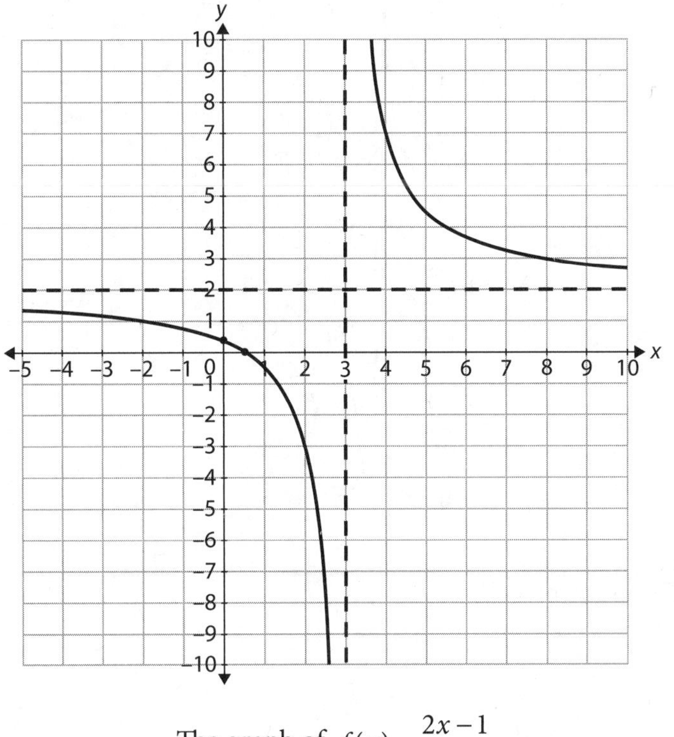

The graph of $f(x) = \frac{2x-1}{x-3}$

Deal with Variations

More complicated rational functions, such as those with more than one vertical asymptote, tend to behave as $y=\frac{1}{x}$ or $y=\frac{1}{x^2}$ on the ends, but there is a variety of behaviors possible between the vertical asymptotes, so you may want to plot a few points.

The graph of $g(x)=\frac{3x^2+6x+3}{x^2-4x+3}=\frac{3(x+1)^2}{(x-1)(x-3)}$ has vertical asymptotes at $x=1$ and $x=3$ and a horizontal asymptote of $y=3$. The y-intercept is (0,1), and the x-intercept is (–1,0). Because the denominator is second degree, you can expect the two outermost sections to both be above the vertical asymptote, but you'll need to hunt a bit for that middle piece.

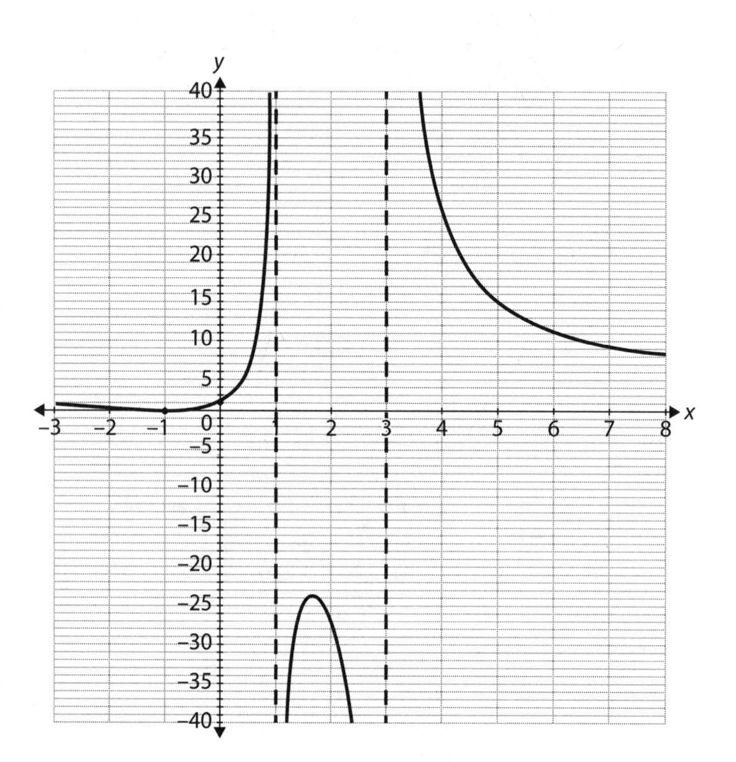

The graph of $g(x)=\frac{3x^2+6x+3}{x^2-4x+3}=\frac{3(x+1)^2}{(x-1)(x-3)}$ is in three sections

Exercise 5.5

Identify the asymptotes and intercepts of the graph of each function, and sketch a graph.

1. $f(x)=\dfrac{3}{x-1}$

2. $f(x)=\dfrac{-1}{x+5}$

3. $f(x)=\dfrac{x+3}{x-2}$

4. $f(x)=\dfrac{x}{x+3}$

5. $f(x)=\dfrac{5x-1}{x+1}$

6. $f(x)=\dfrac{4}{x^2-9}$

7. $f(x)=\dfrac{-3}{x^2+3x+2}$

8. $f(x)=\dfrac{x}{x^2+4x+4}$

9. $f(x)=\dfrac{3x+4}{x^2-5x+6}$

10. $f(x)=\dfrac{x^2-4}{2x-1}$

6

Conic Sections

The conic sections are so named because each is a geometric figure that can be created by slicing through a cone. The difference among them is the direction of the slice. The circle is created by a slice parallel to the base of the cone, the ellipse and the parabola by slices on various angles, and the hyperbola by a cut perpendicular to the base.

Unlike most of the equations you study in pre-calculus, not all of these are functions. They are relations, but the key requirement in the definition of a function, that each element of the domain be paired with only one element of the range, is not met by most of the conics. There are parabolas that are functions but other parabolas that are not, and circles, ellipses, and hyperbolas are not functions.

> To graph relations on a graphing calculator, you'll need to first solve the equation for y as a function of x. That will involve taking the square root of both sides, and that square root will be preceded by a ± sign. That means you'll actually have to enter two equations into your calculator to produce one graph. One equation will use the positive square root and produce the top half of the graph. The other will use the negative square root and produce the bottom half of the graph.

Step 1. Analyze and Graph Parabolas

When you studied quadratic functions, you looked at parabolas with equations of the form $f(x) = a(x - h)^2 + k$. These parabolas have vertex at (h,k) and open up or down, depending on the sign of a; they are functions. Other parabolas, those of the form $x - h = a(y - k)^2$, are not functions. These have vertex (h,k) but open to the right or to the left and fail the vertical line test. The equation $x - 3 = -2(y - 5)^2$ is a parabola with vertex $(3,5)$ that opens to the left. The equation $y - 4 = 3(x - 1)^2$ is a parabola with vertex (1,4) that opens up.

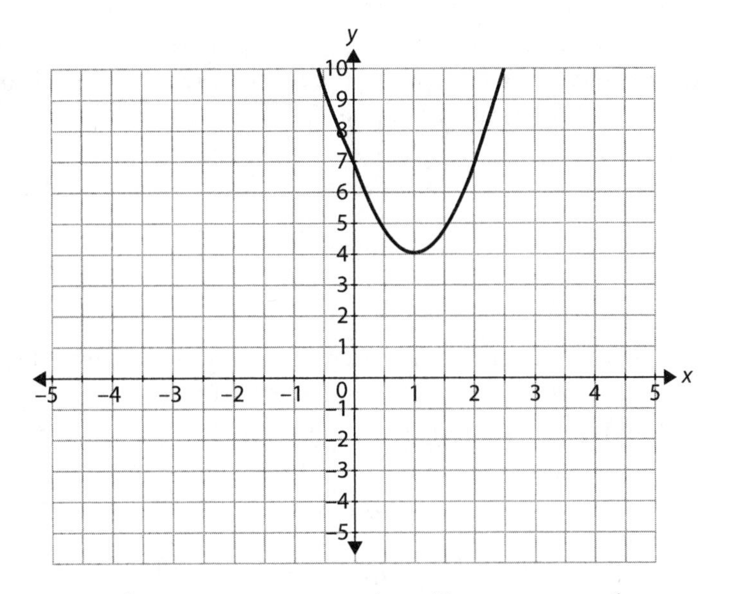

The equation $y - 4 = 3(x - 1)^2$ opens upward

A parabola is the set of all points equidistant from a fixed point, called the focus, and a fixed line, called the directrix. The parabola curves around the focus, turning away from the directrix. The distance between the vertex and the focus, called the focal length, is $\frac{1}{4a}$. The directrix is located an equal distance from the vertex, but on the other side. The parabola $x - 3 = -2(y - 5)^2$ has a focal length of $-\frac{1}{8}$, so, from the vertex of (3,5), the focus will be $\frac{1}{8}$ of a unit to the left, at $\left(3-\frac{1}{8}, 5\right)=\left(2\frac{7}{8}, 5\right)$, and the directrix is the line $x=3\frac{1}{8}$. For the parabola $y - 4 = 3(x - 1)^2$, the focal length is $\frac{1}{12}$, so the focus is $\frac{1}{12}$ of a unit above the vertex, at $\left(1, 4\frac{1}{12}\right)$, and the directrix is the line $y=3\frac{11}{12}$.

If you know the focal length, the vertex, and the direction of opening, you can write the equation of the parabola. Choose the form by the direction of opening, replace h and k with the coordinates of the vertex, and replace a with 1 divided by four times the focal length.

If the equation is not in $x - h = a(y - k)^2$ or $y - k = a(x - h)^2$ form, you can use the technique of completing the square to transform the equation. To put the equation $2y^2 + 16y - x + 35 = 0$ in vertex form, first move the variables to opposite sides of the equation. Move the constant term to the side of the first power variable, in this case, x. Factor out the coefficient of the squared term.

$$2y^2+16y-x+35=0$$
$$2y^2+16y=x-35$$
$$2(y^2+8y)=x-35$$

Complete the square in the parentheses, but remember, when you add to the other side, that the multiplier in front of the parentheses will also affect the constant you add to complete the square.

$$2(y^2+8y)=x-35$$
$$2(y^2+8y+16)=x-35+2\cdot 16$$
$$2(y+4)^2=x-3$$

This parabola has its vertex at (3,–4), opens to the right, and has a focal length of $\frac{1}{4\cdot 2}=\frac{1}{8}$. That puts its focus at $\left(3\frac{1}{8},-4\right)$, and the directrix is the vertical line $x=2\frac{7}{8}$.

The eccentricity of any conic is a number that characterizes the shape and is defined as the ratio of two distances: from a point to the focus and from the point to the directrix. In a parabola, those distances are equal, so every parabola has an eccentricity of 1.

Exercise 6.1

Complete the following exercises.

1. Find the vertex, focus, and directrix of $y + 4 = 8(x - 2)^2$.

2. Find the vertex, focus, and directrix of $x-3=\frac{1}{2}(y+2)^2$.

3. Put the equation in vertex form: $x^2 + 8x + 2y + 3 = 0$.
4. Put the equation in vertex form: $y^2 - 12y + x + 2 = 0$.
5. Sketch the graph of $x^2 - 6x - 4y = 0$.
6. Sketch the graph of $y^2 - 2y + 12x - 3 = 0$.
7. Sketch the graph of $y^2 + 6y + 8x + 25 = 0$.
8. Sketch the graph of $x^2 - 2x + 4y + 9 = 0$.
9. Write the equation of a parabola with focus (6,–3) and directrix $x = 2$.
10. Write the equation of a parabola with focus (–7,1) and directrix $y = 5$.

Step 2. Analyze and Graph Ellipses

While a circle is the set of points at a fixed distance from a single point, an ellipse is a set of points for which the sum of the distances from two focal points is constant. Rather than a single radius, the ellipse has a major axis and a minor axis. The ellipse with center (h,k), major axis of length $2a$, and minor axis of length $2b$ has an equation of one of these forms:

$$\frac{(x-h)^2}{a^2}+\frac{(y-k)^2}{b^2}=1$$

$$\frac{(x-h)^2}{b^2}+\frac{(y-k)^2}{a^2}=1$$

The equation $\frac{(x-4)^2}{25}+\frac{(y+3)^2}{4}=1$ is an ellipse centered at (4,–3). Its major axis is horizontal and 10 units long. The vertices of the ellipse are (4 – 5,–3) = (–1,–3) and (4 + 5,–3) = (9,–3). The minor axis is vertical and 4 units long, with the co-vertices at (4,–3 – 2) = (4,–5) and (4,–3 + 2) = (4,–1).

If you change the equation to $\frac{(x-4)^2}{4}+\frac{(y+3)^2}{25}=1$, the center remains at (4,–3), but the major axis is now vertical, with a length of 10, and the minor axis horizontal with a length of 4. The vertices are (4,–3 – 5) = (4,–8) and (4,–3 + 5) = (4,2), and the co-vertices are (4 – 2,–3) = (2,–3) and (4 + 2,–3) = (6,–3).

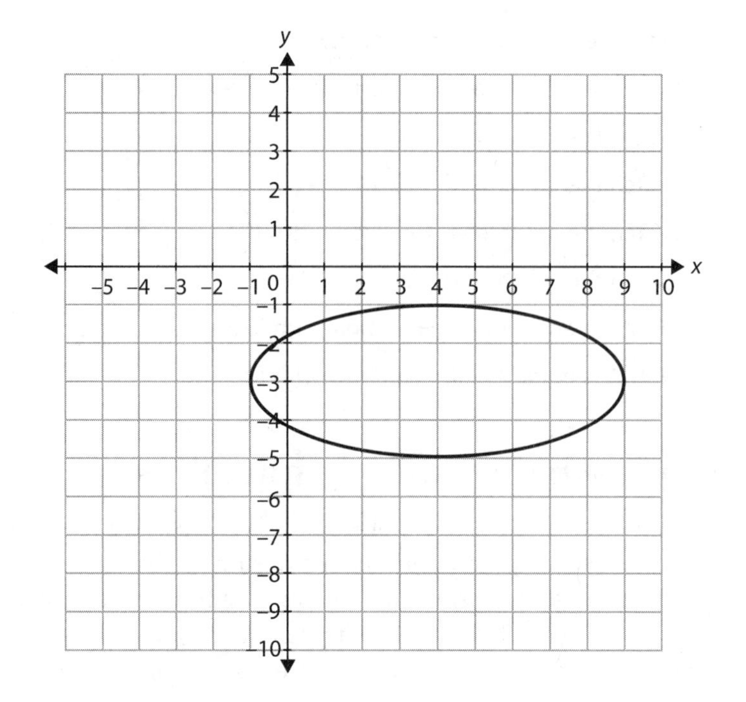

The graph of $\frac{(x-4)^2}{25}+\frac{(y+3)^2}{4}=1$ is an ellipse centered at (4,–3)

To find the foci of the ellipse, first calculate the focal length, c, using the formula $c^2 = a^2 - b^2$. For the equation $\frac{(x-4)^2}{25}+\frac{(y+3)^2}{4}=1$, $c^2 = 25 - 4 = 21$, so $c=\pm\sqrt{21}$. The foci are always located on the major axis, equidistant from the center. So the foci of $\frac{(x-4)^2}{25}+\frac{(y+3)^2}{4}=1$ are located at $(4\pm\sqrt{21},-3)$. The ellipse $\frac{(x-4)^2}{4}+\frac{(y+3)^2}{25}=1$ has foci at $(4,-3\pm\sqrt{21})$.

The eccentricity of an ellipse is $e=\frac{c}{a}$. Because $c=\sqrt{a^2-b^2}<a$, the eccentricity of an ellipse will always be a value between 0 and 1. The closer the lengths of the major and minor axes are, the smaller the eccentricity of the ellipse. As the difference between the lengths of the axes increases, so does the eccentricity. The eccentricity of an ellipse can never reach 1, however, because the ratio $\frac{c}{a}$ will never be 1.

Exercise 6.2

Complete the following exercises.

1. Write the equation of an ellipse with center (2,1), vertex (7,1), and focus (5,1).
2. Write the equation of an ellipse with center (–3,2), co-vertex (–8,2), and focus (–3,–10).
3. Write the equation of an ellipse with vertices (6,–7) and (2,–7) and co-vertices (4,–6) and (4,–8).
4. Find the center, vertices, co-vertices, and foci of the ellipse $\frac{(x-2)^2}{16}+\frac{(y+1)^2}{9}=1$.
5. Find the center, vertices, co-vertices, and foci of the ellipse $16x^2 + 25y^2 - 50y = 375$.
6. Find the center, vertices, co-vertices, and foci of the ellipse $25x^2 + 144y^2 = 3600$.
7. Sketch a graph of $(x - 3)^2 + 9(y - 2)^2 = 9$.
8. Sketch a graph of $9x^2 + 25y^2 - 36x - 50y - 164 = 0$.
9. Sketch a graph of $9x^2 + 4y^2 - 36x + 8y + 31 = 0$.
10. Sketch a graph of $12x^2 + 9y^2 + 48x - 36y + 3 = 0$.

Step 3. Analyze and Graph Circles

Circles can be viewed as a special case of the ellipse but generally are treated as a category of their own. The circle is the set of all points at a fixed distance, called the radius, from a fixed point, called the center. If the center is the point (h,k) and the radius is r, the equation of the circle is $(x - h)^2 + (y - k)^2 = r^2$. The circle with center at $(-4,2)$ and radius 3 has the equation $(x + 4)^2 + (y - 2)^2 = 9$, or

$$(x+4)^2+(y-2)^2=9$$
$$x^2+8x+16+y^2-4y+4=9$$
$$x^2+y^2+8x-4y+11=0$$

If you're presented with an equation in this second form, you can complete the square to transform the equation to a form that shows the center and radius. To convert $x^2 + y^2 - 6x + 10y + 18 = 0$ to standard form, move the constant to the other side, and group the x-terms and the y-terms. Complete the square for the x-terms and then for the y-terms.

$$x^2+y^2-6x+10y+18=0$$
$$x^2-6x+y^2+10y=-18$$
$$x^2-6x+9+y^2+10y+25=-18+9+25$$
$$(x-3)^2+(y+5)^2=16$$

The center of the circle is (3,–5) and the radius is 4.

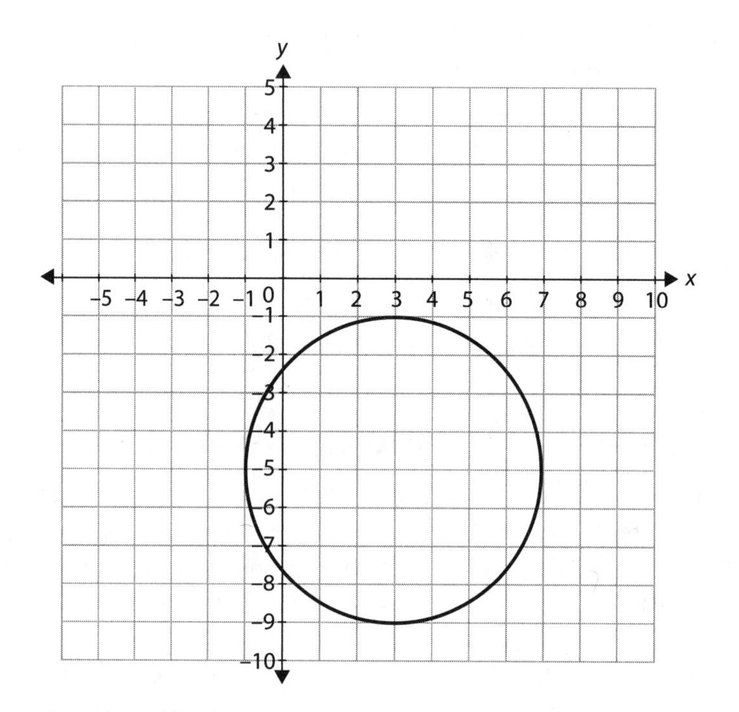

The graph of $(x + 4)^2 + (y - 2)^2 = 9$ is a circle of radius 4 centered at (3,–5)

For ellipses, the eccentricity is defined as $e=\frac{c}{a}$. You can think of a circle as a special case of the ellipse, $c=\sqrt{a^2-b^2}$, but both a and b are the radius, so $c = 0$. As a result, every circle has an eccentricity of 0.

Exercise 6.3

Complete the following exercises.

1. Write the equation in standard form: $x^2 + y^2 - 4x + 2y - 20 = 0$.
2. Write the equation in standard form: $x^2 + y^2 + 8x - 8y = 112$.
3. Write the equation in standard form: $16x^2 + 16y^2 - 32x - 8y - 15 = 0$.
4. Sketch the graph of $x^2 + (y - 3)^2 = 4$.
5. Sketch the graph of $(x - 5)^2 + (y + 2)^2 = 16$.
6. Sketch the graph of $x^2 + y^2 - 8x - 9 = 0$.
7. Sketch the graph of $x^2 + y^2 - 12x + 4y = 81$.
8. Write the equation of a circle with center (3,5) and radius 7.
9. Write the equation of a circle with center (–3,1) and radius 13.
10. Write the equation of a circle with center (0,–4) and radius 8.

Step 4. Analyze and Graph Hyperbolas

The hyperbola, like the ellipse, has two foci, and the shape of the graph is determined by the distances between a point and each of the foci. But while those distances add to a constant for an ellipse, for the hyperbola they subtract to a constant. The difference of the distances from a point on the hyperbola to the two foci is constant.

To give the form of the equation of a hyperbola, you need to discuss several possible variations. The center of the hyperbola is some point (h,k). The vertices of the hyperbola lie on the transverse axis, which can be vertical or horizontal. If the transverse axis is horizontal, the hyperbola opens to the left and the right. If the transverse axis is vertical, the hyperbola opens up and down. If the length of the transverse axis is $2a$, the equation of the hyperbola takes one of two forms. The first version is a hyperbola that opens left and right, and the second opens up and down.

$$\frac{(x-h)^2}{a^2} - \frac{(y-k)^2}{b^2} = 1$$

$$\frac{(y-k)^2}{a^2} - \frac{(x-h)^2}{b^2} = 1$$

The equation $\frac{(x-1)^2}{16}-\frac{(y-4)^2}{9}=1$ produces a hyperbola that opens left and right. Its center is (1,4), and its vertices are at (–3,4) and (5,4).

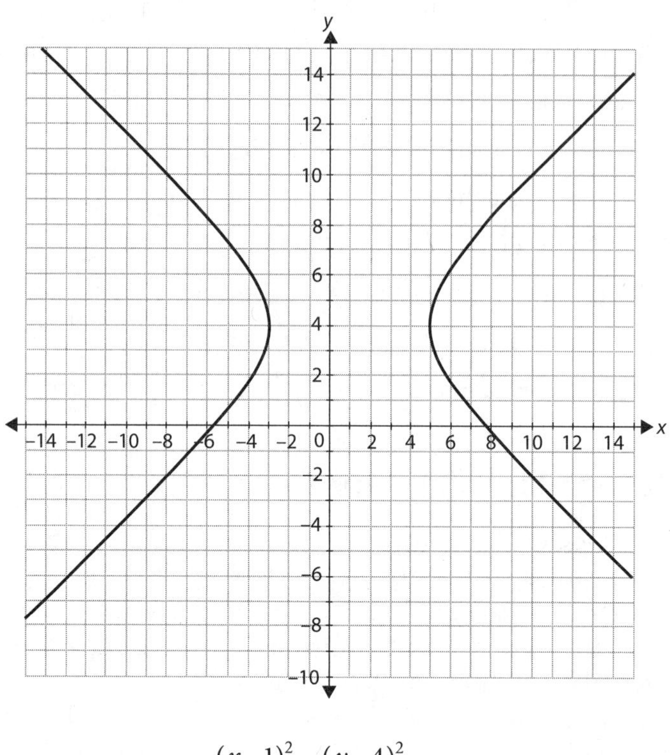

The graph of $\frac{(x-1)^2}{16}-\frac{(y-4)^2}{9}=1$ is a hyperbola

The equation $\frac{(y+2)^2}{4}-\frac{(x+3)^2}{25}=1$ produces a hyperbola (see figure on page 97) that opens up and down, with center at (–3,–2). Its vertices are on the vertical axis, at (–3,–4) and (–3,0).

If the equation of the hyperbola is not in standard form, complete the square to transform it, but be aware of the negative sign that will need to be factored out before you complete the square and will affect the constant added to the other side. To transform $16x^2 - 4y^2 - 160x - 24y + 300 = 0$,

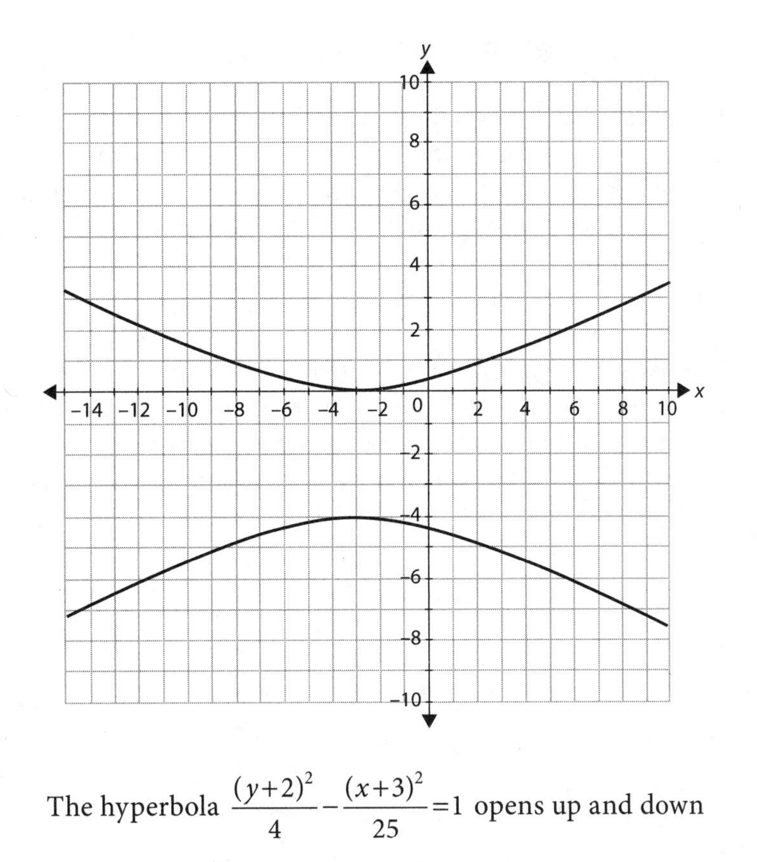

The hyperbola $\frac{(y+2)^2}{4}-\frac{(x+3)^2}{25}=1$ opens up and down

group the x-terms and the y-terms, and move the constant to the other side. Factor 16 out of the x-terms and -4 out of the y-terms.

$$16x^2-4y^2-160x-24y+300=0$$
$$16x^2-160x-4y^2-24y=-300$$
$$16(x^2-10x)-4(y^2+6y)=-300$$

Complete the square in each parenthesis, and remember that the factors of 16 and -4 multiply the constants to be added to the other side.

$$16(x^2-10x+25)-4(y^2+6y+9)=-300+16\cdot 25+-4\cdot 9$$
$$16(x-5)^2-4(y+3)^2=64$$
$$\frac{(x-5)^2}{4}-\frac{(y+3)^2}{16}=1$$

To sketch the graph of a hyperbola, begin as you would with the ellipse. Locate the center, and from the center, count out the axes. In the hyperbola, the transverse axis, which contains the vertices, will not always be the longer axis. It may be shorter, longer, or equal to the other axis. The square root of the denominator in the $(x - h)^2$ term is half of the horizontal axis, and the square root of the denominator of the $(y - k)^2$ term is half of the vertical axis. Instead of sketching an ellipse, lightly sketch a rectangle with its sides passing through the vertices and co-vertices.

The hyperbola has asymptotes that contain the two curves and that happen to contain the diagonals of the rectangle you just sketched. So draw the diagonals and extend them. The equations of these asymptotes can be found because both pass through the center, and their slopes are the ratio of the vertical to the horizontal axis and its opposite. So the equations of the asymptotes would have the form

$$y-k=\pm\frac{b}{a}(x-h).$$

The two curves of the hyperbola will just touch the rectangle at the vertices on the transverse axis and will curve out toward the asymptotes.

The graph of $\frac{(x-1)^2}{16}-\frac{(y-4)^2}{9}=1$ begins with a rectangle that passes through the points (–3,4), (5,4), (1,1), and (1,7). It has asymptotes $y-4=\pm\frac{3}{4}(x-1)$ and passes through the vertices (–3,4) and (5,4), curving out toward the asymptotes, opening left and right.

The graph of $\frac{(y+2)^2}{4}-\frac{(x+3)^2}{25}=1$ begins with a rectangle that passes through the points (–3,–4), (–3,0), (–8,–2), and (2,–2). The asymptotes of the hyperbola are $y+2=\pm\frac{2}{5}(x+3)$. The hyperbola sits above and below the rectangle, just touching it at the vertices (–3,–4) and (–3,0) and curving toward the asymptotes.

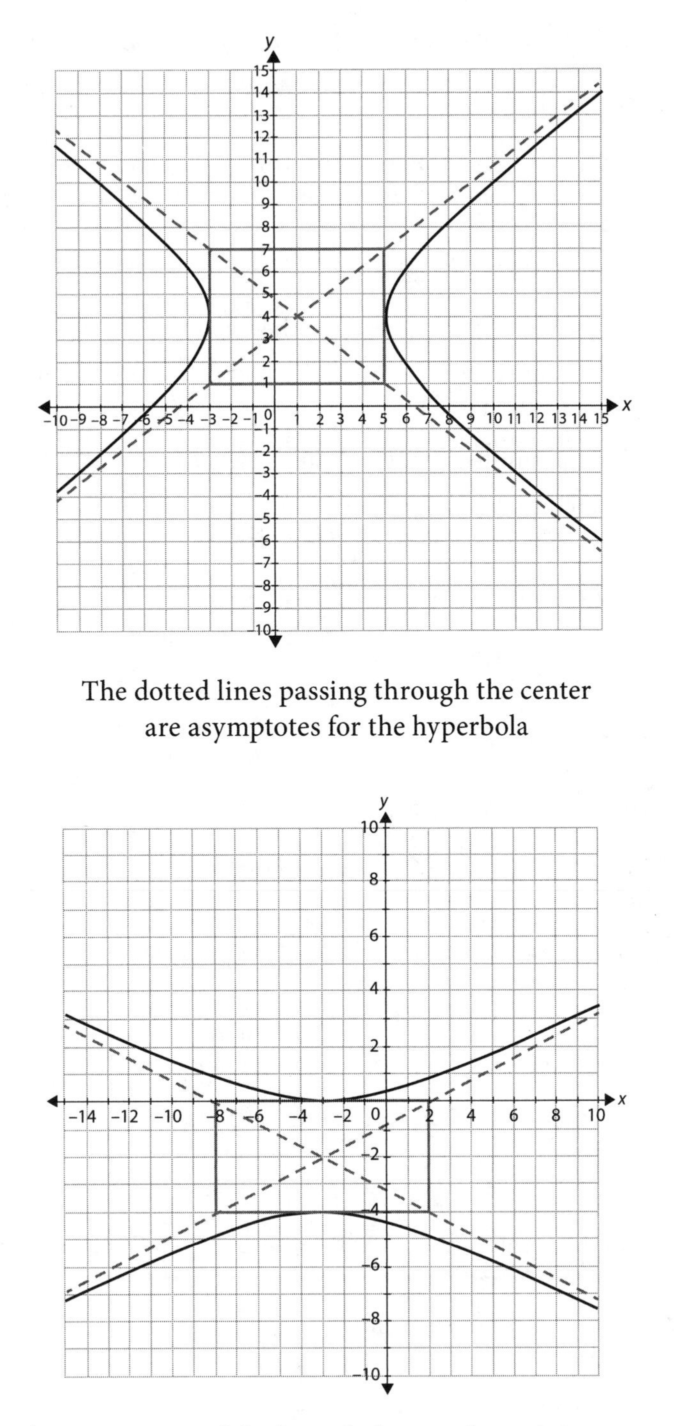

The dotted lines passing through the center are asymptotes for the hyperbola

The asymptotes of the hyperbola pass through its center

To find the focal length, c, for the hyperbola, use the formula $c^2 = a^2 + b^2$. For the hyperbola $\frac{(x-1)^2}{16}-\frac{(y-4)^2}{9}=1$, the focal length $c=\pm\sqrt{16+9}=\pm 5$. The foci of the hyperbola lie on the transverse axis, in this case, the horizontal axis. The foci are the points (–4,4) and (6,4). For the hyperbola $\frac{(y+2)^2}{4}-\frac{(x+3)^2}{25}=1$, the foci will lie on the vertical axis. The focal length will be $c=\pm\sqrt{4+25}=\pm\sqrt{29}$, and the foci will be $(-3,-2\pm\sqrt{29})$.

The eccentricity of the hyperbola is $e=\frac{c}{a}$, where c is the focal length and a is half the length of the transverse axis. The eccentricity of a hyperbola will always be a number greater than 1, because $c^2 = a^2 + b^2$. The eccentricity of $\frac{(x-1)^2}{16}-\frac{(y-4)^2}{9}=1$ is $e=\frac{5}{4}$, and the eccentricity of $\frac{(y+2)^2}{4}-\frac{(x+3)^2}{25}=1$ is $e=\frac{\sqrt{29}}{2}$.

Exercise 6.4

Complete the following exercises.

1. Write the equation of a hyperbola with center (5,1), vertex (3,1), and focus (2,1).
2. Write the equation of a hyperbola with center (–3,0), vertex (–3,2), and focus (–3,5).
3. Write the equation of a hyperbola with center (2,–7), vertex (2,–6), and focus (2,0).
4. Find the center, foci, and asymptotes of the hyperbola $\frac{(y+1)^2}{9}-\frac{(x-2)^2}{16}=1$.
5. Find the center, foci, and asymptotes of the hyperbola $16x^2 - 25y^2 - 50y = 425$.
6. Find the center, foci, and asymptotes of the hyperbola $144x^2 - 25y^2 = 3600$.
7. Sketch a graph of $(y-2)^2 - 9(x-3)^2 = 9$.

8. Sketch a graph of $9x^2 - 25y^2 - 36x - 50y - 214 = 0$.
9. Sketch a graph of $9y^2 - 4x^2 - 36y + 8x - 68 = 0$.
10. Sketch a graph of $16x^2 - 9y^2 + 64x - 36y - 116 = 0$.

Step 5. Graph and Solve Quadratic Systems

A system of quadratic relations may have multiple solutions, a single solution, or no solution, and sketching a graph of the system before you begin an algebraic solution will allow you to know how many solutions you're looking for.

While it's possible to isolate a variable in one equation and substitute into the other equation, the squares often make that difficult. It's usually easier to eliminate a variable by adding or subtracting multiples of the equations.

Begin the solution of a quadratic system by sketching a graph of the equations:

$$9x^2 + 25y^2 = 225$$
$$x^2 + y^2 = 16$$

From the graph, you can anticipate that there will be four solutions. To find those solutions, multiply the equation of the circle by −9, and add.

$$\begin{aligned} 9x^2 + 25y^2 &= 225 \\ -9x^2 - 9y^2 &= -144 \\ \hline 16y^2 &= 81 \\ y^2 &= \frac{81}{16} \\ y &= \pm\frac{9}{4} \end{aligned}$$

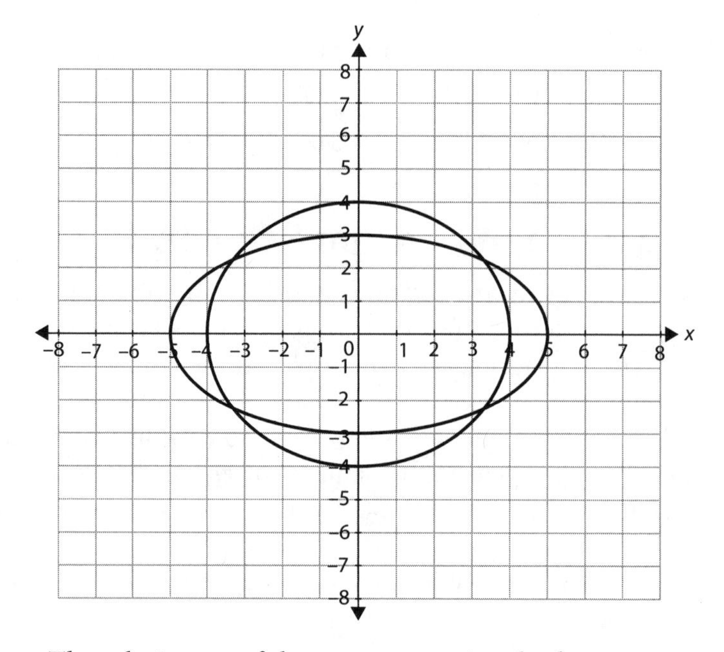

The solution set of the system contains the four points at which the ellipses intersect

Substitute for y in one of the original equations, and solve for x.

$$x^2 + y^2 = 16$$

$$x^2 + \left(\pm\frac{9}{4}\right)^2 = 16$$

$$x^2 + \frac{81}{16} = 16$$

$$x^2 = 16 - \frac{81}{16}$$

$$x^2 = \frac{175}{16}$$

$$x = \pm\sqrt{\frac{175}{16}} = \pm\frac{5\sqrt{7}}{4}$$

The four solutions are $\left(\frac{5\sqrt{7}}{4}, \frac{9}{4}\right)$, $\left(\frac{5\sqrt{7}}{4}, -\frac{9}{4}\right)$, $\left(-\frac{5\sqrt{7}}{4}, \frac{9}{4}\right)$, and $\left(-\frac{5\sqrt{7}}{4}, -\frac{9}{4}\right)$.

Exercise 6.5

Sketch a graph of the system, and identify the number of solutions.

1. $x^2 + y^2 = 9$
   $x + y + 1 = 0$
2. $x^2 + 4y^2 = 4$
   $x^2 + y^2 = 1$
3. $x^2 + 4y^2 = 4$
   $x^2 + y^2 = 4$
4. $x^2 - 4y^2 = 4$
   $16x^2 + 9y^2 = 144$
5. $y = x^2 - 4$
   $x = 4 - y^2$

Solve each system algebraically.

6. $x^2 + y^2 = 25$
   $x - y = -1$
7. $x^2 - y^2 = 4$
   $2x - 3y = 0$
8. $x^2 + y^2 = 32$
   $x = y^2$
9. $x + 1 = y^2$
   $x^2 + 3y^2 = 3$
10. $x^2 + 9y^2 = 9$
    $9x^2 + y^2 = 9$

7

Exponential and Logarithmic Functions

Logarithms originally developed as computational tools, but when calculators came into common usage, that function was no longer critical. Still, logarithmic functions are the inverses of exponential functions, and each has important applications.

Step 1. Get to Know the Exponential Function

Exponential functions are functions of the form $f(x) = b^x$, where the base b is a constant and b is greater than 0. The variable appears in the exponent position, distinguishing the exponential function from the power function, where the variable is the base and the exponent is a real number. The basic exponential function has a domain of all real numbers and a range of positive real numbers. The exponential function with the natural base, $f(x) = e^x$, is frequently used in applications.

> The natural base, e, is defined as $\lim_{x \to \infty}\left(1+\frac{1}{x}\right)^x$.
> An irrational number, e is approximately 2.71828... and is called Euler's number. It takes its name, and the letter e, from Leonhard Euler.

Graphs of exponential functions have a characteristic shape, almost flat on one end and very steep on the other. The flat end approaches a horizontal asymptote. The graph of an exponential function with a base greater than 1 rises sharply on the right and approaches a horizontal asymptote on the left. If the base is a number greater than 0 but less than 1, the graph is reflected, reversed left to right, so it falls steeply from left to right and flattens out on the right end. The graph of the parent exponential function has a horizontal asymptote of $y = 0$, and the graph approaches 0 as x approaches $-\infty$ if $b > 1$ and approaches 0 as x approaches ∞ if $0 < b < 1$.

An asymptote is a line that the graph comes very close to but doesn't touch.

Plotting a few key points can help you shape the graph of an exponential equation. Always look for the y-intercept. If $y = ab^x$, when $x = 0$, $b^0 = 1$, so the y-intercept will be a. Plugging in 1 and –1 for x will give you two more points that will easily set the shape: $(1,ab)$ and $\left(-1,\frac{a}{b}\right)$. The graph of $y = 3(2)^x$ has a y-intercept of $(0,3)$ and passes through the points $(1,6)$ and $\left(-1,\frac{3}{2}\right)$.

The key points that form the parent function $y = b^x$ are $(0,1)$, $(1,b)$, and $\left(-1,\frac{1}{b}\right)$.

The rules for transformations can be applied to exponential functions to help you sketch the graph quickly. Changes to the exponent cause a horizontal translation, reflection, stretch, or compression. The graph of $y = e^{x+5}$ is the graph of $y = e^x$ shifted five units left, and the graph of $y = e^{-x}$ is reflected across the y-axis. Changing the exponent to $y = e^{3x}$ compresses the graph, pushing the sharply increasing side closer to the y-axis, while changing to $y = e^{\frac{x}{2}}$ will stretch the graph horizontally. Changes elsewhere in the equation will cause vertical changes in the graph. The multiplier a in $y = ab^x$ is the factor of stretch or compression. If a is negative, the graph is reflected over the x-axis. Constants added or subtracted translate the graph up or down.

Exercise 7.1

Sketch a graph of each function.

1. $y = 2e^x$
2. $y = 4(.75)^x$
3. $y = 2^{x-3}$
4. $y = e^x + 4$
5. $y = 3(2)^x$
6. $y = -e^{x-2}$
7. $y = 3\left(\frac{1}{2}\right)^x - 1$
8. $y = -2(3)^x + 4$
9. $y = 5(0.1)^x + 2$
10. $y = 3(2)^x - 5$

Step 2. Define the Logarithmic Function

Logarithmic functions are inverses of exponential functions. Logarithms originated as an attempt to simplify multiplication and division by seeing all numbers as powers of a common base and adding or subtracting exponents. When you see an expression such as $\log_2 32 = 5$, you should think "the exponent I'd put on 2 to make 32 is 5." Each log statement is equivalent to an exponential statement.

$$\log_b n = x \text{ is equivalent to } b^x = n.$$

To find the inverse of an exponential function $y = b^x$, you swap the variables to get $x = b^y$. Trying to solve for y again means using that exponential-logarithmic equivalence: $y = \log_b x$.

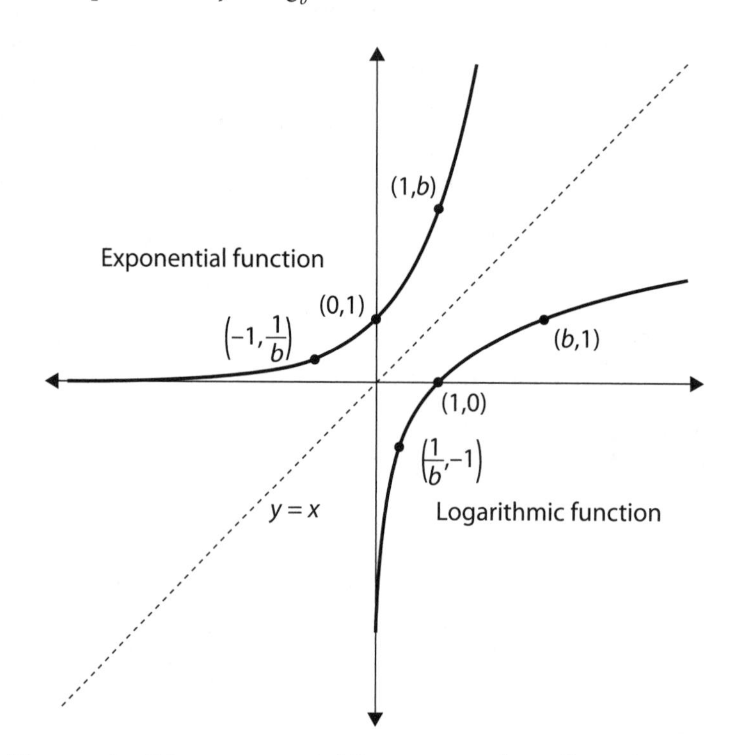

The graph of the exponential function $y = b^x$, reflected over the line $y = x$, forms the graph of the inverse function $y = \log_b x$

Because you swap the variables when you form an inverse, the domain of a function is the range of its inverse, and the range of the function is the domain of the inverse. The function $f(x) = \log_b x$, the inverse of $f(x) = b^x$, has a domain of positive real numbers and a range of all real numbers. The graph of the parent log function is a reflection, over the line $y = x$, of the parent exponential function, and it has a vertical asymptote of $x = 0$. The key points for the parent log function are (1,0), (b,1), and $\left(\frac{1}{b},-1\right)$. All the transformations you applied to aid in graphing other functions can be applied to the graph of the logarithmic function as well.

The graph of $f(x) = -2\log_3 (x + 4) + 1$ is a variant of the graph of $f(x) = \log_3 x$. The parent graph shifts four units left, is stretched vertically by a factor of two, is reflected across the x-axis, and is finally moved up one unit. The graph of $f(x) = -2\log_3 (x + 4) + 1$ is shown in the following figure. It has a domain of $(-4,\infty)$ and a range of $(-\infty,\infty)$.

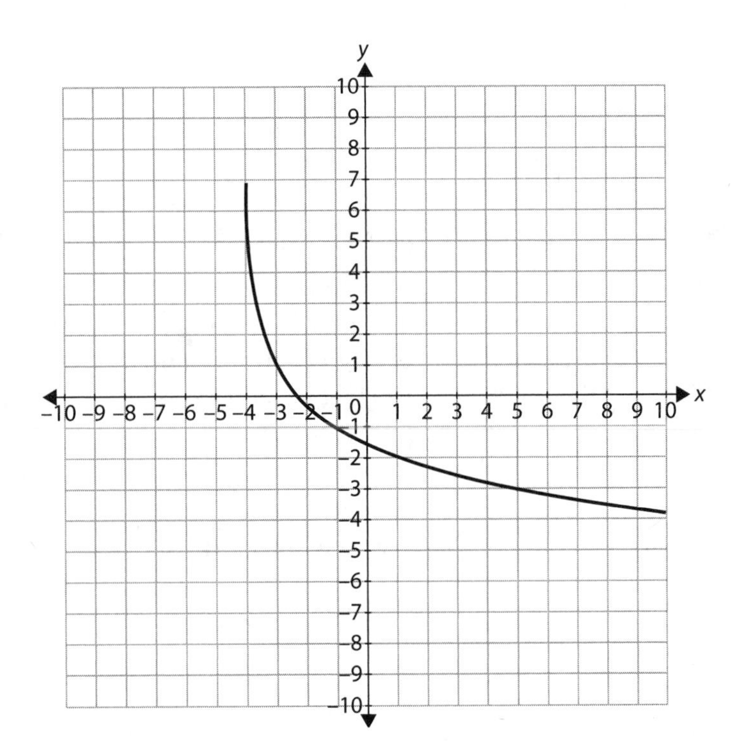

The graph of $f(x) = -2\log_3(x + 4) + 1$

The function $y = \log_{10} x$, the inverse of the function $y = 10^x$, is called the common logarithm and is generally written without the base shown, as just $y = \log x$. The inverse of the function $y = e^x$ is the function $y = \log_e x$, which is referred to as the natural log and given its own symbol, $y = \ln x$.

The "ln" comes from the French *log naturel*.

Exercise 7.2

Find the inverse of each function, and give the domain and range of the inverse.

1. $f(x) = 10^{x+1}$
2. $f(x) = e^x + 5$
3. $f(x) = 5 \cdot 2^{x-3}$
4. $f(x) = 100\left(\dfrac{1}{2}\right)^{\frac{x}{4}}$
5. $f(x) = 5 - 4e^{2x+1}$

Find the exact value of each expression without a calculator.

6. $\log_3 27$
7. $\log_b \sqrt{b}$
8. $\log_8 \dfrac{1}{64}$
9. $\ln e^{4x}$
10. $e^{(\ln e^x)}$

Step 3. Learn to Use Properties of Logarithms

While logarithms are no longer commonly used for calculations, the properties that originally made them useful for computation can be helpful in simplifying expressions and solving equations. The three principal properties of logs parallel the three principal properties of exponents: to multiply, add the exponents; to divide, subtract the exponents; to raise a power to a power, multiply the exponents. In logarithmic terms, this becomes:

$$\log_b xy = \log_b x + \log_b y$$

$$\log_b \frac{x}{y} = \log_b x - \log_b y$$

$$\log_b x^n = n \log_b x$$

In addition, because any nonzero number to the zero power is one, $\log_b 1 = 0$ for any base b.

Applied to numbers, these properties tell you that $\log_b 35 = \log_b (5 \cdot 7) = \log_b 5 + \log_b 7$, that $\log_b \frac{1}{2} = \log_b 1 - \log_b 2 = 0 - \log_b 2 = -\log_b 2$, and that $\log_b 8 = \log_b 2^3 = 3\log_b 2$. When you're solving equations, you'll want to use these properties to simplify variable expressions. Sometimes you'll want to break the log of a complex expression into an expression involving several simpler logs. Other times, you'll need to consolidate an expression involving several logs into a single log.

Notice that it doesn't matter what the base is.

Expand

The log of a complex expression can be expanded into sums, differences, and products of the logs of single numbers and variables by applying the basic rules. To expand a log, first note the overall structure of the expression. The $\log(x^5 y^3)$, although it involves exponents, is basically a product, but $\log\sqrt[3]{x(y+3)} = \log[x(y+3)]^{\frac{1}{3}}$ is first a power, although the radicand is a product. Apply the appropriate property of logs to make the first rewrite, and then continue expanding the remaining logs until each log involves the simplest possible expression.

$$\log(x^5 y^3) = \log x^5 + \log y^3 = 5\log x + 3\log y$$

$$\begin{aligned}\log\sqrt[3]{x(y+3)} &= \log[x(y+3)]^{\frac{1}{3}} \\ &= \frac{1}{3}\log[x(y+3)] \\ &= \frac{1}{3}[\log x + \log(y+3)] \\ &= \frac{1}{3}\log x + \frac{1}{3}\log(y+3)\end{aligned}$$

Remember there is no way to expand the log of a sum or difference: $\log(x+y) \neq \log x + \log y$

Condense

To write an expression involving several logs as a single log, first simplify the expression as much as you can by removing parentheses and combining like terms where possible. Apply the log of a power rule to move any coefficients back to the exponent position, and then work left to right, changing the sum of logs into the log of a product and the difference of logs into the log of a quotient.

To write $3\ln x - 2\ln y + \frac{1}{2}(3\ln z - \ln w)$ as a single log, start by clearing the parentheses.

$$3\ln x - 2\ln y + \frac{1}{2}(3\ln z - \ln w)$$

$$3\ln x - 2\ln y + \frac{3}{2}\ln z - \frac{1}{2}\ln w$$

Move the coefficients to the exponent position.

$$3\ln x - 2\ln y + \frac{3}{2}\ln z - \frac{1}{2}\ln w$$

$$\ln x^3 - \ln y^2 + \ln z^{3/2} - \ln 2^{1/2}$$

Then work from left to right, consolidating the sums and differences.

$$(\ln x^3 - \ln y^2) + \ln z^{3/2} - \ln w^{1/2}$$

$$\left[\ln\left(\frac{x^3}{y^2}\right) + \ln z^{3/2}\right] - \ln w^{1/2}$$

$$\ln\left(\frac{x^3 z^{3/2}}{y^2}\right) - \ln w^{1/2}$$

$$\ln\left(\frac{x^3 z^{3/2}}{y^2 w^{1/2}}\right)$$

Change Base

All the properties of logs depend upon the logs having the same base, so simplifying an expression involving logs with different bases will require you to change base. If you try to evaluate a log using a calculator, you'll find that the calculator has keys for only the common and the natural log. You can solve those problems with the change of base rule:

$$\log_b N = \frac{\log N}{\log b} = \frac{\ln N}{\ln b}$$

The log base b of a number N can be found by finding the log of N and dividing by the log of b. You can use the common log or the natural log.

> The change of base rule would hold true using any base, for example, $\log_3 8 = \frac{\log_7 8}{\log_7 3} = \frac{\log_4 8}{\log_4 3}$, but switching to bases other than 10 or e is generally not helpful.

To evaluate $\log_6 39$, apply the change of base rule $\log_6 39 = \frac{\ln 39}{\ln 6} \approx \frac{3.6636}{1.7918} \approx$ 2.0447.

If you encounter an equation involving logs with different bases, you can use the change of base rule to rewrite the equation in a more convenient form:

$$\log_4(x+3) = \log_2(x+1)$$
$$\frac{\log_2(x+3)}{\log_2 4} = \log_2(x+1)$$
$$\frac{\log_2(x+3)}{2} = \log_2(x+1)$$
$$\log_2(x+3) = 2\log_2(x+1)$$
$$\log_2(x+3) = \log_2(x+1)^2$$
$$x+3 = (x+1)^2$$
$$x+3 = x^2+2x+1$$
$$x^2+x-2=0$$

Solve by factoring, but keep the domain of the log function in mind when you solve.

$$(x+2)(x-1)=0$$
$$x+2=0 \qquad x-1=0$$
$$x=-2 \qquad x=1$$

The solution of $x = -2$ is extraneous, but the solution of $x = 1$ is valid.

Exercise 7.3

Write as a sum and/or difference of logs. Express exponents as coefficients.

1. $\ln[x(x+1)]$
2. $\ln\left(\frac{x}{x-3}\right)$
3. $\log_b\left(\frac{xy}{z}\right)^2$
4. $\ln\left(\frac{1}{x^2}\sqrt{x-2}\right)$

Write as a single logarithm.

5. $4 \ln x + 3 \ln y$
6. $\frac{1}{2}\log_b(x+5) - \log_b x$
7. $\frac{1}{5}[\ln x + 3\ln y - 4\ln z]$

Use the given information and the change of base rule to evaluate each expression.

$$\ln 2 \approx 0.6931$$

$$\ln 3 \approx 1.0986$$

$$\ln 5 \approx 1.6094$$

$$\ln 7 \approx 1.9459$$

8. $\log_3 5$
9. $\log_3 \frac{4}{5}$

10. $\log_7 1.5$

Step 4. Solve Exponential and Logarithmic Equations

When you first learned to solve equations, you were taught to use inverse operations: add to undo subtraction, divide to undo multiplication. That fundamental rule still holds for exponential and logarithmic equations. To solve an exponential equation, take the log of both sides, and to solve a log equation, reexpress it as an equivalent exponential equation.

Exponential Equations

The basic technique for solving exponential equations is to take the log of both sides. The key to successful solutions is in knowing when to do that. In the case of simple equations, where you can transform the equation to equal powers of the same base, "taking the log" will just mean looking at the exponents.

Method 1

1. Express each side of the equation as a single power of the same base. (If this is not possible, go to method 2.)
2. Equate the exponents.
3. Solve the resulting equation and check.

To solve $25^x \cdot 5^{3x-1} = \left(\frac{1}{5}\right)^{x+4}$, first express everything as a power of five.

$$\begin{aligned} 25^x \cdot 5^{3x-1} &= \left(\frac{1}{5}\right)^{x+4} \\ (5^2)^x \cdot 5^{3x-1} &= (5^{-1})^{x+4} \\ 5^{2x} \cdot 5^{3x-1} &= 5^{-x-4} \end{aligned}$$

Combine the powers on the left side by adding the exponents, so that you have a single power on each side.

$$\begin{aligned} 5^{2x} \cdot 5^{3x-1} &= 5^{-x-4} \\ 5^{5x-1} &= 5^{-x-4} \end{aligned}$$

Your next step is, technically, to take the log of both sides, but if you take the log using a base of five, $\log_5 5^{5x-1} = 5x - 1$ and $\log_5 5^{-x-4} = -x - 4$, so taking the log is just taking the exponents and setting them equal.

$$\begin{aligned} 5^{5x-1} &= 5^{-x-4} \\ 5x-1 &= -x-4 \\ 6x &= -3 \\ x &= -\frac{1}{2} \end{aligned}$$

If it's not possible to express both sides as single powers of the same base, you can still follow the same basic steps, but the process will look a bit more complicated.

Method 2

1. Express each side as a single power, even if you must use different bases.
2. Take the log of both sides, using any convenient base.
3. Isolate the variable.
4. Evaluate the logs.
5. Solve and check.

"Any convenient base" generally means base 10 or base *e*, because you have keys on your calculator for the common logarithm (base 10) and the natural log (base *e*).

To solve $\frac{5^{3x-1}}{7^{2x}}=1$, first express the equation as equal powers, and then take the log of both sides.

$$\frac{5^{3x-1}}{7^{2x}}=1$$
$$5^{3x-1}=7^{2x}$$
$$\ln(5^{3x-1})=\ln(7^{2x})$$

Use properties of logs to simplify both sides.

$$\ln(5^{3x-1})=\ln(7^{2x})$$
$$(3x-1)\ln 5=2x\ln 7$$
$$3x\ln 5-\ln 5=2x\ln 7$$

Move all terms involving the variable to one side and constant terms to the other. Factor out the variable and divide. Don't rush to evaluate the logs. The introduction of all those decimals, even rounded to a reasonable number of places, will only make things look worse.

$$3x\ln 5-2x\ln 7=\ln 5$$
$$x(3\ln 5-2\ln 7)=\ln 5$$
$$x=\frac{\ln 5}{3\ln 5-2\ln 7}$$

Once the variable is isolated, use a calculator to evaluate the log expression.

$$x=\frac{\ln 5}{3\ln 5-2\ln 7}$$
$$x\approx\frac{1.6094}{3(1.6094)-2(1.9459)}$$
$$x\approx 1.7186$$

Quadratic Forms

You will sometimes find an exponential equation that is quadratic in form. An equation such as $e^{2x}+5e^x-14=0$ can be seen as $(e^x)^2+5e^x-14=0$. First solve the quadratic form, letting $y=e^x$ and imagining the equation as $y^2+5y-14=0$. By factoring, you can find that $(y+7)(y-2)=0$ and $y=-7$ or

$y = 2$. Return the exponential expression and consider $e^x = -7$, which has no solution, and $e^x = 2$, which gives you the solution $x = \ln 2$.

Logarithmic Equations

To solve a logarithmic equation, you need to translate it to an equivalent exponential equation. How you accomplish that goal will depend upon the complexity of the log equation. If logs appear on only one side of the equation, you can rewrite the equation in its equivalent exponential form. If both sides have logs (with the same base), you can exponentiate, which actually isn't as complicated as it sounds.

Method 1

1. Simplify each side of the equation until you have a single log equal to a constant.
2. Rewrite the log equation as an equivalent exponential equation.
3. Evaluate the powers.
4. Solve the resulting equation.
5. Check the solution(s), keeping the domain of the log function in mind.

To solve $\log_3 x - \log_3 (x - 5) = -2$, first condense the left side to a single log.

$$\log_3 x - \log_3(x-5) = -2$$
$$\log_3\left(\frac{x}{x-5}\right) = -2$$

This log equation has an equivalent exponential form, and once you've written that you can evaluate the power and solve the resulting equation.

$$\log_3\left(\frac{x}{x-5}\right) = -2$$
$$3^{-2} = \frac{x}{x-5}$$
$$\frac{1}{9} = \frac{x}{x-5}$$
$$9x = x-5$$
$$8x = -5$$
$$x = -\frac{5}{8}$$

When you check this solution, unfortunately you find that it must be rejected as extraneous, because the domain of the log function is nonnegative real numbers. You're forced to conclude that this equation has no solution.

Method 2

1. Simplify each side of the equation until you have a single log equal to a single log.
2. Exponentiate. In simpler terms, equate the exponents. If $\log_b x = \log_b y$, then $x = y$.
3. Solve the equation.
4. Check the solution(s), keeping the domain of the log function in mind.

To solve $\ln(x - 3) + \ln(x + 5) = \ln(3x + 5)$, first condense the left side to a single logarithm.

$$\ln(x-3)+\ln(x+5)=\ln(3x+5)$$
$$\ln[(x-3)(x+5)]=\ln(3x+5)$$
$$\ln(x^2+2x-15)=\ln(3x+5)$$

Technically, your next step is to say that $e^{\ln(x^2+2x-15)} = e^{\ln(3x+5)}$, but that simply means $x^2 + 2x - 15 = 3x + 5$. Solve the equation $x^2 + 2x - 15 = 3x + 5$ by factoring.

$$\ln(x^2+2x-15)=\ln(3x+5)$$
$$x^2+2x-15=3x+5$$
$$x^2-x-20=0$$
$$(x-5)(x+4)=0$$

The quadratic equation has two solutions, $x = 5$ and $x = -4$, but only one of them is in the domain of the log function. The solution of the equation $\ln(x - 3) + \ln(x + 5) = \ln(3x + 5)$ is $x = 5$.

Exercise 7.4

Solve each equation. Be sure to check for extraneous solutions.

1. $9^{x-3} = 27^{4-x}$
2. $9^{x-3} = 7^{2x}$
3. $25e^{0.38x} = 350$
4. $4500(1.02)^{15r} = 10{,}000$
5. $e^{2x} - 3e^{x} - 40 = 0$
6. $\log_2 (3x - 5) = 6$
7. $\ln x + \ln(x + 1) = \ln 30$
8. $\ln(x+1) - \ln(x+5) = \ln\frac{1}{2}$
9. $3 \ln x - \ln(x + 2) = 2 \ln(x - 1)$
10. $\frac{1}{2}\log_5 x + \frac{1}{2}\log_5(x+3) = \log_5(x-1)$

Step 5. Explore Exponential Growth and Decay

When $b > 1$, the exponential function $f(x) = b^x$ represents exponential growth. As x increases, $f(x)$ increases sharply. In contrast, when b is less than 1 but greater than 0, the equation represents exponential decay, a decreasing quantity. If $0 < b < 1$, as x increases, the function decreases. The model of exponential growth and the model of exponential decay both have important applications.

Exponential Growth

Compound Interest. Compound interest is a common example of exponential growth. To calculate the value of an investment receiving compound interest, use the formula $A = P\left(1+\frac{r}{n}\right)^{nt}$, where P is the principal or original investment, r is the rate of interest per year (converted to a decimal), n is the number of times per year interest is compounded, and t is the number of years. If you invest \$1000 at 5 percent per year for two years compounded quarterly, $P = \$1000$, $r = 0.05$, $t = 2$, and $n = 4$.

$$A = P\left(1+\frac{r}{n}\right)^{nt}$$

$$A = 1000\left(1+\frac{.05}{4}\right)^{4\cdot 2}$$

$$= 1000(1.0125)^8$$

$$= 1000(1.104486101)$$

$$= \$1104.49$$

After two years you will have earned \$104.49 of interest, and your investment will be worth \$1104.49.

When the interest is compounded continuously, the formula becomes $A = Pe^{rt}$. If the same \$1000 is invested at 5 percent compounded continuously for two years,

$$A = 1000e^{0.05(2)} = 1000e^{0.1} \approx \$1105.17$$

This $A = Pe^{rt}$ formula, commonly referred to as the Pert formula, can model many exponential growth or decay situations, with $r > 0$ for growth and $r < 0$ for decay.

With continuous compounding, this investment results in \$1105.17, or interest of \$5.17.

Population Growth. Under the right conditions, the population of a particular organism may grow exponentially, if only for a period of time. In such cases, the models $P = P_0\,(1 + r)^t$ or $A = Pe^{kt}$ can be used to calculate the population at a given time.

If a colony of bacteria is created with 200 bacteria and the population increases by 50 percent every hour, the population 24 hours later can be found by

$$P = P_0(1+r)^t$$

$$P = 200(1+0.5)^{24}$$

$$P \approx 200(16834.1122)$$

$$P \approx 3{,}366{,}822$$

Alternately, you could use $A = Pe^{kt}$ with $P = 200$ and $A = 300$, the population after one hour. Find k.

$$\begin{aligned} A &= Pe^{kt} \\ 300 &= 200e^{k\cdot 1} \\ 1.5 &= e^{k} \\ \ln 1.5 &= k \end{aligned}$$

Then you can evaluate the population after 24 hours.

$$\begin{aligned} A &= 200e^{24\ln(1.5)} \\ &\approx 200(16834.1122) \\ &\approx 3{,}366{,}822 \end{aligned}$$

Exponential Decay

When the base of a power is greater than 1, the power will grow as the exponent increases, so equations of the form $y = ab^x$ represent exponential growth when b is greater than 1. In contrast, when b is less than 1 but greater than 0, the equation represents exponential decay, a pattern of decrease.

Radioactive Decay. The classic example of exponential decrease is the decay of radioactive substances over time. The information about the rate of decay is generally given in the form of the half-life of the element, the time it takes for a quantity of the element to decay to half its mass. Using that information, you can calculate the rate of decay and use $A = Pe^{rt}$ to model the change.

The half-life of the radioisotope sodium-24 is 15 hours. Using $A = Pe^{rt}$, and substituting $P = 1$, $A = \frac{1}{2}$, and $t = 15$, you can find the rate of decay.

$$\begin{aligned} A &= Pe^{rt} \\ \frac{1}{2} &= 1\cdot e^{r\cdot 15} \\ \ln\left(\frac{1}{2}\right) &= 15r \\ r &= \frac{1}{15}\ln\left(\frac{1}{2}\right) \\ r &\approx -0.0462 \end{aligned}$$

The rate of decay will always be the natural log of one-half, divided by the half-life.

If a sample of 10 grams of sodium-24 is present at time $t = 0$ hours, then two days later, when $t = 48$ hours, the remaining sodium-24 is

$$A = Pe^{rt}$$
$$A \approx 10e^{(-0.0462)(48)}$$
$$A \approx 1.0887$$

Only 1.0887 grams of sodium-24 remain.

Depreciation. Major purchases, such as factory equipment or vehicles, lose value, or depreciate, as they age. The pattern of this decline in value is generally exponential, with a steep decline in the first few years of the product life and a slower decrease over time. Models of exponential decay, such as $A = P(1 - r)^t$ or $A = Pe^{rt}$ with $r < 0$, often describe this depreciation well.

If a new car is purchased for \$24,000 and depreciates at 12 percent per year, the value of the car three years later can be found using

$$A = P(1-r)^t$$
$$A = 24{,}000(1-0.12)^3$$
$$= 24{,}000(0.88)^3$$
$$\approx 16355.33$$

Exercise 7.5

Identify each situation as growth or decay, and evaluate the result.

1. A patient is given an injection of 250 mg of a drug. Each hour, as the body metabolizes the drug, the level in the bloodstream is reduced by 20 percent. What is the level in the bloodstream 8 hours later to the nearest milligram?
2. A city had a population of 250,000 in 2008, and the population was increasing by 11 percent per year. What was the population in 2010?
3. A county had 45,000 acres of forested land in 1996, but that acreage was decreasing at 5 percent per year. To the nearest acre, how many acres of forested land remained in the county in 2000?
4. If a culture is initiated with 100 bacteria and doubles every 4 hours, when will the culture contain 10,000 bacteria, to the nearest tenth of an hour?

5. A truck purchased for \$35,000 depreciates at a rate of 8 percent per year. When will its value drop below \$2,000? (Round to the nearest year.)
6. The population of Texas grows at approximately 2 percent per year. If the population was 20,851,820 in the 2000 census ($t = 0$), in what year should the population of Texas have reached 25 million?
7. The half-life of thorium-234 is 24.1 days. If a sample of 2 g of thorium-234 is present on January 1, 2012, when will less than 0.001 g remain?

Calculate the value of each investment after the specified time, when invested as described.

8. An investment of \$7,500, at 2.5 percent per year, compounded continuously, for 5 years
9. An investment of \$15,000, at 12 percent per year, compounded quarterly, for 6 years
10. An investment of \$25,000, at 6 percent per year, compounded continuously, for 15 years

8

Radical Functions

The word *radical* comes from the Latin meaning "root," and it is used to describe the symbol that indicates a root as well as all expressions involving that symbol. Radical functions are the inverses of power functions. The square root function is the inverse of the squaring function. The cube root function is the inverse of the cubic function. In general, the inverse of $f(x) = x^n$ is $f^{-1}(x) = \sqrt[n]{x}$.

If n is an odd number, the domain of the nth root function is all real numbers, and the range is all reals as well. If n is an even number, the domain and the range are nonnegative real numbers.

Before you analyze radical functions, you'll want to review rules for exponents and a bit about power functions.

Step 1. Review Rules for Exponents

To multiply powers of the same base, keep the base and add the exponents.

$$a^2 \cdot a^3 = a^{2+3} = a^5$$

To divide powers of the same base, keep the base and subtract the exponents.

$$\frac{t^7}{t^3} = t^{7-3} = t^4$$

To raise a power to a power, keep the base and multiply the exponents.

$$(b^3)^2 = (b^3)(b^3) = b^{3+3} = b^{2\times 3} = b^6$$

When a product is raised to a power, each factor is raised to that power. When a quotient is raised to a power, both the numerator and the denominator are raised to that power.

Zero Exponent

If you follow the rule for dividing powers to evaluate $\frac{x^5}{x^5}$, you'll conclude that $\frac{x^5}{x^5} = x^0$. You know from arithmetic, however, that any number divided by itself equals 1, so $\frac{x^5}{x^5} = 1$. If $a \neq 0$, $a^0 = 1$.

Negative Exponents

According to the rules for exponents, $\frac{x^5}{x^6} = x^{-1}$ and $\frac{y^3}{y^7} = y^{-4}$. If you simplify $\frac{x^5}{x^6}$, you get $\frac{1}{x}$ and if you simplify $\frac{y^3}{y^7}$, you get $\frac{1}{y^4}$. So $x^{-1} = \frac{1}{x}$ and $y^{-n} = \frac{1}{y^n}$.

Rational Exponents

From working with radicals, you know that $(\sqrt{x})^2 = x$ and $\sqrt{x^2} = |x|$. Because $\left(x^{\frac{1}{2}}\right)^2 = x$ and $(x^2)^{\frac{1}{2}} = |x|$, you can define an exponent to represent the square root: $\sqrt{x} = x^{\frac{1}{2}}$. You can generalize that definition for any root: $\sqrt[n]{x} = x^{\frac{1}{n}}$.

Exercise 8.1

Rewrite each expression in simplest form, using only positive exponents.

1. $(2x)^4(x^{-3})\left(\frac{x}{2}\right)^{-4}\left(\frac{2}{x}\right)^3$
2. $(x^2y)^3(x^5y^2)^{-3}$
3. $\left(\frac{a^2b^{-3}c^4}{a^{-3}b^6c^{-1}}\right)^{-2}$
4. $\frac{(3xy^4z^{-3})^2}{(9x^8y^{12}z^{-4})^{\frac{1}{2}}}$
5. $\left[\frac{(2x^3)^{-2}(18xy^{-4})^3}{8x^5y^{-2}}\right]^{\frac{1}{2}}$
6. $(4x^5y)^{\frac{3}{4}}(4(xy)^5)^{\frac{1}{4}}$
7. $\left(\frac{9x^7y^{-4}}{x^{-3}y^6}\right)^{\frac{5}{2}}$
8. $\left(\frac{2187x^2y^5}{81x^8y^{-1}}\right)^{\frac{1}{3}}$
9. $\frac{x^{-1}+y^{-1}}{(x+y)^{-1}}$
10. $\frac{(x+y)^3}{(x-y)^{-2}}$

Step 2. Simplify Radicals and Rationalize Denominators

It is common practice to put radical expressions in simplest radical form and to rationalize denominators when the radicand, the expression under the radical, is a number. When you work with radical functions, you'll want to simplify radical expressions involving variables as well.

- Simplify square roots by factoring the radicand and taking the square root of any perfect squares.

$$\sqrt{x^4-4x^3+4x^2}=\sqrt{x^2(x^2-4x+4)}=\sqrt{x^2(x-2)^2}=|x(x-2)|$$

- Simplify other roots by factoring the radicand and taking roots whenever a factor is raised to a power that is a multiple of the index of the radical. This process can be made simpler by using fractional exponents.

$$\begin{aligned}\sqrt[4]{x^6y^7z^9} &= (x^6y^7z^9)^{\frac{1}{4}}\\ &= x^{\frac{6}{4}}y^{\frac{7}{4}}z^{\frac{9}{4}}\\ &= x\cdot x^{\frac{2}{4}}\cdot y\cdot y^{\frac{3}{4}}\cdot z^2\cdot z^{\frac{1}{4}}\\ &= xyz^2\cdot x^{\frac{2}{4}}y^{\frac{3}{4}}z^{\frac{1}{4}}\\ &= xyz^2(x^2y^3z)^{\frac{1}{4}}\\ &= xyz^2\sqrt[4]{x^2y^3z}\end{aligned}$$

- Find the product of two roots with the same index by finding the root of the product of the radicands.

$$\begin{aligned}\sqrt{x^2-5x}\cdot\sqrt{2x^2+x} &= \sqrt{(x^2-5x)(2x^2+x)}\\ &= \sqrt{x(x-5)x(2x+1)}\\ &= \sqrt{x^2(x-5)(2x+1)}\\ &= |x|\sqrt{2x^2-9x-5}\end{aligned}$$

- If a denominator is a single square root, rationalize by multiplying by the square root over itself.

$$\begin{aligned}\frac{x-3}{\sqrt{x^2-x-6}} &= \frac{x-3}{\sqrt{x^2-x-6}}\cdot\frac{\sqrt{x^2-x-6}}{\sqrt{x^2-x-6}}\\ &= \frac{(x-3)\sqrt{x^2-x-6}}{x^2-x-6}\\ &= \frac{(x-3)\sqrt{x^2-x-6}}{(x-3)(x+2)}\\ &= \frac{\sqrt{x^2-x-6}}{x+2}\end{aligned}$$

- If the denominator is a single root of index n, multiply the numerator and denominator by the radical raised to the n–1 power to rationalize the denominator.

$$\frac{3}{\sqrt[5]{x+3}}\cdot\frac{(\sqrt[5]{x+3})^{5-1}}{(\sqrt[5]{x+3})^{5-1}} = \frac{3(\sqrt[5]{x+3})^4}{(\sqrt[5]{x+3})^5} = \frac{3(\sqrt[5]{x+3})^4}{x+3}$$

- If the denominator is a sum or difference involving a radical, multiply the numerator and denominator by the conjugate of the denominator.

$$\frac{3x}{5x-\sqrt{x-3}}\cdot\frac{5x+\sqrt{x-3}}{5x+\sqrt{x-3}}=\frac{3x(5x+\sqrt{x-3})}{(5x-\sqrt{x-3})(5x+\sqrt{x-3})}$$
$$=\frac{15x^2+3x\sqrt{x-3}}{25x^2-(x-3)}$$
$$=\frac{15x^2+3x\sqrt{x-3}}{25x^2-x+3}$$

Exercise 8.2

Write each expression in simplest radical form.

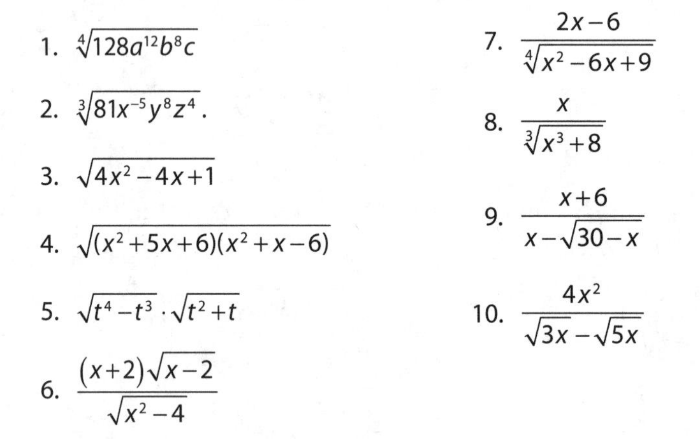

1. $\sqrt[4]{128a^{12}b^8c}$
2. $\sqrt[3]{81x^{-5}y^8z^4}$.
3. $\sqrt{4x^2-4x+1}$
4. $\sqrt{(x^2+5x+6)(x^2+x-6)}$
5. $\sqrt{t^4-t^3}\cdot\sqrt{t^2+t}$
6. $\dfrac{(x+2)\sqrt{x-2}}{\sqrt{x^2-4}}$
7. $\dfrac{2x-6}{\sqrt[4]{x^2-6x+9}}$
8. $\dfrac{x}{\sqrt[3]{x^3+8}}$
9. $\dfrac{x+6}{x-\sqrt{30-x}}$
10. $\dfrac{4x^2}{\sqrt{3x}-\sqrt{5x}}$

Step 3. Explore the Square Root and Cube Root Functions

The square root function is the inverse of the squaring function, but the squaring function, $f(x) = x^2$, defined on $(-\infty,\infty)$, is not one-to-one. In order for the inverse to be a function, the domain must be restricted. The function $f(x)=\sqrt{x}$, the parent function for all square root functions, is the inverse of $f(x) = x^2$, restricted to $[0,\infty)$. The function $f(x)=\sqrt{x}$ has a domain of non-negative real numbers and a range of nonnegative reals.

The graph of $f(x)=\sqrt{x}$ has the shape of half of a parabola. Remember that the graph of the inverse of any function is the reflection of the original graph over the line $y = x$. Restrict the graph of $f(x) = x^2$ to the nonnegative numbers, and you have half a parabola. Reflect that half parabola $f(x) = x^2$ across the line $y = x$ to produce the graph of $f(x)=\sqrt{x}$.

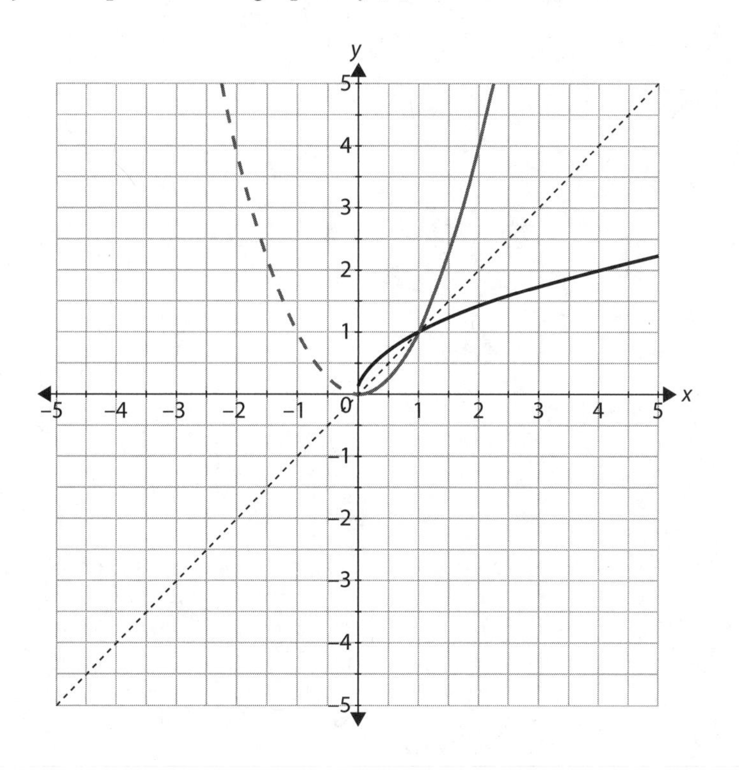

The graph of $f(x) = x^2$ for $x \geq 0$, reflected across the line $y = x$, produces the graph of the inverse $f(x)=\sqrt{x}$

The cube root function $f(x)=\sqrt[3]{x}$ is the inverse of the cubing function $f(x) = x^3$. Both the cubing function and the cube root function have a domain of all real numbers and a range of all real numbers. The graph of $f(x)=\sqrt[3]{x}$ is a reflection of the graph of $f(x) = x^3$ over the line $y = x$.

All of the transformations you were able to make on the graphs of quadratic and polynomial functions are possible with radical functions. The cube root function $f(x)=2\sqrt[3]{x-4}+1$ is translated four units right and one unit up and stretched vertically by a factor of two.

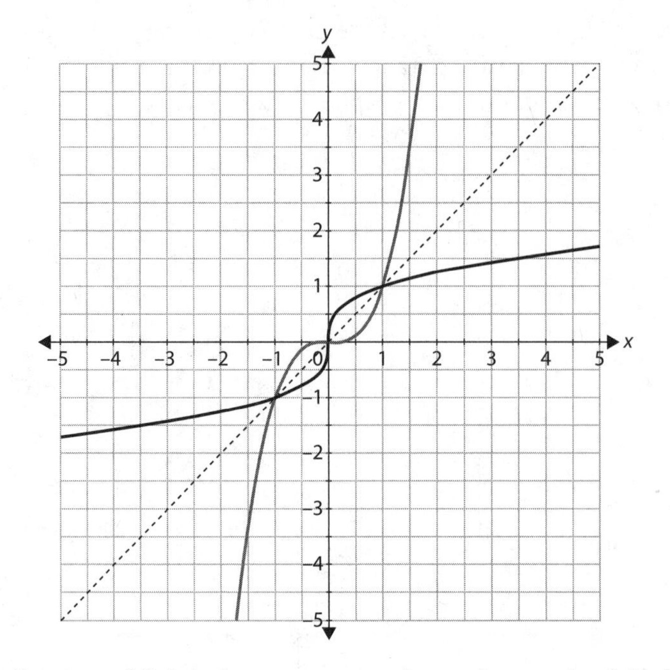

The reflection of $f(x) = x^3$ over $y = x$ produces the graph of $f(x)=\sqrt[3]{x}$

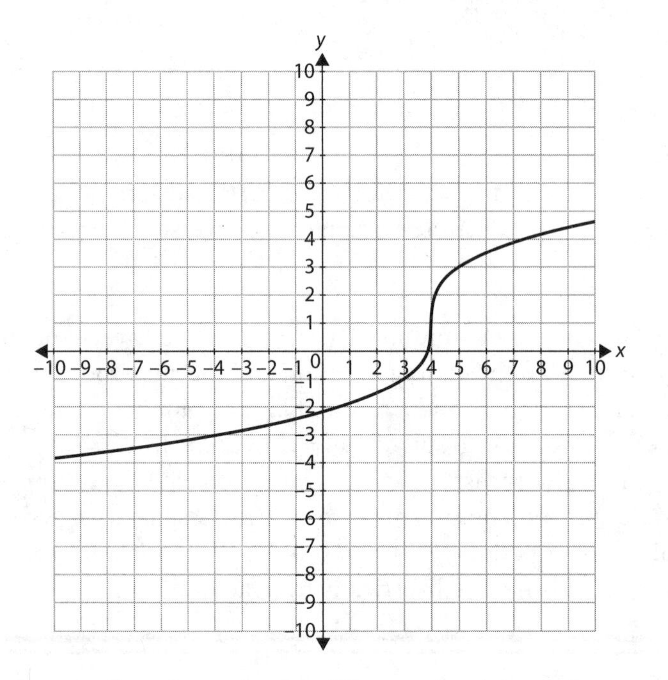

The cube root function, translated four units right and one unit up and stretched vertically by a factor of two

Exercise 8.3

Sketch a graph of each function and give the domain and range.

1. $f(x) = 2 + \sqrt{x-3}$
2. $f(x) = 4 - 2\sqrt[3]{x}$
3. $f(x) = \sqrt{\dfrac{x}{2}}$
4. $f(x) = \sqrt[3]{2(x-3)} + 4$
5. $f(x) = \dfrac{1}{2}\sqrt{-3x} - 4$

9

Systems of Equations

A single equation containing more than one variable has an infinite number of solutions, but it is possible to find a unique solution for a collection of equations, each of which involves those variables, if you have as many equations as variables. In algebra, you learned to solve systems of two linear equations with two variables, but you'll want to extend your skills to larger systems and other types of equations.

Step 1. Review Two-Variable Linear Systems

Algebra courses generally include three methods for solving systems of equations. Graphing clarifies the process but is not always practical for actually finding the solution. Substitution is a useful algebraic method if one of the equations is simple enough to allow you to isolate a variable. Elimination, sometimes called the method of linear combination, is a powerful and flexible method that can be used in diverse circumstances.

Graphing

A system of two linear equations in two variables can be solved by graphing each of the equations and locating the point of intersection. The coordinates of the point of intersection are the values of x and y that solve the system.

With the help of a graphing calculator, equations—linear or other—can be graphed easily, and the point of intersection of two graphs can be found using the intersect feature on the calculator.

By graphing $2x - 3y = 6$ and $y = 3x + 5$ on the same set of axes, you can see that they cross at the point (−3,−4). That means that $x = -3$, and $y = -4$ will solve both equations. Each equation has an infinite number of solutions, but this one is the only one they have in common.

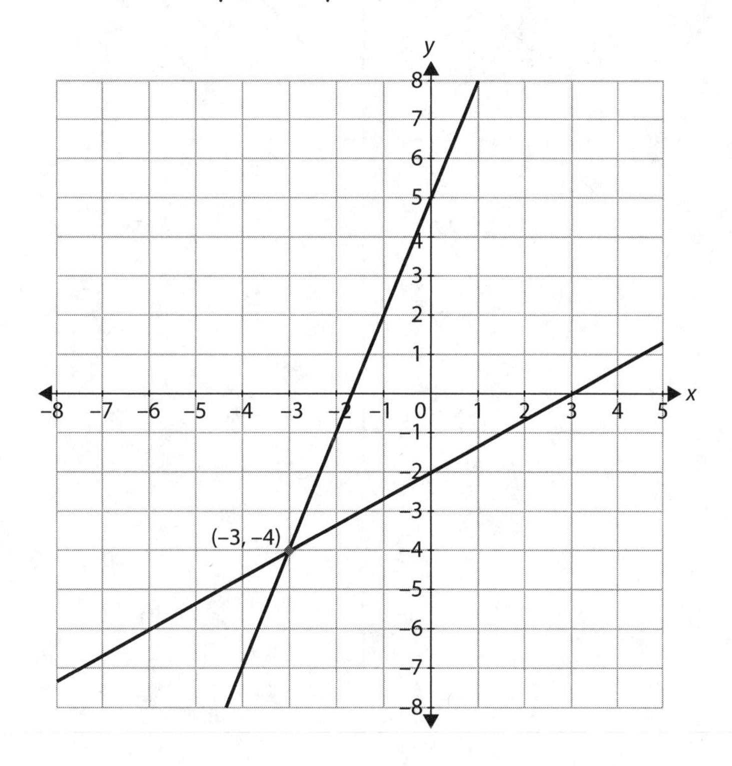

The solution of the system is the point of intersection of the graphs

If the lines are parallel, the system has no solution and is called an inconsistent system. If the two graphs prove to be the same line, the system is dependent and has infinite solutions.

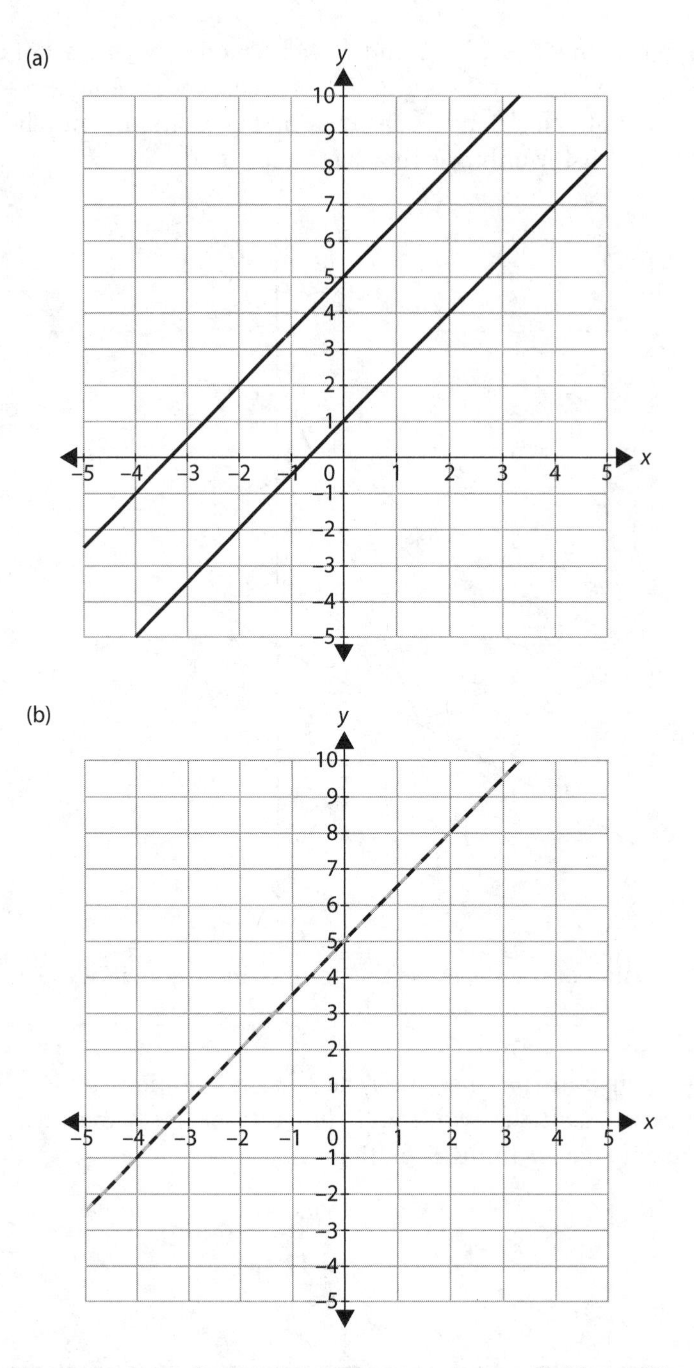

An inconsistent linear system appears as parallel lines (a), but the graphs coincide in a dependent system (b)

Substitution

When one equation in a system expresses one variable as a function of the other, such as $y = 3x - 1$ or $x = 4 + y$, it can be used to substitute into the other equation. That substitution produces an equation with only one variable, which can be easily solved. To solve the system $\begin{cases} y = 3x - 1 \\ 2x + 5y = 12 \end{cases}$ by substitution, replace the y in the second equation with $3x - 1$.

$$\begin{aligned} 2x + 5y &= 12 \\ 2x + 5(3x - 1) &= 12 \\ 2x + 15x - 5 &= 12 \\ 17x - 5 &= 12 \\ 17x &= 17 \\ x &= 1 \end{aligned}$$

Once you know the value of one variable, substitute that into one of the original equations to find the value of the other variable.

$$\begin{aligned} y &= 3x - 1 \\ y &= 3(1) - 1 \\ y &= 2 \end{aligned}$$

Elimination

If both equations are in standard form (or can easily be transformed to standard form), then the method of linear combinations, or elimination, allows you to add or subtract the equations, or multiples of the equations, in a way that will make one variable drop out. Make sure both equations are in standard form and that like terms are aligned under one another before you begin.

To solve the system $\begin{cases} 2x + 3y = 17 \\ 5x - 2y = 14 \end{cases}$, multiply the first equation by 2 and the second equation by 3. That will create an equivalent system in which the coefficients of the variable y in the equations are opposites, $\begin{cases} 4x + 6y = 34 \\ 15x - 6y = 42 \end{cases}$. This system has the same solution as the original, but in this version, you can eliminate y by adding the equations. If the coefficients were identical rather than opposite,

Be careful to multiply through the entire equation by the constant. Multiplying the variable terms but not the constant is a common error. Checking your solution in both of the original equations will help you catch your errors.

you could eliminate a variable by subtracting rather than adding. When the value of one variable is known, substitute it back into one of the original equations to find the other.

Exercise 9.1

Solve the following systems.

1. $3a+2b=15$
   $2a+b=8$
2. $3x+2y=23$
   $x-y=-9$
3. $8x+2y=16$
   $2x-y=7$
4. $5x-2y=29$
   $3x+4y=33$
5. $4x-3y=14$
   $3x+2y=19$
6. $2x+3y=4$
   $3x-8y=-9$
7. $3x-7y=30$
   $5x+y=12$
8. $7x-3y=1$
   $2x-y=1$
9. $2x+4y=5$
   $4x+5y=6$
10. $2x+3y=21$
    $4x+y=9$

Step 2. Solve Systems of Linear Equations in Three Variables

Because the graph of an equation with three variables is a three-dimensional surface, solving three-variable systems by graphing becomes impractical. For systems with three (or more) variables, algebraic methods are the better choice, although even those can be cumbersome.

There are more convenient methods, some of which will be covered in the next chapter.

Substitution

It is possible to solve a three-variable system by substitution if you can isolate a variable in one of the equations. You can substitute for that variable in the two remaining equations, leaving you with a two-variable system to solve. You'll need to substitute the solutions you find for that system back into one of the original equations to find the third variable.

To solve the system $\begin{cases} 2x - 3y + z = 27 \\ 3x + 5y - 2z = -2 \\ x + 3y - 5z = -22 \end{cases}$ by substitution, isolate z in the first equation to find $z = 27 - 2x + 3y$. Substitute $27 - 2x + 3y$ for z in the other two equations.

$$\begin{aligned} 3x + 5y - 2(27 - 2x + 3y) &= -2 \\ x + 3y - 5(27 - 2x + 3y) &= -22 \end{aligned}$$

Simplify each equation and then solve the system for x and y.

$$\begin{aligned} 3x + 5y - 54 + 4x - 6y = -2 \quad &\Rightarrow \quad 7x - y = 52 \\ x + 3y - 135 + 10x - 15y = -22 \quad &\Rightarrow \quad 11x - 12y = 113 \end{aligned}$$

$$\begin{aligned} 12(7x - y) &= (52) \cdot 12 \\ 11x - 12y &= 113 \\ \hline 84x - 12y &= 624 \\ 11x - 12y &= 113 \\ \hline 73x &= 511 \\ x &= 7 \\ 7x - y &= 52 \\ 7 \cdot 7 - y &= 52 \\ 49 - y &= 52 \\ y &= -3 \end{aligned}$$

Go back to one of the original equations to find z:

$$z = 27 - 2(7) + 3(-3) = 27 - 14 - 9 = 4$$

Elimination

If solving for one variable produces an expression that would be awkward to substitute, the elimination method may be a better choice. Elimination involves the same steps used in solving two-variable systems, but a careful, organized approach is essential.

Using two of the three equations, eliminate one variable. Then choose a different pair of equations, and eliminate the same variable. This gives you two equations with the same two variables.

To solve $\begin{cases} 2x - 3y + z = 27 \\ 3x + 5y - 2z = -2 \\ x + 3y - 5z = -22 \end{cases}$, first reduce it to a two-variable system.

Eliminate z in the first pair of equations.

$$\begin{array}{r} 2(2x - 3y + z) = (27)\cdot 2 \\ 3x + 5y - 2z = -2 \\ \hline 4x - 6y + 2z = 54 \\ 3x + 5y - 2z = -2 \\ \hline 7x - y = 52 \end{array}$$

Then eliminate the same variable using a different pair of equations.

$$\begin{array}{r} 5(3x + 5y - 2z) = (-2)\cdot 5 \\ 2(x + 3y - 5z) = (-22)\cdot 2 \\ \hline 15x + 25y - 10z = -10 \\ 2x + 6y - 10z = -44 \\ \hline 13x + 19y = 34 \end{array}$$

Solve the two-variable system you've created by any convenient method. Determine the value of either of the variables in the system. In this example, the reduced system is made up of the equations $7x - y = 52$ and $13x + 19y = 34$.

$$\begin{array}{r} 19(7x - y) = (52)\cdot 19 \\ 13x + 19y = 34 \\ \hline 133x - 19y = 988 \\ 13x + 19y = 34 \\ \hline 146x = 1022 \\ x = 7 \end{array}$$

Substitute the found value into one of the equations from the two-variable system to find the value of a second variable. Then substitute the two known values into one of the original equations to find the value of the third variable.

$$\begin{array}{r} 7x - y = 52 \\ 7\cdot 7 - y = 52 \\ 49 - y = 52 \\ y = -3 \end{array}$$

$$\begin{aligned} 2x - 3y + z &= 27 \\ 2 \cdot 7 - 3(-3) + z &= 27 \\ 14 + 9 + z &= 27 \\ 23 + z &= 27 \\ z &= 4 \end{aligned}$$

Exercise 9.2

Solve each system.

1. $4x - 2y + 3z = -20$
   $2x + 2y - z = 28$
   $x - 2y + 3z = -32$
2. $x - 7y - 3z = 6$
   $3x + y - z = 8$
   $x - y + 5z = 12$
3. $2x - y - 3z = 10$
   $x - 2y + 3z = -22$
   $3x + 5y - z = 63$
4. $x + y - z = -6$
   $x - y + z = 14$
   $x - y - z = 8$
5. $2x + 3y = 5$
   $2y + z = 5$
   $x + 3z = -5$
6. $3x - y + 4z = 36$
   $5x - 2y - 3z = -30$
   $5x - 6y + z = 42$
7. $3x - y + 4z = 16$
   $5x + 2y - 3z = 23$
   $5x - 6y + z = 31$
8. $5x - 7y + 2z = 3$
   $3x + 2y - z = 5$
   $4x - y + 3z = 2$
9. $5x - 7y + 2z = 40$
   $3x + 2y - z = 5$
   $4x - y + 3z = 27$
10. $x + 3y - z = 25$
    $3x + y + z = -5$
    $x - y + 3z = -23$

Step 3. Solve Nonlinear Systems

When systems of equations involve forms other than linear equations, the solution process can be more difficult—or at least less predictable. Each system will make different demands, and you'll need to adapt your strategies to the problem.

Graphing

For systems with two variables, sketching a graph of the system by hand or, when possible, examining the graphs on a graphing utility is often a good first move. The graph will tell you the number of solutions and will allow you to estimate the solutions. Your graphing utility should have a feature that locates the intersection of two graphs. If an algebraic solution is not possible, that feature will provide a good estimate.

Graphing the system of conics $\begin{cases} x^2+4x+4y^2-12=0 \\ x^2-2x-y^2+4y-7=0 \end{cases}$ does require taking the time to put the equations in standard form, but it will help you identify the number of solutions. In standard form, the equations become $\begin{cases} \dfrac{(x+2)^2}{16}+\dfrac{y^2}{4}=1 \\ \dfrac{(x-1)^2}{4}-\dfrac{(y-2)^2}{4}=1 \end{cases}$. The graph will show an ellipse centered at (–2,0) and a hyperbola centered at (1,2), which will intersect at two points. From a hand-drawn sketch of the graph, you can estimate that one point of intersection occurs at approximately (–1,2) and the other at about (–3.5,–2). A graphing utility would identify the points of intersection as (–1,1.94) and (–3.36,–1.88).

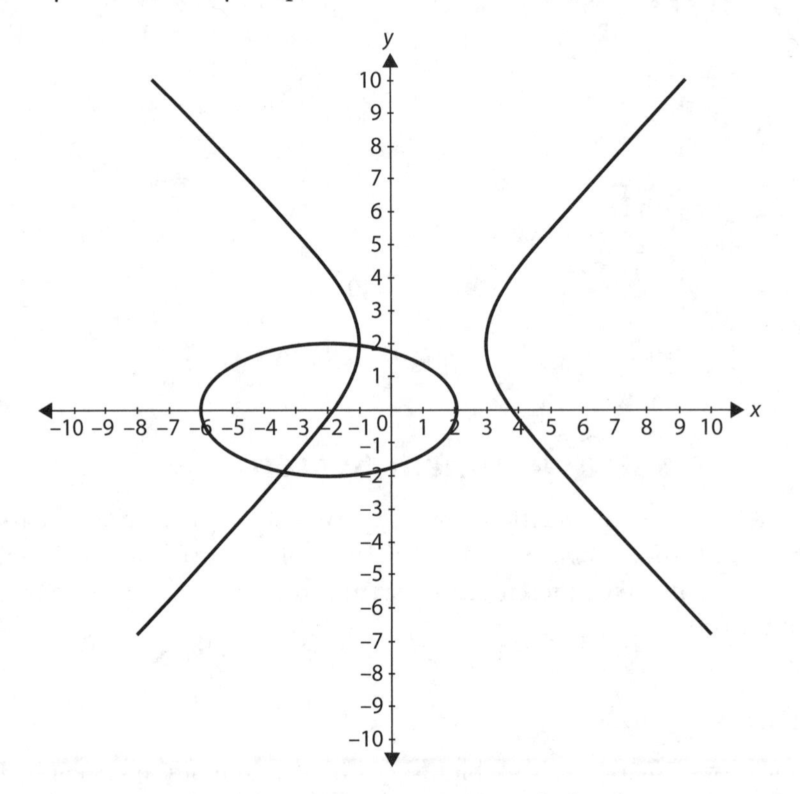

The solution set of this nonlinear system includes the two points at which the left wing of the hyperbola intersects the ellipse

Substitution

If you can isolate a variable in one of the equations, you can substitute into the other equation (or equations) and solve for one of the variables. Once you have the value of one variable, you can use substitution to find the other(s).

The system of conics in the previous example would be tedious to solve by substitution. Isolating y in the first equation would give you $y = \frac{\pm\sqrt{12-4x-x^2}}{2}$, and substituting that into the second equation would leave a lot of algebra between you and a solution.

On the other hand, the system $\begin{cases} y^2 - 9x + 5 = 0 \\ y^2 - 9x - 6y + 14 = 0 \end{cases}$ could be solved by substitution without much trouble. In the first equation, isolate $9x = y^2 + 5$, and substitute for $9x$ in the second equation.

$$\begin{aligned} y^2 - 9x - 6y + 14 &= 0 \\ y^2 - (y^2 + 5) - 6y + 14 &= 0 \\ -5 - 6y + 14 &= 0 \\ -6y &= -9 \\ y &= \frac{3}{2} \end{aligned}$$

You can substitute for y in the original equation to find the value of x.

$$\begin{aligned} y^2 - 9x + 5 &= 0 \\ \left(\frac{3}{2}\right)^2 - 9x + 5 &= 0 \\ \frac{9}{4} - 9x + \frac{20}{4} &= 0 \\ 9x &= \frac{29}{4} \\ x &= \frac{29}{36} \end{aligned}$$

Elimination

If the equations in the system involve like terms, you may be able to solve by elimination. Arrange the equations so that like terms are aligned under one another, multiply one or both equations by a constant as necessary to match coefficients, and add or subtract to eliminate a variable.

The system $\begin{cases} 4x^2-16x-9y^2=20 \\ 9x^2+54x+4y^2=45 \end{cases}$ can be solved by elimination if you align like terms and subtract.

$$\begin{aligned} 4(4x^2-16x-9y^2) &= 20\cdot 4 \\ 9(9x^2+54x+4y^2) &= -45\cdot 9 \\ \hline 16x^2-64x-36y^2 &= 80 \\ 81x^2+486x+36y^2 &= -405 \\ \hline 97x^2+422x \qquad &= -325 \\ 97x^2+422x+325 &= 0 \end{aligned}$$

The quadratic formula gives solutions of $x\approx -3.35$ and $x=-1$. Plugging $x=-1$ back into one of the equations gives you a single y-value of $y=0$.

$$\begin{aligned} 4x^2-16x-9y^2 &= 20 \\ 4(-1)^2-16(-1)-9y^2 &= 20 \\ 4+16-9y^2 &= 20 \\ -9y^2 &= 0 \\ y &= 0 \end{aligned}$$

Substituting $x\approx -3.35$ produces two values.

$$\begin{aligned} 4x^2-16x-9y^2 &= 20 \\ 4(-3.35)^2-16(-3.35)-9y^2 &= 20 \\ 98.49-9y^2 &= 20 \\ -9y^2 &= -78.49 \\ y^2 &= 8.72 \\ y &= \pm 2.95 \end{aligned}$$

The ellipse and hyperbola intersect at three points. They are tangent at (–1,0) and intersect at approximately (–3.35,2.95) and (–3.35,–2.95).

Exercise 9.3

Solve each system.

1. $4x + 3y = 24$
   $x^2 + y^2 = 25$

2. $4x^2 + 9y^2 = 36$
   $x^2 + y^2 - 2y = 9$

3. $x + 4 = y^2$
   $3 - x = y^2$

4. $xy = 8$
   $x^2 + y^2 = 16$

5. $9y^2 - 2x^2 = 16$
   $2x^2 + y^2 = 4$

6. $9x^2 + 4y^2 = 36$
   $9x^2 - 4y^2 = 36$

7. $y = \ln(x + 3)$
   $y = \ln(4 - x)$

8. $x^2 + 4y^2 = 16$
   $y = \sqrt{x + 1}$

9. $y = \ln(x - 2)$
   $e^{2y} + 3e^y - 10 = 0$

10. $y = 2^x$
    $y = 3^{x-1}$

10

Matrices and Determinants

In mathematics, a matrix is a rectangular arrangement of numbers, and the individual numbers within the matrix are called elements. Every matrix is organized into rows and columns. The dimension or order of a matrix is a description of its size, giving first the number of rows and then the number of columns. The matrix $\begin{bmatrix} 2 & 8 & 3 & 6 & 1 \\ 8 & 4 & 2 & 6 & 5 \end{bmatrix}$ is a 2×5 matrix, meaning that it has two rows and five columns. If the number of rows and columns are the same, the matrix is square. A matrix with only one row is called a row matrix, and a matrix with only one column is called a column matrix. Row matrices and column matrices are sometimes referred to as vectors.

Step 1: Master Matrix Arithmetic

Only matrices with identical dimensions can be added or subtracted. If two matrices have the same dimension, they can be added by adding the corresponding elements or subtracted by subtracting corresponding elements.

Your graphing calculator is probably capable of performing most matrix arithmetic operations.

To add the matrix $\begin{bmatrix} 8 & 6 & 2 & 9 \\ 7 & 1 & 5 & 0 \end{bmatrix}$ to the matrix $\begin{bmatrix} 9 & 3 & 7 & 2 \\ 0 & 4 & 1 & 8 \end{bmatrix}$, create a new matrix of the same dimension, and fill it with the sums of the corresponding elements.

$$\begin{bmatrix} 8 & 6 & 2 & 9 \\ 7 & 1 & 5 & 0 \end{bmatrix} + \begin{bmatrix} 9 & 3 & 7 & 2 \\ 0 & 4 & 1 & 8 \end{bmatrix} = \begin{bmatrix} 8+9 & 6+3 & 2+7 & 9+2 \\ 7+0 & 1+4 & 5+1 & 0+8 \end{bmatrix} = \begin{bmatrix} 17 & 9 & 9 & 11 \\ 7 & 5 & 6 & 8 \end{bmatrix}$$

To subtract the matrices $\begin{bmatrix} 8 & 6 & 2 & 9 \\ 7 & 1 & 5 & 0 \end{bmatrix} - \begin{bmatrix} 9 & 3 & 7 & 2 \\ 0 & 4 & 1 & 8 \end{bmatrix}$, form a new matrix of the same dimensions and fill it with the differences of the corresponding elements.

$$\begin{bmatrix} 8 & 6 & 2 & 9 \\ 7 & 1 & 5 & 0 \end{bmatrix} - \begin{bmatrix} 9 & 3 & 7 & 2 \\ 0 & 4 & 1 & 8 \end{bmatrix} = \begin{bmatrix} 8-9 & 6-3 & 2-7 & 9-2 \\ 7-0 & 1-4 & 5-1 & 0-8 \end{bmatrix} = \begin{bmatrix} -1 & 3 & -5 & 7 \\ 7 & -3 & 4 & -8 \end{bmatrix}$$

In matrix arithmetic, there are two types of multiplication, scalar multiplication and matrix multiplication. Scalar multiplication is the multiplication of a single number times a matrix. The single number is called a scalar.

In the scalar multiplication $3\begin{bmatrix} 1 & 7 & -2 \\ 0 & 5 & -1 \end{bmatrix}$, multiplying the 2×3 matrix by the scalar 3 has the effect of adding three copies of the matrix. Because scalar multiplication represents repeated addition, to multiply a matrix by a scalar, multiply each element of the matrix by the scalar.

Two matrices can be multiplied only if the number of elements in each row of the first matrix is equal to the number of elements in each column of the second. If $A = \begin{bmatrix} 1 & 3 & 2 \\ -2 & 0 & 4 \end{bmatrix}$ and $B = \begin{bmatrix} 1 & 0 & 3 & -2 \\ 2 & 1 & -4 & 3 \\ 5 & 0 & -3 & 1 \end{bmatrix}$, you can multiply $A \cdot B$ because there are three elements of each row of A and three elements of each column of B. In other words, the number of columns in the first matrix must be equal to the number of rows in the second. Matrix A has two rows and three columns and matrix B has three rows and four columns. If you write the dimensions of matrix A and then the dimensions of matrix B, and the dimensions in the middle match, the multiplication is possible. If the dimensions do not match, the multiplication cannot be performed.

$$\begin{matrix} \dim(A) & \dim(B) \\ 2 \times 3 & 3 \times 4 \end{matrix}$$

The middle dimensions must match for multiplication to be possible.

Looking at the dimensions can give you another piece of useful information as well. The remaining numbers tell you the dimension of the product matrix that will result. When you multiply $A \cdot B$, the dimensions tell you not only that the multiplication is possible but also that the product matrix will have dimension 2 × 4.

$$\begin{array}{ll} \dim(A) & \dim(B) \\ 2\times 3 & 3\times 4 \end{array}$$

The outer numbers are the dimensions of the product matrix.

The need for matching dimensions makes the order of multiplication important. While we have seen that it is possible to multiply $A \cdot B$, it is not possible to perform the multiplication $B \cdot A$.

$$\begin{array}{ll} \dim(B) & \dim(A) \\ 3\times 4 & 2\times 3 \end{array}$$

The dimensions show that the product *BA* is not possible even though *AB* is possible.

You find the product of two matrices by repeating a process of multiplying one row by one column. Multiply the first element in the row by the first element in the column, then the second element in the row by the second element in the column, and so on. Add these products to form the single element in the product matrix. The result of multiplying a row times a column is a single element.

$$\begin{bmatrix} 3 & 2 & 5 & 1 \end{bmatrix} \cdot \begin{bmatrix} -2 \\ 1 \\ 4 \\ -3 \end{bmatrix} = [3(-2)+2(1)+5(4)+1(-3)] = [-6+2+20+-3] = [13]$$

The process of multiplying two larger matrices repeats these same steps, with each row of the first matrix producing a row of the product matrix. To multiply two matrices, begin by multiplying the first row of the first matrix by each column of the second matrix, placing the results in the first row of the product matrix. Repeat the process using each row of the first matrix, and place the results in the corresponding row of the product matrix.

It's wise to practice problems involving a row matrix times a column matrix until you're confident with the multiplication process. Multiplying larger matrices is just a matter of repeating that operation until you've multiplied each row of the first matrix by each column of the second.

The 4 × 2 matrix $\begin{bmatrix} 1 & 3 \\ 2 & 4 \\ 5 & -1 \\ -2 & -3 \end{bmatrix}$ multiplied by the 2 × 3 matrix $\begin{bmatrix} -2 & 1 & 0 \\ 3 & -1 & 4 \end{bmatrix}$

will yield a 4 × 3 product.

Multiply the first row by each column.

$$\begin{bmatrix} 1 & 3 \\ 2 & 4 \\ 5 & -1 \\ -2 & -3 \end{bmatrix} \times \begin{bmatrix} -2 & 1 & 0 \\ 3 & -1 & 4 \end{bmatrix} = \begin{bmatrix} (1\cdot -2)+(3\cdot 3) & (1\cdot 1)+(3\cdot -1) & (1\cdot 0)+(3\cdot 4) \\ ? & ? & ? \\ ? & ? & ? \\ ? & ? & ? \end{bmatrix}$$

$$= \begin{bmatrix} 7 & -2 & 12 \\ ? & ? & ? \\ ? & ? & ? \\ ? & ? & ? \end{bmatrix}$$

Repeat for each row.

$$\begin{bmatrix} 1 & 3 \\ 2 & 4 \\ 5 & -1 \\ -2 & -3 \end{bmatrix} \times \begin{bmatrix} -2 & 1 & 0 \\ 3 & -1 & 4 \end{bmatrix} = \begin{bmatrix} 7 & -2 & 12 \\ 8 & -2 & 16 \\ ? & ? & ? \\ ? & ? & ? \end{bmatrix}$$

$$\begin{bmatrix} 1 & 3 \\ 2 & 4 \\ 5 & -1 \\ -2 & -3 \end{bmatrix} \times \begin{bmatrix} -2 & 1 & 0 \\ 3 & -1 & 4 \end{bmatrix} = \begin{bmatrix} 7 & -2 & 12 \\ 8 & -2 & 16 \\ -13 & 6 & -4 \\ ? & ? & ? \end{bmatrix}$$

$$\begin{bmatrix} 1 & 3 \\ 2 & 4 \\ 5 & -1 \\ -2 & -3 \end{bmatrix} \times \begin{bmatrix} -2 & 1 & 0 \\ 3 & -1 & 4 \end{bmatrix} = \begin{bmatrix} 7 & -2 & 12 \\ 8 & -2 & 16 \\ -13 & 6 & -4 \\ -5 & 1 & -12 \end{bmatrix}$$

Exercise 10.1

Perform each operation, or explain why it is not possible.

1. $4\begin{bmatrix} 2 & -3 \\ 1 & 4 \end{bmatrix}$

2. $\begin{bmatrix} 1 & 2 & 3 \end{bmatrix}\begin{bmatrix} 2 \\ -1 \\ 4 \end{bmatrix}$

3. $\begin{bmatrix} 2 \\ -1 \\ 4 \end{bmatrix}\begin{bmatrix} 1 & 2 & 3 \end{bmatrix}$

4. $\begin{bmatrix} 7 & 0 & 8 \\ -5 & 1 & 4 \end{bmatrix}+\begin{bmatrix} -3 & 5 & 0 \\ 7 & 3 & 6 \end{bmatrix}$

5. $\begin{bmatrix} 4 & 0 & -3 \\ -5 & 2 & 0 \end{bmatrix}+\begin{bmatrix} -4 & 1 \\ 3 & 7 \end{bmatrix}$

6. $\begin{bmatrix} 2 & -1 & 7 \\ 5 & 8 & 2 \end{bmatrix}-\begin{bmatrix} 2 & 9 & 3 \\ -5 & 2 & -2 \end{bmatrix}$

7. $\left(\begin{bmatrix} 4 & 0 \\ -1 & 5 \end{bmatrix}-\begin{bmatrix} 6 & -2 \\ 3 & 4 \end{bmatrix}\right)+\begin{bmatrix} 3 & 5 \\ -1 & 8 \end{bmatrix}$

8. $\begin{bmatrix} 1 & -6 \\ 9 & 7 \\ 5 & 3 \end{bmatrix}+\begin{bmatrix} 0 & 3 & 6 \\ 1 & 4 & 7 \\ 2 & 5 & 8 \end{bmatrix}$

9. $\begin{bmatrix} 4 & 1 & 0 \\ -3 & 5 & 2 \end{bmatrix}\begin{bmatrix} 9 & -3 \\ -6 & 4 \\ 0 & 1 \end{bmatrix}$

10. $\begin{bmatrix} 9 & -3 \\ -6 & 4 \\ 0 & 1 \end{bmatrix}\begin{bmatrix} 4 & 1 & 0 \\ -3 & 5 & 2 \end{bmatrix}$

Step 2. Find Determinants

The determinant of a matrix is a single number associated with the matrix. The determinant of a matrix $A=\begin{bmatrix} 4 & 8 \\ 3 & 1 \end{bmatrix}$ is indicated as $|A|$ or by $\begin{vmatrix} 4 & 8 \\ 3 & 1 \end{vmatrix}$.

Only square matrices have determinants. In a square matrix, the diagonal path from upper left to lower right is called the major diagonal. The diagonal from upper right to lower left is the minor diagonal. In the matrix $\begin{bmatrix} 4 & 8 \\ 3 & 1 \end{bmatrix}$, the major diagonal contains the elements 4 and 1, while the minor diagonal contains 8 and 3.

The determinant of a 2×2 matrix is equal to the product of the elements on the major diagonal minus the product of the elements on the minor diagonal. The determinant $\begin{vmatrix} 4 & 8 \\ 3 & 1 \end{vmatrix} = 4\times 1-8\times 3=4-24=-20$. In symbolic terms,

$\begin{vmatrix} a & b \\ c & d \end{vmatrix} = ad - bc$. The determinant of the square matrix $M = \begin{bmatrix} -2 & 3 \\ 1 & -4 \end{bmatrix}$ is $|M| = \begin{vmatrix} -2 & 3 \\ 1 & -4 \end{vmatrix} = -2 \times -4 - 3 \times 1 = 8 - 3 = 5$.

To find the determinant of a 3 × 3 (or larger) matrix, most people turn to technology. The paper and pencil method is tedious, but if you don't have the technology available to you, just work carefully.

Your graphing calculator should be able to find a determinant. Look in the menu of matrix math for a command such as det(*A*).

Begin with a general expression for the determinant of a 3 × 3 matrix: $\begin{vmatrix} a & b & c \\ d & e & f \\ g & h & i \end{vmatrix}$. Imagine three copies of this determinant, but in each copy, circle one element of the top row and then cross out the rest of the row and the rest of the column containing the circled element. The three versions should look like this:

$$\begin{vmatrix} \boxed{a} & \not{b} & \not{c} \\ \not{d} & e & f \\ \not{g} & h & i \end{vmatrix} \qquad \begin{vmatrix} \not{a} & \boxed{b} & \not{c} \\ d & \not{e} & f \\ g & \not{h} & i \end{vmatrix} \qquad \begin{vmatrix} \not{a} & \not{b} & \boxed{c} \\ d & e & \not{f} \\ g & h & \not{i} \end{vmatrix}$$

In each version, you see four elements untouched and forming a 2 × 2 matrix. Calculate the determinant of each of the remaining 2 × 2 matrices. The determinant of the 3 × 3 matrix is:

$$\begin{vmatrix} a & b & c \\ d & e & f \\ g & h & i \end{vmatrix} = a\begin{vmatrix} e & f \\ h & i \end{vmatrix} - b\begin{vmatrix} d & f \\ g & i \end{vmatrix} + c\begin{vmatrix} d & e \\ g & h \end{vmatrix}$$

Notice that the signs alternate positive and negative. If you expand across an odd-numbered row (or down an odd-numbered column), start with a positive and alternate. If you expand across an even-numbered row or column, start with a negative and alternate.

To find the determinant of the 3 × 3 matrix $\begin{vmatrix} 4 & 1 & 3 \\ -2 & 2 & -1 \\ 0 & -3 & 1 \end{vmatrix}$, expand across the first row.

$$\begin{vmatrix} 4 & 1 & 3 \\ -2 & 2 & -1 \\ 0 & -3 & 1 \end{vmatrix} = 4\begin{vmatrix} 2 & -1 \\ -3 & 1 \end{vmatrix} - 1\begin{vmatrix} -2 & -1 \\ 0 & 1 \end{vmatrix} + 3\begin{vmatrix} -2 & 2 \\ 0 & -3 \end{vmatrix}$$

$$= 4[2\cdot 1-(-3)(-1)]-1[-2\cdot 1-0(-1)]+3[(-2)(-3)-0\cdot 2]$$

$$= 4[2-3]-1[-2-0]+3[6-0]$$

$$= 4[-1]-1[-2]+3[6]$$

$$= -4+2+18$$

Exercise 10.2

Solve the following problems.

1. $\begin{vmatrix} 4 & 2 \\ 9 & 6 \end{vmatrix}$

2. $\begin{vmatrix} -5 & 6 \\ 5 & 4 \end{vmatrix}$

3. $\begin{vmatrix} -3 & 1 \\ -5 & 8 \end{vmatrix}$

4. $\begin{vmatrix} 12 & 0 \\ -3 & \frac{1}{2} \end{vmatrix}$

5. $\begin{vmatrix} 5 & 3 \\ 2 & -9 \end{vmatrix}$

6. $\begin{vmatrix} 11 & 10 \\ 2 & 1 \end{vmatrix}$

7. $\begin{vmatrix} 3 & 2 \\ 1 & 0 \end{vmatrix}$

8. $\begin{vmatrix} 4 & 1 & -2 \\ 3 & 0 & 1 \\ -1 & -2 & 5 \end{vmatrix}$

9. $\begin{vmatrix} -3 & 7 & 2 \\ 0 & 4 & -1 \\ 3 & 2 & 0 \end{vmatrix}$

10. $\begin{vmatrix} -1 & 2 & 1 \\ 2 & 0 & 0 \\ 3 & -4 & 2 \end{vmatrix}$

Step 3. Apply Cramer's Rule

Cramer's rule is a method for determining the solution of a system of equations by means of determinants. In the elimination method of solving a system of equations, one or both equations are multiplied by a constant, and the equations are added or subtracted to eliminate one of the variables. That process often leads to a pattern of multiplication that is replicated by the process of finding determinants.

To eliminate x from the system $\begin{cases} 2x+3y=19 \\ 3x-4y=3 \end{cases}$, the first equation can be multiplied by 3 and the second equation multiplied by −2.

$$\begin{aligned} 3(2x+3y) &= (19)\cdot 3 \\ -2(3x-4y) &= (3)\cdot -2 \end{aligned} \quad \text{becomes} \quad \begin{aligned} 6x+9y &= 57 \\ -6x+8y &= -6 \end{aligned}$$

Adding the equations eliminates x.

$$\begin{aligned} 6x+9y &= 57 \\ -6x+8y &= -6 \\ \hline 17y &= 51 \\ y &= \frac{51}{17} = 3 \end{aligned}$$

If instead we choose to eliminate y, the first equation is multiplied by 4 and the second by 3.

$$\begin{aligned} 4(2x+3y) &= (19)\cdot 4 \\ 3(3x-4y) &= (3)\cdot 3 \end{aligned} \quad \text{becomes} \quad \begin{aligned} 8x+12y &= 76 \\ 9x-12y &= 9 \end{aligned}$$

Then the equations are added.

$$\begin{aligned} 8x + \cancel{12y} &= 76 \\ 9x - \cancel{12y} &= 9 \\ \hline 17x \quad\quad &= 85 \\ x &= \frac{85}{17} = 5 \end{aligned}$$

The denominator of 17, common to both $x=\frac{85}{17}=5$ and $y=\frac{51}{17}=3$, is equal to 2 · 4 + 3 · 3, the opposite of the determinant of the coefficient matrix: $\begin{vmatrix} 2 & 3 \\ 3 & -4 \end{vmatrix} = 2(-4)-3(3) = -8-9 = -17$. The 85 in $x=\frac{85}{17}=5$ can be produced from a determinant involving the coefficients of the y-terms and the constants and the 51 in $y=\frac{51}{17}=3$ from the coefficients of the x-terms and the constants. Cramer's rule recognizes this and uses determinants to arrive at the solution of the system quickly and easily.

To use Cramer's rule to solve a system, first form a matrix of the coefficients of the variables, placing the coefficients of each variable in a column. Find the determinant of that matrix. For each variable in the system, modify the coefficient matrix by replacing the values in the column that corresponds to that variable with the constants. Find the determinant of each matrix.

For the system $\begin{cases} 2x+3y=19 \\ 3x-4y=3 \end{cases}$, the matrix of coefficients has the determinant $D=\begin{vmatrix} 2 & 3 \\ 3 & -4 \end{vmatrix}=2(-4)-3(3)=-8-9=-17$. When the first column is replaced with the constants, the matrix has determinant $N_x=\begin{vmatrix} 19 & 3 \\ 3 & -4 \end{vmatrix}=19(-4)-3(3)=-76-9=-85$. When the y column is replaced with the constants, the determinant of the resulting matrix is $N_y=\begin{vmatrix} 2 & 19 \\ 3 & 3 \end{vmatrix}=2(3)-3(19)=6-57=-51$.

To find the values of the variables, divide the determinant created when the constants are inserted by the determinant of the coefficient matrix.

$$x=\frac{N_x}{D}=\frac{-85}{-17}=5$$

$$y=\frac{N_y}{D}=\frac{-51}{-17}=3$$

Exercise 10.3

Solve each system.

1. $2x+y=4$
   $4x-3y=13$
2. $2x+3y=4$
   $3x-8y=-9$
3. $2x+3y=-8$
   $x+2y=-3$
4. $3x-7y=30$
   $5x+y=12$
5. $2x-y=5$
   $-x+y=-3$
6. $5x-2y=29$
   $3x+4y=33$
7. $-3x+4y=-6$
   $5x-3z=-22$
   $3y+2z=-1$
8. $x+2y-3z=5$
   $x-y+2z=-3$
   $x+y-z=2$
9. $x+2y+3z=17$
   $-x-5y+4z=2$
   $2x+7y-5z=-1$
10. $5x+10y=70$
    $5x+25z=270$
    $10y+25z=300$

Step 4. Find Inverse Matrices

In arithmetic, you learned that every nonzero number has a reciprocal and that the product of the number and its reciprocal is 1. The reciprocal of $\frac{3}{5}$, for example, is $\frac{5}{3}$, and the product $\frac{3}{5} \cdot \frac{5}{3} = 1$.

What is commonly called the reciprocal is formally called the multiplicative inverse. Two numbers are multiplicative inverses if their product is 1, the identity element for multiplication.

To transfer the concept of reciprocal (multiplicative inverse) to matrix arithmetic, remember that the identity element for matrix multiplication is a square matrix composed of ones on the major diagonal and zeros elsewhere. In order for two matrices to be called inverses, their product must be such an identity matrix. Matrix multiplication is not generally commutative, so add the condition that to be called inverses the two matrices must produce the same identity when multiplied in either order, which requires that the two matrices both be square. If they were not square, then the product $A \times B$ would be of a different size than the product $B \times A$, even if both were identities.

If A and B are square matrices and I is an identity matrix of the same dimension, and if $A \times B = B \times A = I$, then A and B are inverse matrices. We denote the inverse of matrix M as M^{-1}.

Just as only square matrices have determinants, so only square matrices have inverses. Not all square matrices actually do have inverses, however, and those matrices that have determinants of 0 have no inverse. Such a matrix is not invertible.

If you try to find the inverse of a matrix with a 0 determinant, the calculator error will probably say "singular matrix."

To determine whether two matrices are inverses, check both possible products. To determine if $A = \begin{bmatrix} 4 & 3 \\ 3 & 2 \end{bmatrix}$ and $B = \begin{bmatrix} -2 & 3 \\ 3 & -4 \end{bmatrix}$ are inverses, calculate both the products $A \times B$ and $B \times A$.

$$A \times B = \begin{bmatrix} 4 & 3 \\ 3 & 2 \end{bmatrix} \times \begin{bmatrix} -2 & 3 \\ 3 & -4 \end{bmatrix} = \begin{bmatrix} -8+9 & 12-12 \\ -6+6 & 9-8 \end{bmatrix} = \begin{bmatrix} 1 & 0 \\ 0 & 1 \end{bmatrix}$$

$$B \times A = \begin{bmatrix} -2 & 3 \\ 3 & -4 \end{bmatrix} \times \begin{bmatrix} 4 & 3 \\ 3 & 2 \end{bmatrix} = \begin{bmatrix} -8+9 & -6+6 \\ 12-12 & 9-8 \end{bmatrix} = \begin{bmatrix} 1 & 0 \\ 0 & 1 \end{bmatrix}$$

It is important to check both products. It is possible to find two matrices that produce an identity when multiplied in one order but not in the other. Such matrices are referred to as one-sided inverses. One failure is enough to tell you that the two matrices are not inverses, however.

$$M \times N = \begin{bmatrix} 1 & 4 \\ 0 & 2 \end{bmatrix} \times \begin{bmatrix} 1 & -4 \\ 0 & 0.5 \end{bmatrix} = \begin{bmatrix} 1+0 & -4+4 \\ 0+0 & 0+1 \end{bmatrix} = \begin{bmatrix} 1 & 0 \\ 0 & 1 \end{bmatrix}$$

$$N \times M = \begin{bmatrix} 1 & -4 \\ 0 & 0.5 \end{bmatrix} \times \begin{bmatrix} 1 & 4 \\ 0 & 2 \end{bmatrix} = \begin{bmatrix} 1+0 & 4-8 \\ 0+0 & 0+1 \end{bmatrix} = \begin{bmatrix} 1 & -4 \\ 0 & 1 \end{bmatrix}$$

Finding the inverse of a 2 × 2 matrix is a relatively simple process.

1. Find the determinant of the matrix.
2. Exchange the elements on the major diagonal.
3. Change the signs of the elements on the minor diagonal.
4. Multiply by the reciprocal of the determinant.

To find the inverse of the matrix $\begin{bmatrix} 1 & 4 \\ 0 & 2 \end{bmatrix}$ we first find the determinant of the matrix. Matrices with determinants of 0 have no inverse.

$$\begin{vmatrix} 1 & 4 \\ 0 & 2 \end{vmatrix} = 2 - 4 = -2$$

Next swap the elements on the major diagonal, 1 and 2, and change the signs of the elements on the minor diagonal. Because 0 is neither positive nor negative, it remains 0.

$$\begin{bmatrix} 1 & 4 \\ 0 & 2 \end{bmatrix} \text{ becomes } \begin{bmatrix} 2 & -4 \\ 0 & 1 \end{bmatrix}.$$

Finally, perform a scalar multiplication, multiplying the matrix just formed by the reciprocal of the determinant of the original matrix, $\frac{1}{-2}$. In cases when the determinant is 0, it is impossible to find a reciprocal and the process is stopped.

$$\frac{1}{-2} \cdot \begin{bmatrix} 2 & -4 \\ 0 & 1 \end{bmatrix} = \begin{bmatrix} -1 & 2 \\ 0 & -\frac{1}{2} \end{bmatrix}$$

The inverse of the matrix $\begin{bmatrix} 1 & 4 \\ 0 & 2 \end{bmatrix}$ is the matrix $\begin{bmatrix} -1 & 2 \\ 0 & -\frac{1}{2} \end{bmatrix}$.

For larger matrices, it's helpful to have a calculator with matrix operations because the process of finding an inverse gets complicated quickly as the matrix size increases. If you must find the inverse of a larger matrix, follow these steps.

1. Find the determinant of the matrix. If the determinant is 0, the matrix is not invertible, so stop right there.
2. Fill a matrix of the same size with position signs. Start with a plus in row 1, column 1, and alternate plus and minus across the row. Start the next row with a minus, and continue alternating.
3. Determine the cofactor of each element, and fill the matrix of cofactors. The cofactor of an element is found by eliminating the row and column that contain the element and calculating the determinant of the remaining matrix. For a 3×3 matrix, crossing out a row and column will leave you with the determinant of a 2×2, but for larger matrices, this step is where it gets complicated.
4. Transpose the matrix of cofactors to form the adjoint. Form the adjoint by placing the elements that had been the first row of the original matrix in the first column of the adjoint, the second row in the second column, and so on.
5. Multiply the adjoint by the reciprocal of the determinant.

This should make it clear why most people prefer to use technology when it's available. Just enter the matrix, and then type the matrix name and the inverse (x^{-1}) key.

To find the inverse of the matrix $\begin{bmatrix} 1 & 3 & -2 \\ 5 & 0 & -1 \\ 4 & 2 & -5 \end{bmatrix}$, begin by finding the determinant.

$$\begin{aligned} \begin{vmatrix} 1 & 3 & -2 \\ 5 & 0 & -1 \\ 4 & 2 & -5 \end{vmatrix} &= 1\begin{vmatrix} 0 & -1 \\ 2 & -5 \end{vmatrix} - 3\begin{vmatrix} 5 & -1 \\ 4 & -5 \end{vmatrix} + {}^{-}2\begin{vmatrix} 5 & 0 \\ 4 & 2 \end{vmatrix} \\ &= 1(0+2) - 3(-25+4) + -2(10-0) \\ &= 2 + 63 - 20 \\ &= 45 \end{aligned}$$

Because the matrix is square and the determinant is nonzero, the inverse exists. Create a matrix of the same size, and fill it with an alternating pattern of pluses and minuses, the position signs.

$$\begin{bmatrix} + & - & + \\ - & + & - \\ + & - & + \end{bmatrix}$$

Return to the original matrix to find the cofactor of each element. In the original matrix, find the cofactor of the 1 in row 1, column 1 by crossing out the first row and first column, and find the determinant of the remaining matrix.

$$\begin{bmatrix} \cancel{1} & \cancel{3} & \cancel{2} \\ \cancel{5} & 0 & -1 \\ \cancel{4} & 2 & -5 \end{bmatrix}$$

$$\begin{vmatrix} 0 & -1 \\ 2 & -5 \end{vmatrix} = 0 - -2 = 2$$

The cofactor of the first element is 2. Place this cofactor in the first row, first column of the matrix of position signs: $\begin{bmatrix} +2 & - & + \\ - & + & - \\ + & - & + \end{bmatrix}$. Repeat this process for each element in the original matrix, placing the cofactor in the corresponding place in the matrix of position signs. To find the cofactor of the 3 in row 1, column 2, eliminate row 1 and column 2, and find the determinant $\begin{vmatrix} 5 & -1 \\ 4 & -5 \end{vmatrix} = -25 + 4 = -21$, and place it with its position sign. Because the sign for this position is a minus, the element becomes positive.

$$\begin{bmatrix} +2 & -{}^{-}21 & + \\ - & + & - \\ + & - & + \end{bmatrix} = \begin{bmatrix} +2 & 21 & + \\ - & + & - \\ + & - & + \end{bmatrix}$$

Repeat the process, moving from element to element until the new matrix is full. The matrix of cofactors is $\begin{bmatrix} +2 & 21 & +10 \\ -{}^{-}11 & +3 & -{}^{-}10 \\ +{}^{-}3 & -9 & +{}^{-}15 \end{bmatrix} = \begin{bmatrix} 2 & 21 & 10 \\ 11 & 3 & 10 \\ -3 & -9 & -15 \end{bmatrix}$.

Transpose the matrix of cofactors $\begin{bmatrix} 2 & 21 & 10 \\ 11 & 3 & 10 \\ -3 & -9 & -15 \end{bmatrix}$ to $\begin{bmatrix} 2 & 11 & -3 \\ 21 & 3 & -9 \\ 10 & 10 & -15 \end{bmatrix}$. The transposed matrix of cofactors is called the adjoint. Finally, multiply the adjoint by the reciprocal of the determinant. The determinant, calculated earlier, was 45, so multiply by $\frac{1}{45}$.

$$\frac{1}{45}\cdot\begin{bmatrix} 2 & 11 & -3 \\ 21 & 3 & -9 \\ 10 & 10 & -15 \end{bmatrix} = \begin{bmatrix} \frac{2}{45} & \frac{11}{45} & -\frac{3}{45} \\ \frac{21}{45} & \frac{3}{45} & -\frac{9}{45} \\ \frac{10}{45} & \frac{10}{45} & -\frac{15}{45} \end{bmatrix} = \begin{bmatrix} \frac{2}{45} & \frac{11}{45} & -\frac{1}{15} \\ \frac{7}{15} & \frac{1}{15} & -\frac{1}{5} \\ \frac{2}{9} & \frac{2}{9} & -\frac{1}{3} \end{bmatrix}$$

Your calculator will generally display decimal equivalents, but look in the math menu for a function that converts back to fraction form.

Exercise 10.4

Determine whether the given matrices are inverses.

1. $\begin{bmatrix} 4 & 3 \\ 3 & 2 \end{bmatrix}$ and $\begin{bmatrix} -2 & 3 \\ 3 & -4 \end{bmatrix}$

2. $\begin{bmatrix} 1 & 4 \\ 0 & 2 \end{bmatrix}$ and $\begin{bmatrix} 1 & -4 \\ 0 & \frac{1}{2} \end{bmatrix}$

3. $\begin{bmatrix} 1 & 2 & 3 \\ 0 & 2 & 1 \\ 0 & 0 & 3 \end{bmatrix}$ and $\begin{bmatrix} 1 & -1 & -\frac{2}{3} \\ 0 & \frac{1}{2} & -\frac{1}{6} \\ 0 & 0 & \frac{1}{3} \end{bmatrix}$

4. $\begin{bmatrix} 2 & 1 & 3 \\ 5 & 1 & -5 \\ -3 & 0 & -2 \end{bmatrix}$ and $\begin{bmatrix} -\frac{1}{15} & \frac{1}{15} & -\frac{4}{15} \\ \frac{5}{6} & \frac{1}{6} & \frac{5}{6} \\ \frac{1}{10} & -\frac{1}{10} & -\frac{1}{10} \end{bmatrix}$

5. $\begin{bmatrix} 1 & 3 & 8 \\ -4 & 0 & 2 \\ -5 & 6 & 1 \end{bmatrix}$ and $\begin{bmatrix} 1 & -4 & -5 \\ 3 & 0 & 6 \\ 8 & 2 & 1 \end{bmatrix}$

Find the inverse of each matrix, if possible.

6. $\begin{bmatrix} 3 & -1 \\ -5 & 2 \end{bmatrix}$

7. $\begin{bmatrix} 8 & 4 \\ 5 & 3 \end{bmatrix}$

8. $\begin{bmatrix} -1 & 4 & 2 \\ -2 & 2 & -3 \end{bmatrix}$

9. $\begin{bmatrix} 2 & 1 & 0 \\ 0 & 1 & 2 \\ 1 & 0 & 2 \end{bmatrix}$

10. $\begin{bmatrix} 2 & 8 & 1 \\ 0 & 5 & 4 \\ 0 & 0 & 0 \end{bmatrix}$

Step 5. Solve Systems by Inverse Matrices

Cramer's rule allows you to solve systems of equations by means of determinants formed from the coefficients and constants of the system. With the help of inverses, you can solve systems even more quickly.

If you organize the information in the system into matrices—one for the coefficients of the variables, one for the variables themselves, and one for the constants—the system can be represented by a single matrix multiplication. The coefficients form a square matrix, which can be multiplied by a column matrix containing the variables, with the result equal to a column matrix containing the constants. The system $\begin{cases} 3x - 2y = 13 \\ x + 3y = 8 \end{cases}$ breaks down into the coefficient matrix $\begin{bmatrix} 3 & -2 \\ 1 & 3 \end{bmatrix}$, the variable matrix $\begin{bmatrix} x \\ y \end{bmatrix}$, and the constant matrix $\begin{bmatrix} 13 \\ 8 \end{bmatrix}$, and because $\begin{bmatrix} 3 & -2 \\ 1 & 3 \end{bmatrix} \cdot \begin{bmatrix} x \\ y \end{bmatrix} = \begin{bmatrix} 3x - 2y \\ x + 3y \end{bmatrix}$, the matrix equation $\begin{bmatrix} 3 & -2 \\ 1 & 3 \end{bmatrix} \cdot \begin{bmatrix} x \\ y \end{bmatrix} = \begin{bmatrix} 13 \\ 8 \end{bmatrix}$ is equivalent to the system.

The algebra of matrices allows you to multiply both sides of this equation by the same matrix, just as standard algebra allows you to multiply both sides of an equation by the same number. If you multiply both sides of this equation by the inverse of the coefficient matrix, you'll have a solution, because $[A]^{-1} \cdot [A] \cdot [X] = [A]^{-1} \cdot [B]$ simplifies to $[X] = [A]^{-1} \cdot [B]$.

The coefficient matrix from the previous example is $\begin{bmatrix} 3 & -2 \\ 1 & 3 \end{bmatrix}$, and its inverse is $\begin{bmatrix} \frac{3}{11} & \frac{2}{11} \\ \frac{-1}{11} & \frac{3}{11} \end{bmatrix}$. Multiplying both sides of the equation by the inverse gives you

$$\begin{bmatrix} 3 & -2 \\ 1 & 3 \end{bmatrix}^{-1} \cdot \begin{bmatrix} 3 & -2 \\ 1 & 3 \end{bmatrix} \cdot \begin{bmatrix} x \\ y \end{bmatrix} = \begin{bmatrix} 3 & -2 \\ 1 & 3 \end{bmatrix}^{-1} \cdot \begin{bmatrix} 13 \\ 8 \end{bmatrix}$$

$$\underbrace{\begin{bmatrix} \frac{3}{11} & \frac{2}{11} \\ \frac{-1}{11} & \frac{3}{11} \end{bmatrix} \cdot \begin{bmatrix} 3 & -2 \\ 1 & 3 \end{bmatrix}}_{\begin{bmatrix} 1 & 0 \\ 0 & 1 \end{bmatrix}} \cdot \begin{bmatrix} x \\ y \end{bmatrix} = \begin{bmatrix} \frac{3}{11} & \frac{2}{11} \\ \frac{-1}{11} & \frac{3}{11} \end{bmatrix} \cdot \begin{bmatrix} 13 \\ 8 \end{bmatrix}$$

$$\begin{bmatrix} 1 & 0 \\ 0 & 1 \end{bmatrix} \begin{bmatrix} x \\ y \end{bmatrix} = \begin{bmatrix} \frac{3}{11} & \frac{2}{11} \\ \frac{-1}{11} & \frac{3}{11} \end{bmatrix} \cdot \begin{bmatrix} 13 \\ 8 \end{bmatrix}$$

$$\begin{bmatrix} x \\ y \end{bmatrix} = \begin{bmatrix} \frac{39}{11} + \frac{16}{11} \\ \frac{-13}{11} + \frac{24}{11} \end{bmatrix} = \begin{bmatrix} 5 \\ 1 \end{bmatrix}$$

The logic of this method holds for systems of any size, and because calculators allow you to find inverses and to multiply matrices easily, the technique is an easy way to solve systems of any size.

Except where required to document your work, it's not necessary to find the inverse before solving the system. If the coefficients are in matrix [*A*] and the constants in matrix [*B*], typing $[A]^{-1}[B]$ will produce the solution.

To solve $\begin{cases} 2x - 3y + z = 27 \\ 3x + 5y - 2z = -2 \\ x + 3y - 5z = -22 \end{cases}$ create the equivalent matrix equation.

$$\begin{bmatrix} 2 & -3 & 1 \\ 3 & 5 & -2 \\ 1 & 3 & -5 \end{bmatrix} \cdot \begin{bmatrix} x \\ y \\ z \end{bmatrix} = \begin{bmatrix} 27 \\ -2 \\ -22 \end{bmatrix}$$

Then multiply both sides by the inverse of the coefficient matrix.

$$\begin{bmatrix} x \\ y \\ z \end{bmatrix} = \begin{bmatrix} 2 & -3 & 1 \\ 3 & 5 & -2 \\ 1 & 3 & -5 \end{bmatrix}^{-1} \cdot \begin{bmatrix} 27 \\ -2 \\ -22 \end{bmatrix}$$

Completing the calculation tells you that $x = 7$, $y = -3$, and $z = 4$.

Exercise 10.5

Solve each system by inverse matrices.

1. $4x - 3y = 14$
   $3x + 2y = 19$
2. $2x + 3y = 4$
   $3x - 8y = -9$
3. $3x - 7y = 30$
   $5x + y = 12$
4. $7x - 3y = 1$
   $2x - y = 1$
5. $2x + 4y = 5$
   $4x + 5y = 6$
6. $2x + 3y = 21$
   $4x + y = 9$
7. $3x - y + 4z = 37.5$
   $5x + 2y - 3z = 25.5$
   $5x - 6y + z = 63.5$
8. $5x - 7y + 2z = 44$
   $3x + 2y - z = -14$
   $4x - y + 3z = 17$
9. $5x - 7y + 2z = 40$
   $3x + 2y - z = 5$
   $4x - y + 3z = 27$
10. $x + 3y - z = 25$
    $3x + y + z = -5$
    $x - y + 3z = -23$

11

Triangle Trigonometry

Trigonometry, or "triangle measurement," developed as a means to calculate measurements of sides and angles of triangles. It has applications far beyond the triangle, but it's always a good plan to start at the beginning.

Step 1. Review Right Triangle Trigonometry

If the three sides of the right triangle are labeled as the hypotenuse, the side opposite a particular acute angle A, and the side adjacent to the acute angle A, six different ratios are possible. The six ratios are called the sine (sin), cosine (cos), tangent (tan), cosecant (csc), secant (sec), and cotangent (cot) and are defined as

$$\sin(A) = \frac{\text{opposite}}{\text{hypotenuse}} \qquad \csc(A) = \frac{\text{hypotenuse}}{\text{opposite}}$$

$$\cos(A) = \frac{\text{adjacent}}{\text{hypotenuse}} \qquad \sec(A) = \frac{\text{hypotenuse}}{\text{adjacent}}$$

$$\tan(A) = \frac{\text{opposite}}{\text{adjacent}} \qquad \cot(A) = \frac{\text{adjacent}}{\text{opposite}}$$

Because you know the relationships of the sides in the 45-45-90 and 30-60-90 right triangles, you can easily determine the values of the trigonometric ratios for angles of 30°, 45°, and 60°.

	sin	cos	tan
30°	$\frac{1}{2}$	$\frac{\sqrt{3}}{2}$	$\frac{\sqrt{3}}{3}$
45°	$\frac{\sqrt{2}}{2}$	$\frac{\sqrt{2}}{2}$	1
60°	$\frac{\sqrt{3}}{2}$	$\frac{1}{2}$	$\sqrt{3}$

Find Sides

With these six ratios, it is possible to solve for any unknown side of a right triangle if another side and an acute angle are known. Choose a ratio that incorporates the side you know and the side that you want to find, substitute the values you know, and solve for the unknown.

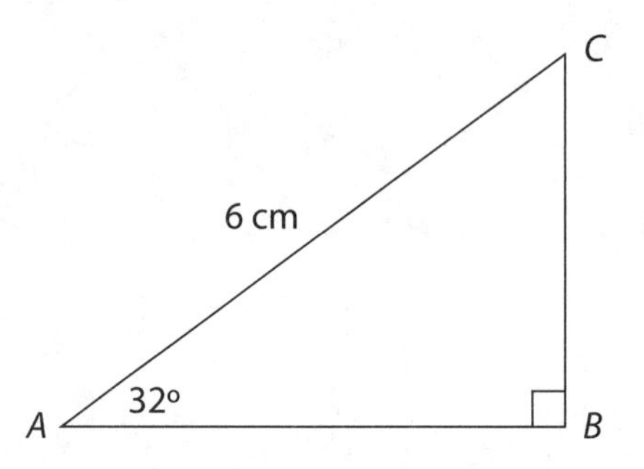

Right ΔABC has an acute angle of 32°

In right ΔABC, hypotenuse $\overline{AC}$ is 6 centimeters long, and $\angle A$ measures 32°. To find the length of the shorter leg, $\overline{BC}$, the side opposite the 32° angle, you need a ratio that talks about the opposite side and the hypotenuse.

$$\sin 32^\circ = \frac{BC}{AC} = \frac{x}{6}$$

$$\sin 32^\circ = \frac{x}{6}$$

$$x = 6\sin 32^\circ$$

$$x \approx 6(0.5299)$$

Find Angles

In addition to finding the other sides of a right triangle, trigonometric ratios can be used to find the measures of the acute angles of the right triangle if you know the lengths of the sides. Calculate one of the trig ratios using the lengths of two sides known, and work backward to the angle.

Whenever you work with trigonometry, double-check to make sure your calculator is in the mode that matches your units of measurement. Generally, you'll use degrees when working with triangles and radians when working with functions, unless the question indicates otherwise.

Working backward means you know the sine (or cosine or tangent) of the angle and want to find the angle that has that sine. The common way to say "the angle whose sine is N" is $\arcsin(N)$ or $\sin^{-1}(N)$. The angle whose cosine is N can be indicated by $\arccos(N)$ or $\cos^{-1}(N)$, and the angle whose tangent is N by $\arctan(N)$ or $\tan^{-1}(N)$.

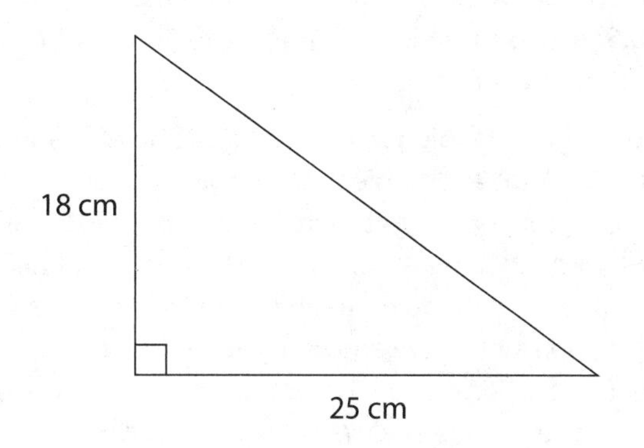

A right triangle with legs measuring 18 cm and 25 cm

If the legs of a right triangle measure 18 centimeters and 25 centimeters, you can use the two known sides to find the tangent of one of the angles. The tangent of the smaller angle will be $\frac{18}{25}$, or the tangent of the larger angle will be $\frac{25}{18}$. To find the measure of the angle, $\tan^{-1}\left(\frac{18}{25}\right) \approx 35.75°$ or $35°45'$. The two acute angles of a right triangle are complementary, so the larger of the acute angles measures approximately $90° - 35°45' = 54°15'$.

Exercise 11.1

Complete the following exercises.

1. Find the missing sides of a 45°–45°–90° triangle ΔARM with hypotenuse $\overline{AM}$ measuring 12 ft.
2. Find the missing sides of a 45°–45°–90° triangle ΔLEG with leg $\overline{EG}$ measuring $5\sqrt{6}$ m.
3. Find the missing sides of a 30°–60°–90° triangle ΔCAT with shorter leg $\overline{CA}$ measuring $14\sqrt{6}$ ft.
4. Find the missing sides of a 30°–60°–90° triangle ΔDOG with hypotenuse $\overline{DG}$ measuring $4\sqrt{21}$ cm.

5. In right ΔXYZ with right angle at Y, $\angle X = 32°$, and side $\overline{YZ}$ is 58 m. Find the lengths of the other two sides.
6. A ladder 28 ft long makes an angle of 15° with the wall of a building. How far from the wall is the foot of the ladder?
7. From the top of the ski slope, Elise sees the lodge at an angle of depression of 18.5°. If the slope is known to have an elevation of 1500 ft, how far is Elise from the lodge?

   If you imagine standing looking straight ahead of you and then raising your eyes to look up at an object, the angle between your original, horizontal gaze and your line of sight to the object above is the angle of elevation. On the other hand, if you're in an elevated position looking straight ahead and shift your gaze down to an object below, the angle between your original, horizontal gaze and your line of sight to the object below is the angle of depression. Because the horizontal lines are parallel, a little basic geometry shows that the angle of elevation is equal to the angle of depression.

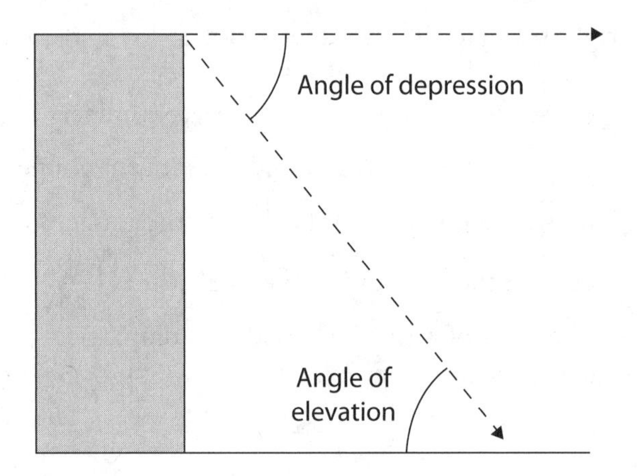

8. If an observer notes that the angle of elevation to the top of a 162-m tower is 38°24′, how far is the observer from the tower?

> 1 degree = 60 minutes, written 60′. You can change 38°24′ to $38\frac{24}{60} = 38.4°$.

9. What is the angle of elevation of the sun at the instant a 68-ft flagpole casts a shadow of 81 ft?
10. If the legs of a right triangle measure 349.2 m and 716.8 m, find the measures of the acute angles of the triangle.

Step 2. Use Trigonometry to Find Areas

In geometry, you learned that the area of a parallelogram was the product of its base and its height, $A = bh$, and that the area of a triangle was half the product of its base and height, $A = \frac{1}{2}bh$, but if you don't know the altitude, or height, the formula isn't helpful. If you know two sides of a triangle and the angle included between them, or two adjacent sides of a parallelogram and the angle included between them, it's possible to use trigonometry to find the height and therefore the area.

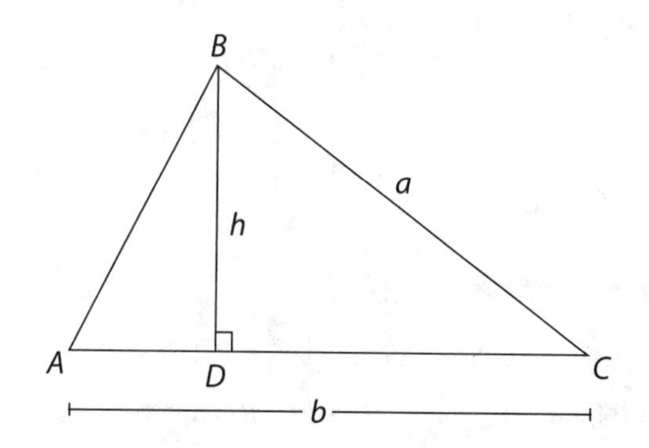

ΔABC with altitude $\overline{BD} \perp \overline{AC}$ has area $A = \frac{1}{2}ab\sin C$

Drop an altitude from one vertex to the opposite side, forming a right triangle. One of the known sides forms the hypotenuse of the right triangle. If the lengths of two adjacent sides, a and b, and the measurement of the angle between them, $\angle C$, are known, the area of the parallelogram is $A = ab\sin C$ and the triangle is $A = \frac{1}{2}ab\sin C$.

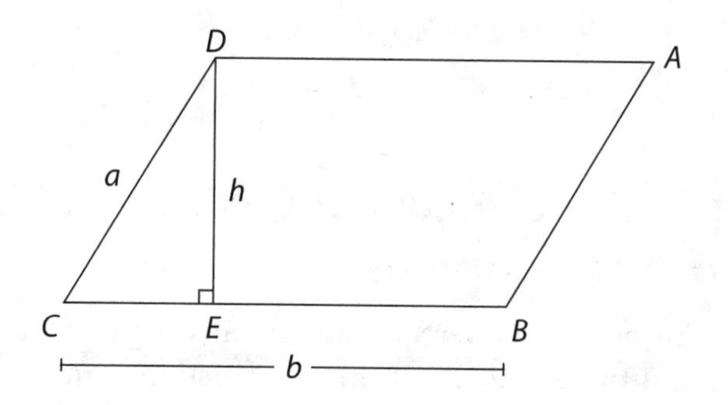

Parallelogram $ABCD$ with altitude $\overline{DE}$ has area $A = ab\sin C$

To find the area of a parallelogram with sides of 18 inches and 22 inches and an included angle of 60°, use the area formula:

$$\begin{aligned} A &= ab\sin C \\ &= 18 \cdot 22\sin 60^\circ \\ &= 396 \cdot \frac{\sqrt{3}}{2} \\ &= 198\sqrt{3}\text{ in}^2 \end{aligned}$$

Exercise 11.2

Find the area of each triangle.

1. In ΔRST, $RS = 10$, $ST = 7.5$, and $\angle S = 121°$.
2. In ΔARM, $AR = 10$, $RM = 7.5$, and $\angle R = 84°$.
3. In ΔLEG, $LE = 10$, $EG = 7.5$, and $\angle E = 30°$.
4. In ΔXYZ, $XY = 12$, $YZ = 4$, and $\angle Y = 121°$.
5. In ΔABC, $AB = 12$, $BC = 4$, and $\angle B = 9°$.

Find the area of each parallelogram.

6. In $\square CHEM$, $CH = 2.8$, $HE = 3.5$, and $\angle H = 54°$.
7. In $\square LOVE$, $LO = 5$, $OV = 7$, and $\angle O = 75°$.
8. In $\square SOAP$, $SO = 1.5$, $OA = 11$, and $\angle O = 67°$.
9. In $\square NEXT$, $NE = 3$, $EX = 9$, and $\angle E = 140°$.
10. In $\square ABCD$, $AB = 3$, $BC = 4$, and $\angle B = 97°$.

Step 3. Use the Law of Cosines to Extend to Nonright Triangles

Up to this point, you were able to find missing information only in right triangles, but the usefulness of the trig functions can be extended to other triangles, nonright triangles, with the help of two rules called the law of sines and the law of cosines.

The Law of Cosines

If the angles of a triangle are A, B, and C, and the sides opposite those angles are a, b, and c, respectively, then the law of cosines says that $c^2 = a^2 + b^2 - 2ab\cos C$.

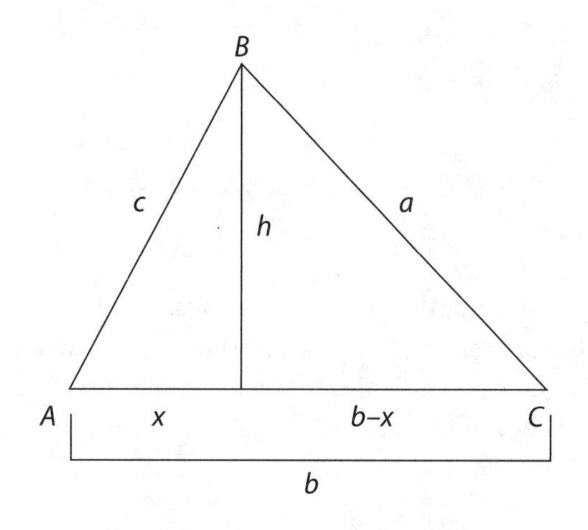

In ΔABC, the altitude to $\overline{AC}$ divides the triangle into two right triangles

The law of cosines may remind you a bit of the Pythagorean theorem, and there's a reason for that. It's derived by dividing the triangle into two right triangles by drawing an altitude. The altitude creates two right triangles. Applying the Pythagorean theorem to each of those right triangles, you get the following.

$$c^2 = h^2 + x^2$$
$$a^2 = h^2 + (b - x)^2$$

Solve the second equation for h^2.

$$h^2 = a^2 - (b - x)^2$$

Substitute for h^2 in the first equation, and simplify.

$$c^2 = a^2 - (b - x)^2 + x^2$$
$$c^2 = a^2 - (b^2 - 2bx + x^2) + x^2$$
$$c^2 = a^2 - b^2 + 2bx - x^2 + x^2$$
$$c^2 = a^2 - b^2 + 2bx$$

Write the ratio for cos*C* and use it to express *x* in terms of *a*, *b*, and cos*C*.

$$\cos C = \frac{b-x}{a}$$
$$a\cos C = b - x$$
$$x = b - a\cos C$$

Replace *x* and simplify.

$$c^2 = a^2 - b^2 + 2b(b - a\cos C)$$
$$c^2 = a^2 - b^2 + 2b^2 - 2ab\cos C$$
$$c^2 = a^2 + b^2 - 2ab\cos C$$

To find a side with the law of cosines, you'll need to know the angle opposite the unknown side. Substitute the two known sides for *a* and *b* and the known angle for $\angle C$. Evaluate the expression to find the value of c^2, and take the square root of both sides.

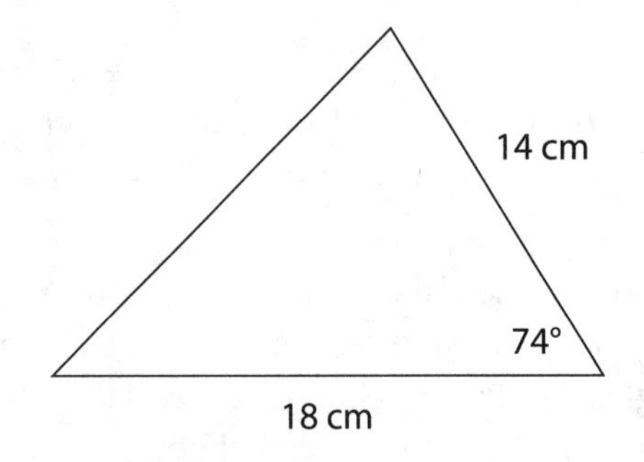

If two sides and the included angle of the triangle are known, the missing side can be found with the law of cosines

In ΔABC, $\angle C$ measures 74°. If side *a* is 14 centimeters and side *b* is 18 centimeters, find the length of side *c*.

Use the law of cosines with $a = 14$, $b = 18$, and $\angle C = 74°$.

$$c^2 = a^2 + b^2 - 2ab\cos C$$
$$c^2 = 14^2 + 18^2 - 2(14)(18)\cos 74°$$

Simplify, observing the order of operations carefully.

$$c^2 = 196 + 324 - 504(0.2756)$$
$$c^2 = 196 + 324 - 138.9024$$
$$c^2 = 381.0976$$

Find the square root.

$$c \approx 19.52$$

To find an angle using the law of cosines, you'll need to know all three sides of the triangle. Replace c with the side opposite the angle and the other two known sides for a and b. Evaluate and solve to find $\cos C$. Then use the inverse function to find the measure of $\angle C$.

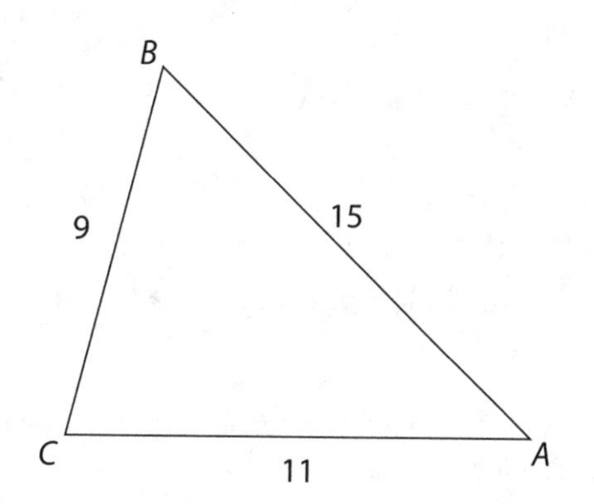

The measures of the angles of ΔABC can be found with the law of cosines

To find the measures of the largest angle of a triangle with sides of 9 inches, 11 inches, and 15 inches, use the law of cosines with $a = 9$, $b = 11$, and $c = 15$.

$$c^2 = a^2 + b^2 - 2ab\cos C$$

$$15^2 = 9^2 + 11^2 - 2(9)(11)\cos C$$

$$225 = 81 + 121 - 198\cos C$$

$$225 = 202 - 198\cos C$$

Solve for $\cos C$.

$$23 = -198\cos C$$

$$-\frac{23}{198} = \cos C$$

$$\cos C = -0.1162$$

Find the angle that has a cosine of -0.1162.

$$\angle C = \cos^{-1}(-0.1162) \approx 96.67°$$

Exercise 11.3

Find the requested side of the triangle to the nearest hundredth by using the law of cosines.

1. In ΔABC, $\angle C$ measures 42°. If side BC is 4 cm and side AC is 8 cm, find the length of side AB.
2. In ΔRST, $\angle T$ measures 163°. If side ST is 81 ft and side RT is 90 ft, find the length of side RS.
3. In ΔTAP, $\angle A$ measures 29°. If side TA is 32 m and side AP is 45 m, find the length of side TP.
4. In ΔDOG, $\angle D$ measures 92°. If side DO is 189 in and side DG is 201 in, find the length of side OG.
5. In ΔRST, $\angle R$ measures 73°. If side RS is 69 ft and side RT is 77 ft, find the length of side ST.

Find the requested angle of the triangle to the nearest hundredth of a degree by using the law of cosines.

6. In ΔABC, $AB = 57$ cm, $BC = 37$ cm, and $AC = 46$ cm. Find the measure of $\angle B$.
7. In ΔXYZ, $XY = 454$ yd, $YZ = 537$ yd, and $XZ = 416.5$ yd. Find the measure of $\angle Z$.
8. In ΔRST, $RS = 13$ in, $ST = 16$ in, and $RT = 25$ in. Find the measure of $\angle T$.
9. In ΔDOG, $DO = 684$ ft, $OG = 932$ ft, and $DG = 841.5$ ft. Find the measure of $\angle D$.
10. In ΔJET, $JE = 29$ in, $ET = 22$ in, and $JT = 27$ in. Find the measure of $\angle E$.

Step 4. Apply the Law of Sines

In geometry, you learned that the longest side of a triangle was opposite the largest angle, and the shortest side opposite the smallest angle, but that lengths of the sides were not proportional to the size of the angles. The law of sines says that the lengths of the sides are proportional to the sines of the angles. If ΔABC is a triangle (not necessarily a right triangle) with side a opposite $\angle A$, side b opposite $\angle B$, and side c opposite $\angle C$, then

$$\frac{a}{\sin A} = \frac{b}{\sin B} = \frac{c}{\sin C}$$

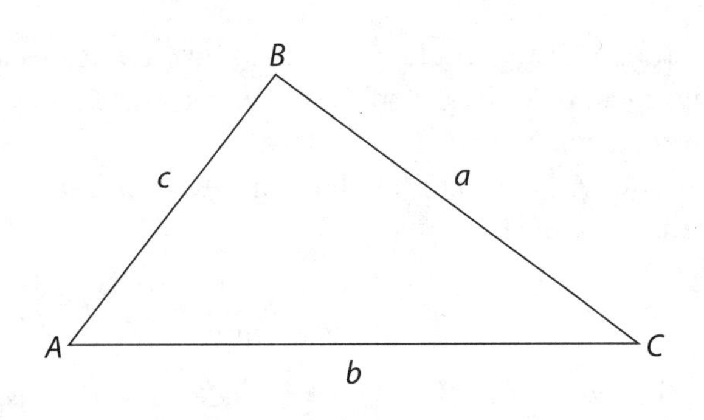

The side opposite $\angle A$ is a, the side opposite $\angle B$ is b, and the side opposite $\angle C$ is c

If you know the measures of two angles of a triangle and the length of one side, you can use a proportion from the law of sines to solve for a missing side. You will need the measurement of the angle opposite the known side and the measure of the angle opposite the side you're trying to find.

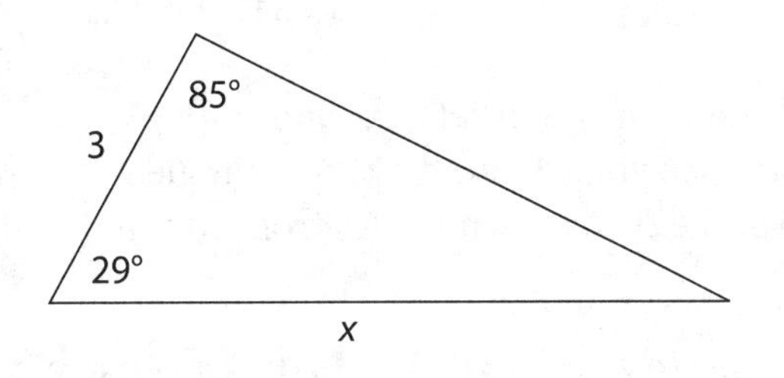

If two angles and the included side of a triangle are known, the other sides can be found with the law of sines

Suppose that in ΔABC, $\angle A$ measures 29° and $\angle B$ measures 85°. You know that $AB = 3$ centimeters, and you want to find AC. Because you know two of the angles, you can find the third. $\angle C$ measures $180° - (29° + 85°) = 66°$. The side labeled as a is $\overline{BC}$, the side opposite $\angle A$, b is side $\overline{AC}$, and c is side $\overline{AB}$. The law of sines becomes

$$\frac{BC}{\sin 29°} = \frac{AC}{\sin 85°} = \frac{3}{\sin 66°}$$

Cross-multiply and solve for AC.

$$\frac{AC}{\sin 85°} = \frac{3}{\sin 66°}$$

$$AC \cdot \sin 66° = 3 \sin 85°$$

$$AC = \frac{3 \sin 85°}{\sin 66°} = \frac{3(0.9962)}{0.9135} = 3.27$$

AC is approximately 3.27 centimeters.

To find a missing angle, you must know two sides and one angle. One of the sides must be opposite the angle you know, and the other must be opposite the angle you're trying to find.

In ΔABC, $AB = 12$ centimeters and $AC = 18$ centimeters. If $\angle B$ measures 77°, use the law of sines.

$$\frac{18}{\sin 77^\circ} = \frac{12}{\sin C}$$

You'll need to solve for the sine of $\angle C$ and then use the inverse sine key on your calculator to find the measure of $\angle C$.

$$\frac{18}{\sin 77^\circ} = \frac{12}{\sin C}$$

$$18 \sin C = 12 \sin 77^\circ$$

$$\sin C = \frac{12 \sin 77^\circ}{18} \approx 0.6496$$

$$m\angle C = \sin^{-1}(0.6496) \approx 40.5^\circ$$

The law of sines will let you find only a side that lies opposite a known angle, but once you've found a second angle in the triangle, you can subtract the total of the two known angles from 180° to find the third angle.

The Ambiguous Case of the Law of Sines

The one difficulty of the law of sines is that you may encounter situations in which there are two triangles that fit your given information or in which no triangle fits your information. This is called the ambiguous case.

In geometry, you learned to prove that a pair of triangles is congruent by SSS (side, side, side), SAS (side, angle, side), ASA (angle, side, angle), and AAS (angle, angle, side), and you learned that SSA (side, side, angle) was not valid. Translating that information over to trigonometry, if your given information is ASA, two angles and the side included between them, or AAS, two angles and a side, you can use the law of sines without any concern. (If your given information is SSS, three sides, or SAS, two sides and the angle included between them, you'll need the law of cosines.)

When you're using the law of sines to find an angle and your given information is SSA—two sides and an angle, but not the angle included between the sides—you're not guaranteed a unique triangle. You're in the ambiguous case. If you call the known angle $\angle A$ and the known sides a and b, the possibilities are summarized in the following table.

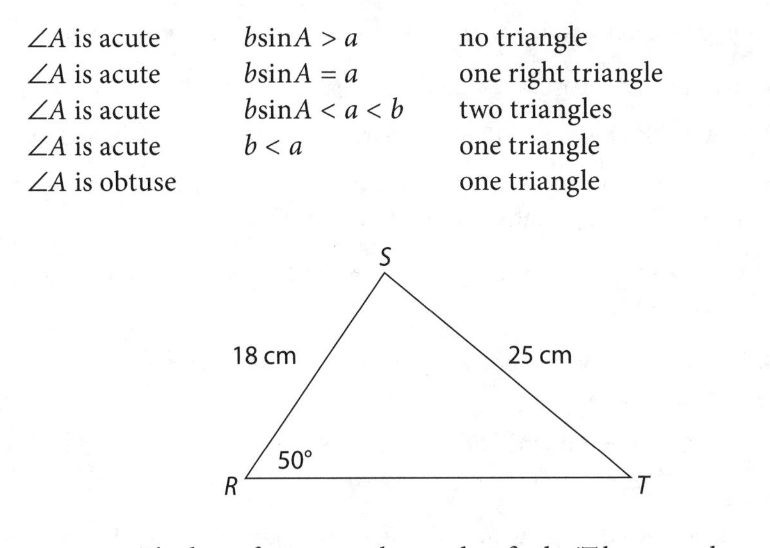

$\angle A$ is acute	$b\sin A > a$	no triangle
$\angle A$ is acute	$b\sin A = a$	one right triangle
$\angle A$ is acute	$b\sin A < a < b$	two triangles
$\angle A$ is acute	$b < a$	one triangle
$\angle A$ is obtuse		one triangle

The law of sines can be used to find $\angle T$ because the opposite side is known

If you know that $\angle R$ measures 50° and that $\overline{RS}$ measures 18 centimeters and $\overline{ST}$ measures 25 centimeters, the product of RS, the side adjacent to $\angle R$, and sin R is $18\sin 50° = 18(0.7660) \approx 13.7888$. That's less than the side opposite the angle, so there is a triangle. The opposite side is larger than the adjacent side, which tells you to expect one solution.

If $\angle X$ measures 50°, $XY = 22$ inches, and $YZ = 14$ inches, $\angle X$ is acute, and $22\sin 50° \approx 16.852$. This is larger than the opposite side of 14 inches, so there is no solution.

If $\angle A$ measures 20°, $AB = 20$ feet, and $BC = 8$ feet, $\angle A$ is acute, and $AB\sin A = 20\sin 20° = 6.8404$. This is less than BC, so there is some solution, but because BC, the opposite side, is less than AB, the adjacent side, there may be two solutions. Solve $\frac{8}{\sin 20°} = \frac{20}{\sin C}$ to find that $\sin C = \frac{20\sin 20°}{8} \approx 0.8551$. Your calculator will say that $\sin^{-1}(0.8551) \approx 58.77°$, so one possible triangle is one with angles of 20°, 58.77°, and 101.23°. But an obtuse angle, $180° - 58.77° = 121.23°$, also has a sine of approximately 0.8551, so a second triangle is possible with angles of 20°, 121.23°, and 38.77°.

Exercise 11.4

Use the given information to solve for the missing side to the nearest hundredth.

1. In ΔABC, $\angle A$ measures 74° and $\angle B$ measures 41°. If $BC = 58$ in, find AC.
2. In ΔRST, $\angle R$ measures 18° and $\angle S$ measures 44°. If $ST = 6$ ft, find RS.
3. If ΔABC is an isosceles triangle, with vertex angle of 12° and a base of 14 cm, find the lengths of the congruent sides.

Use the given information to solve for the missing angle to the nearest hundredth of a degree.

4. In ΔABC, $AB = 342$ yd and $AC = 263$ yd. If $\angle C$ measures 46°, find the measure of $\angle B$.
5. In ΔPQR, $PQ = 7$ m and $QR = 56$ m. If $\angle P$ measures 107°, find the measure of $\angle R$.

Find the requested angle of the triangle to the nearest hundredth of a degree. If more than one triangle is possible, find both. If no triangle is possible, indicate that.

6. In ΔABC, $AB = 51$, $BC = 58$, and $\angle C = 54°$. Find $\angle A$.
7. In ΔXYZ, $XY = 11$, $YZ = 5.4$, and $\angle Z = 154°$. Find $\angle X$.

Solve each problem by the most efficient method. Round to the nearest hundredth. Use the law of sines or the law of cosines, as appropriate. If more than one solution is possible, give both solutions.

8. Find the lengths of the diagonals of a parallelogram whose sides are 32 cm and 48 cm, if the acute angle between the sides is 47°.
9. The angle of elevation from the foot of one building to the roof of a taller building nearby is 48º. The angle of depression from the top of the taller building to the top of the shorter one is 29º. If the shorter building is 65 ft tall, find the distance between the buildings.

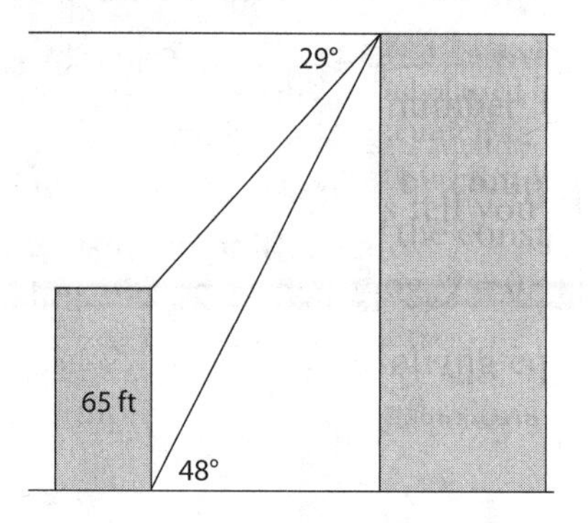

10. From a bench on the shoreline of a lake, a boat is spotted at a bearing of N 53° E. The same boat is spotted from a second bench directly east of the first at N 41° W. If the perpendicular distance from the boat to the shoreline is 60 m, how far apart are the two benches?

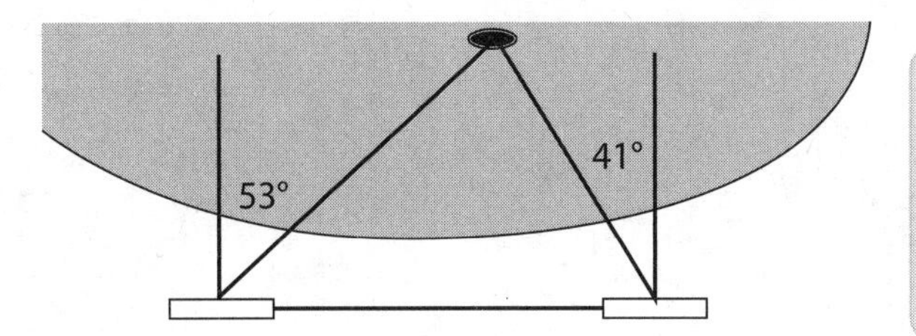

A bearing first specifies a starting direction, usually north or south, then gives a number of degrees to rotate, followed by the direction of rotation. A bearing of N 41° W tells you to start facing north and turn 41° toward the west.

12

Trigonometric Functions

Trigonometry, or triangle measurement, begins in the right triangle but doesn't have to be restricted to triangles. By moving onto the coordinate plane and observing the interaction between the right triangle and a circle centered at the origin, it becomes possible to talk about the sine and cosine of any real number, and you can define six trigonometric functions.

Step 1. Expand Your Concept of Angles

For this expanded view of trigonometry, an angle is in *standard position* if its vertex is at the origin and one of its sides, called the *initial side*, lies on the positive x-axis. The other side of the angle is called the *terminal side*. If the direction of rotation from the initial side to the terminal side is counterclockwise, the measure of the angle between the two sides is considered to be positive. If the direction of the rotation is clockwise, the measure of the angle is negative. Angles may be acute, right, obtuse, straight, or reflex (more than a straight angle). An angle may even be more than a full rotation.

To begin to talk about trigonometry in broader terms, it is helpful to move to a different system of measurement called *radian measure*. A radian is the measure of a central angle whose intercepted arc is equal in length to the radius of the circle. Because the circumference of a circle is 2π times the radius, there are 2π radians in a full rotation, π radians in a half rotation.

Remember to check the mode on your calculator. Whether you're working in radians or degrees, if the calculator is set to the wrong system, your results will be unpredictable.

If you need to convert from degrees to radians or radians to degrees, you can use the proportion $\frac{\text{degrees}}{360^\circ} = \frac{\text{radians}}{2\pi}$. Fill in the known measure, and solve the proportion. To find the radian equivalent of 135°, set up the proportion and put 135 in the degrees position.

$$\frac{\text{degrees}}{360^\circ} = \frac{\text{radians}}{2\pi}$$

$$\frac{135^\circ}{360^\circ} = \frac{r}{2\pi}$$

$$360r = 270\pi$$

$$r = \frac{3\pi}{4}$$

The radian equivalent of 135° is $\frac{3\pi}{4}$ radians.

Common Angles in Radians

Just as you learned the relationships of sides in common right triangles, you'll find it helpful to know the radian equivalents of common angles.

$$0^\circ = 0 \text{ radians} \quad 30^\circ = \frac{\pi}{6} \text{ radians} \quad 45^\circ = \frac{\pi}{4} \text{ radians}$$

$$60^\circ = \frac{\pi}{3} \text{ radians} \quad 90^\circ = \frac{\pi}{2} \text{ radians}$$

Don't become dependent on conversion. Think in radians. Learn to count around the circle, starting from the positive x-axis, by multiples of $\frac{\pi}{2}$, $\frac{\pi}{3}$, $\frac{\pi}{4}$, or $\frac{\pi}{6}$.

Because the terminal side can rotate in different directions and can complete any number of rotations before reaching its final position, two angles in standard position may have different measurements yet share the same terminal side. One may be a positive angle while the other is negative, or one may include one or more full rotations while the other is less than a full rotation. Angles that share the same terminal side are called *coterminal angles*.

Exercise 12.1

Find the radian equivalent of each angle measure.

1. 40°
2. 330°
3. 120°
4. 225°

Find the degree equivalent of each angle measure.

5. $\frac{5\pi}{6}$
6. $\frac{2\pi}{9}$
7. $\frac{\pi}{12}$

Find a positive and a negative angle coterminal with the given angle.

8. $\frac{\pi}{2}$
9. $\frac{11\pi}{6}$
10. $-\frac{5\pi}{3}$

Step 2. Learn the Unit Circle

A circle on the coordinate plane with its center at the origin and a radius of 1 is called a *unit circle*. Choose a point on the circle, connect the origin to the point, and you form an angle in standard position. If θ is an angle in standard position, and the terminal side of θ intersects the unit circle at the point (x,y), then you can define the following:

$$\sin(\theta) = y \quad \csc(\theta) = \frac{1}{y}$$
$$\cos(\theta) = x \quad \sec(\theta) = \frac{1}{x}$$
$$\tan(\theta) = \frac{y}{x} \quad \cot(\theta) = \frac{x}{y}$$

If you know the coordinates of the point where the terminal side of the angle intersects the unit circle, you know the $\cos\theta$ and $\sin\theta$, because they are the coordinates of that point. Although based on the right triangle created

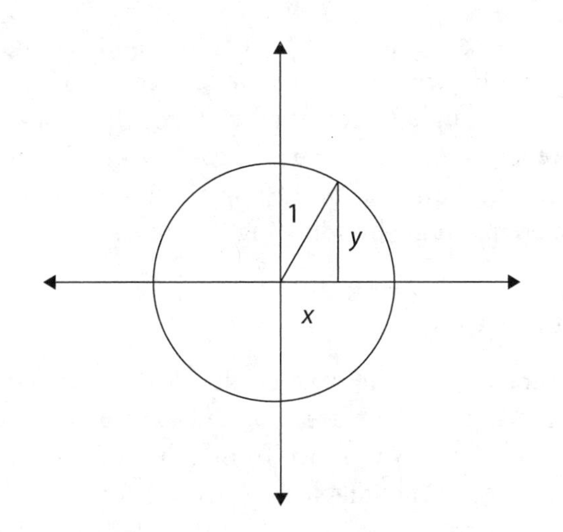

The unit circle with an angle in quadrant I

by dropping a perpendicular to the x-axis from the point on the unit circle, these definitions give you a method of finding the trig values based solely on the point where the terminal side intersects the unit circle. No right triangles are required.

If $\theta = \frac{5\pi}{4}$, the terminal side falls in the third quadrant, intersecting the unit circle at the point $\left(-\frac{\sqrt{2}}{2}, -\frac{\sqrt{2}}{2}\right)$. Using that point, you can determine the following:

$$\sin\frac{5\pi}{4} = -\frac{\sqrt{2}}{2} \qquad \csc\frac{5\pi}{4} = -\frac{2}{\sqrt{2}} = -\sqrt{2}$$

$$\cos\frac{5\pi}{4} = -\frac{\sqrt{2}}{2} \qquad \sec\frac{5\pi}{4} = -\frac{2}{\sqrt{2}} = -\sqrt{2}$$

$$\tan\frac{5\pi}{4} = 1 \qquad \cot\frac{5\pi}{4} = 1$$

Reference Angle

From the point where the terminal side of θ intersects the unit circle, drop a perpendicular to the x-axis, creating a right triangle with hypotenuse of 1.

Designate as α the acute angle of this right triangle that has its vertex at the origin. This acute angle you've labeled α has the same sine, cosine, and tangent as the angle θ, except possibly for the sign. This acute angle is called the reference angle for θ. Every angle has an acute angle that is its reference angle. The trig functions of the angle θ can be found from the trig functions of the reference angle by adjusting the signs.

All-Star Trig Class

The symmetries of the unit circle mean that many of the values of the trigonometric functions are repeated as you move around the circle, with a change of sign as you move from quadrant to quadrant. If you know the values of the six functions for the acute reference angle and you understand how the signs change, you can find the trig functions of any angle with that reference angle.

TERMINAL SIDE FALLS IN	*SIGN OF X*	*SIGN OF Y*	$\sin(\theta)$ $\csc(\theta)$	$\cos(\theta)$ $\sec(\theta)$	$\tan(\theta)$ $\cot(\theta)$
Quadrant I	positive	positive	positive	positive	positive
Quadrant II	negative	positive	positive	negative	negative
Quadrant III	negative	negative	negative	negative	positive
Quadrant IV	positive	negative	negative	positive	negative

There are a variety of mnemonic devices to help you remember those signs. You might try "all-star trig class." The *A* in *all* tells you that in the first quadrant, ALL six trig functions are positive. The *S* in *star* means that, in the second quadrant, the SINE and its reciprocal, the cosecant, are positive, but all others negative. The *T* of *trig* indicates that in the third quadrant the TANGENT and its reciprocal, cotangent, are positive, and the *C* in *class* signals that in quadrant IV the COSINE and its reciprocal, secant, are the only positive functions.

Some people go clockwise from quadrant I and remember ACTS, or go counterclockwise from quadrant IV and remember CAST. Another common mnemonic is "all seniors take calculus"—even if they don't.

Even if you don't know the point where the terminal side intersects the unit circle, if you know a point on the terminal side or the value of one of the trig functions, and you know the quadrant in which the terminal side falls, you can find all six trig functions. Knowing a point on the terminal side allows you to drop a perpendicular, use the Pythagorean theorem to find the hypotenuse, and determine six functions of the reference angle. Knowing one of the functions lets you reconstruct where a point on the terminal side might be.

If $\tan\theta = \frac{3}{4}$, the reference angle is an acute angle of a 3-4-5 right triangle. If θ is a first quadrant angle, all six functions are positive. If θ falls in quadrant III, the other quadrant where tan is positive, only tangent and cotangent are positive, and the rest are negative. If you have the additional piece of information that $\csc\theta < 0$, you know that $\sin\theta < 0$, so the angle must fall in the third quadrant. Once you have that, you can say this:

$$\sin\theta = -\frac{3}{5} \quad \csc\theta = -\frac{5}{3}$$
$$\cos\theta = -\frac{4}{5} \quad \sec\theta = -\frac{5}{4}$$
$$\tan\theta = \frac{3}{4} \quad \cot\theta = \frac{4}{3}$$

Exercise 12.2

Determine the point at which the terminal side of the angle intersects the unit circle. Use special right triangle relationships where helpful.

1. $\frac{7\pi}{6}$
2. $-\frac{5\pi}{4}$

Determine in which quadrant the terminal side of θ falls.

3. $\cos\theta < 0$ and $\csc\theta < 0$
4. $\csc\theta > 0$ and $\cot\theta < 0$

Find the value of all six trig functions of θ from the information given.

5. $\sin\theta = \frac{5}{13}$ and $\tan\theta > 0$
6. $\cos\theta = \frac{3}{5}$ and $\csc\theta < 0$

Find the sine, cosine, and tangent of an angle in standard position if the given point is on the terminal side of the angle.

7. (5, –5)
8. (0, 4)
9. $(\sqrt{3}, -2)$
10. (–2,5)

Step 3. Define Trigonometric Functions

Using the unit circle definitions allows you to define six trig functions on the real numbers. The six trigonometric functions—sine, cosine, tangent, and their reciprocals—are periodic functions. Each of them repeats a certain pattern in a fixed interval, called the period.

The most fundamental sine wave, $y = \sin\theta$, has the following graph. The domain of $y = \sin\theta$ is $(-\infty,\infty)$, and the range of $y = \sin\theta$ is $[-1,1]$. The period of the sine wave, the time it takes to complete one full wave, is 2π.

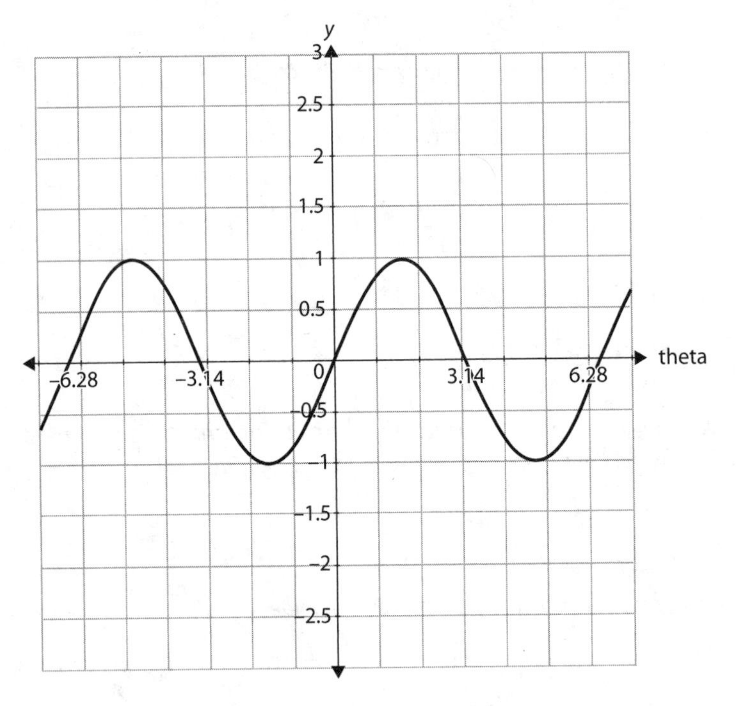

The graph of $y = \sin\theta$

The graph of $y = \cos\theta$ resembles the graph of $y = \sin\theta$ but is shifted, or translated, $\frac{\pi}{2}$ units to the left. The graph of $y = \cos\theta$ is the same graph as $y = \sin\left(\frac{\pi}{2} - \theta\right)$. The cosine wave also has a domain of $(-\infty,\infty)$, a range of $[-1,1]$, and a period of 2π.

The sine and cosine graphs are continuous as well as periodic. They are smooth, connected curves that repeat the wavelike pattern. In different equations, the length or period of the wave may change, its height or

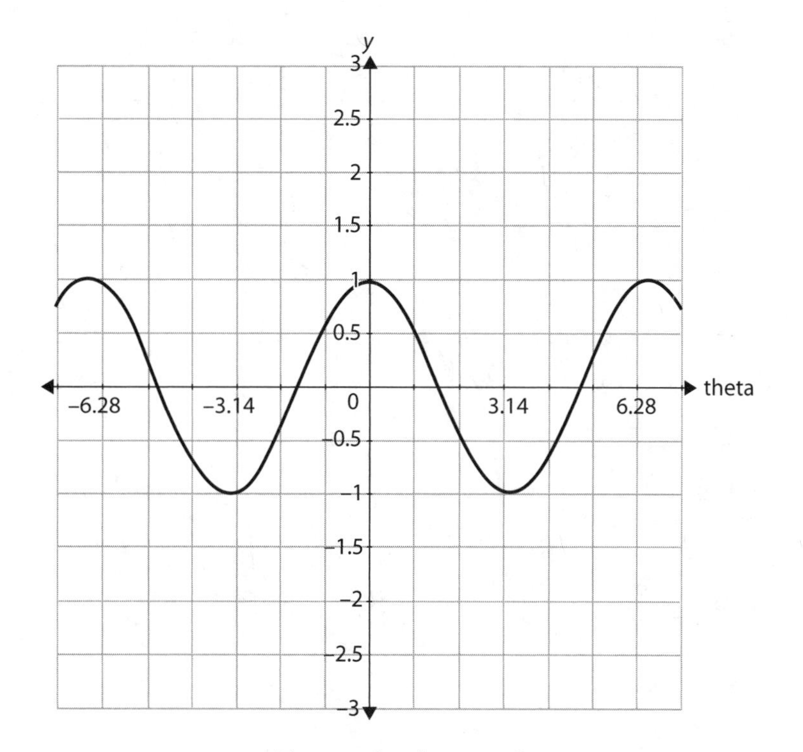

The graph of $y = \cos\theta$

amplitude may change, and it may shift in different directions, but the basic wave form remains for all sine and cosine graphs.

All the other trigonometric functions have graphs that are discontinuous, but as with the sine and cosine, each repeats a pattern. They are discontinuous because each is defined as a quotient, and for each one, there are values that make the denominator 0 and therefore make the function undefined.

The tangent function has a discontinuous graph, because the tangent is defined as the quotient $\frac{y}{x}$ of the coordinates of the point (x,y) at which the terminal side of the angle intersects the unit circle. The graph will have a vertical asymptote at multiples of $\frac{\pi}{2}$, when the x-value of the point on the terminal side is 0. In the space of π units between vertical asymptotes, the same pattern is repeated. The parent graph has its y-intercept at the origin and x-intercepts at all multiples of π. Each wave goes to $-\infty$ on the left and ∞ on the right.

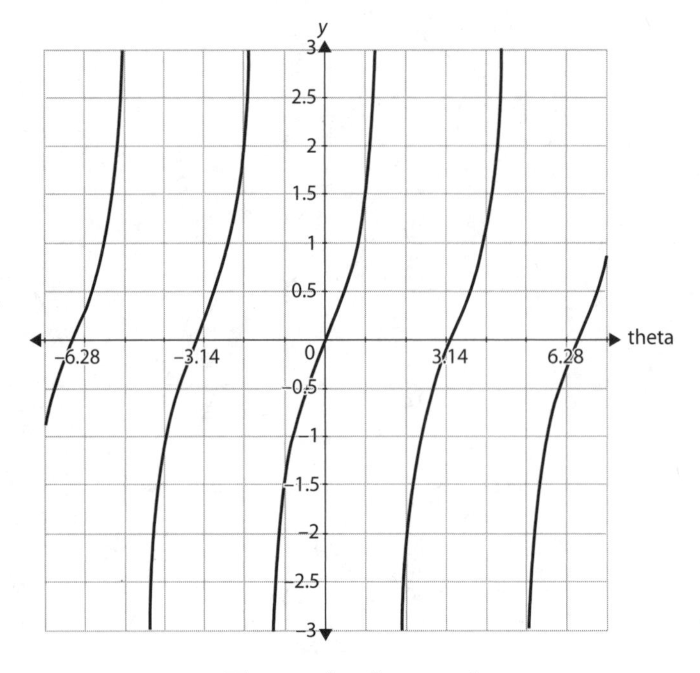

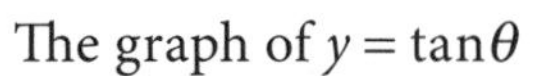
The graph of $y = \tan\theta$

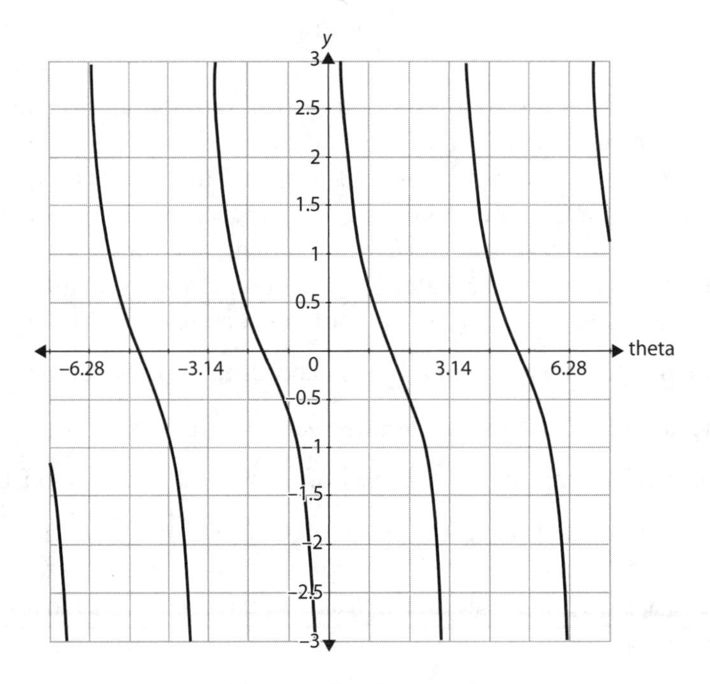

The graph of $y = \cot\theta$

Like the tangent, cotangent is discontinuous, but its vertical asymptotes occur at multiples of π. The x-intercepts of the parent graph fall at odd multiples of $\frac{\pi}{2}$, and each wave goes to ∞ on the left and $-\infty$ on the right.

The secant and cosecant functions are the reciprocals of the cosine and sine functions, respectively, and as with the sine and cosine differ by a shift of $\frac{\pi}{2}$. Each is made up of cup-shaped sections alternating opening up and down, with each section tangent to the graph of its reciprocal. The graphs have vertical asymptotes between the cup-shaped sections, at multiples of π for the cosecant function and at odd multiples of $\frac{\pi}{2}$ for the secant function. The sections of the secant function seem to balance on the peaks and troughs of the cosine graph. In the figures (on page 184), the sine and cosine graphs are shown in gray, with the cosecant and secant graphs in black.

Transformations

There are six possible alterations to the parent graph, but these show up as four numbers in the equations of the trig functions. In the equation $y = a\sin(b(\theta - h)) + k$, the h and k represent translations, or rigid shifts. The number h is the horizontal shift, sometimes called the phase shift, and the k is the vertical shift. The vertical shift defines the midline of the graph. For the parent graph, the midline is the x-axis, and the graph rises above the midline and falls below it by equal distances, called the amplitude.

> If you take a few key points on the parent graph to transform, the parameters b and h, which affect θ before the trig function works on it, will change the x-coordinate of the key point. Their effect is reversed: add h to the x-coordinate, and divide by b. The parameters a and k will change the y-coordinates, and their effect is true. Multiply the y-coordinate by a and add k.

Change $y = \sin\theta$ to $y = \sin(\theta - 3)$ and the graph will move three units to the right. Change $y = \sin\theta$ to $y = \sin(\theta + 2)$ and it will move two units left. Tacking on a constant at the end shifts the graph up or down; $y = \sin(\theta) + 4$ moves the parent graph up four units, and $y = \sin(\theta) - 1$ shifts down one.

The numbers in the a and b positions do double duty, with the sign of the number telling you one thing and the absolute value communicating something else. If a is negative, the graph is reflected across the x-axis. If b is a negative number, the graph is reflected across the y-axis. The absolute value of a, called the *amplitude*, tells you about vertical stretch or compression. When $|a| > 1$, the graph is stretched vertically. If $|a| < 1$, the graph is compressed.

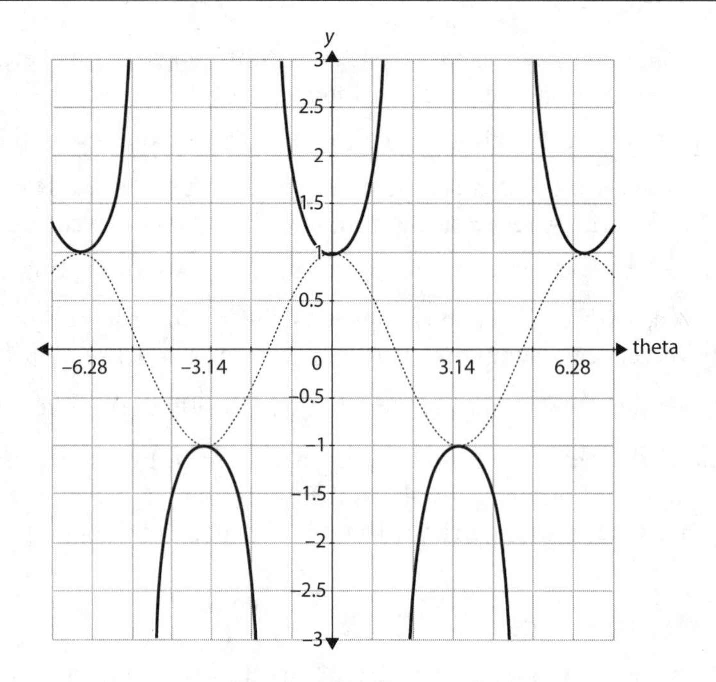

The graph of $y = \sec\theta$

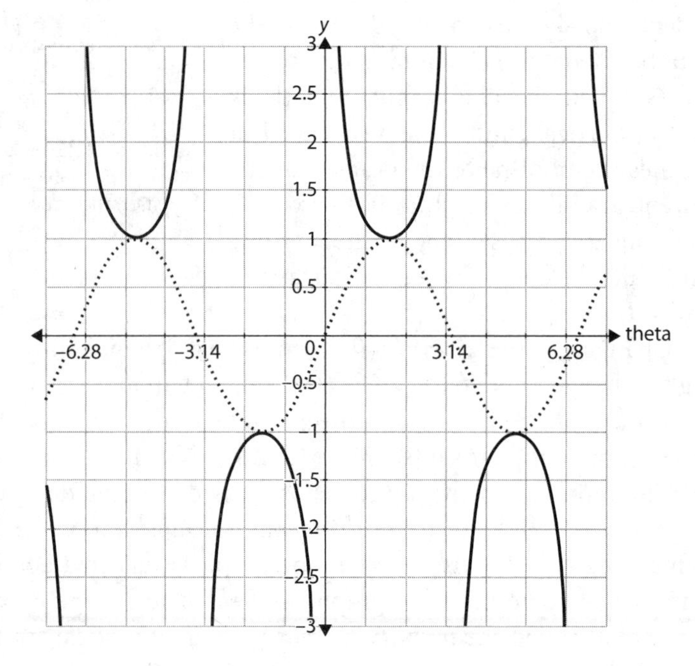

The graph of $y = \csc\theta$

The absolute value of b talks about horizontal stretch and compression. If $|b| > 1$, the graph is compressed horizontally, because more than one full cycle must fit into the space of 2π. When $|b| < 1$, you complete less than one cycle in 2π, so the graph is stretched horizontally. Because stretching or compressing the wave horizontally will change the number of waves that fit in a space of 2π, the number in the b position is called the *frequency*. The product of the frequency and the period will always be 2π.

Here are some examples of these adjustments, one change at a time.

- $y = \cos\left(\theta - \frac{\pi}{4}\right)$. Shift the graph of $y = \cos\theta$ $\frac{\pi}{4}$ units to the right.
- $y = \cos(2\theta)$. Two full waves fit in the space between 0 and 2π. The period of the wave is π units. The wave has been compressed horizontally.
- $y = -\cos\theta$. The graph is reflected over the x-axis.
- $y = 3\cos\theta$. The graph is stretched vertically, so it rises to 3 and falls to -3.
- $y = \cos(\theta) + 1$. Shift the graph up one unit.

If you put all those changes into one equation, you'll have the equation $y = -3\cos\left(2\left(\theta - \frac{\pi}{4}\right)\right) + 1$ or $y = -3\cos\left(2\theta - \frac{\pi}{2}\right) + 1$ with the graph on page 186.

Exercise 12.3

Use transformations to sketch the graph of each of the equations.

1. $y = \cos(4\theta)$
2. $y = 4\sin\theta + 1$
3. $y = \cos\left(x - \frac{\pi}{3}\right)$
4. $y = -\frac{1}{2}\sin(3\theta)$
5. $y = -5\cos(3\theta) + 3$
6. $y = -3\sin\left(\frac{\theta}{2}\right)$
7. $y = \tan(4\theta) + 2$
8. $y = \frac{3}{4}\sec\left(\theta - \frac{\pi}{6}\right)$
9. $y = \csc(4\theta) - 2$
10. $y = -\frac{1}{4}\cot\left(\frac{\theta}{2}\right)$

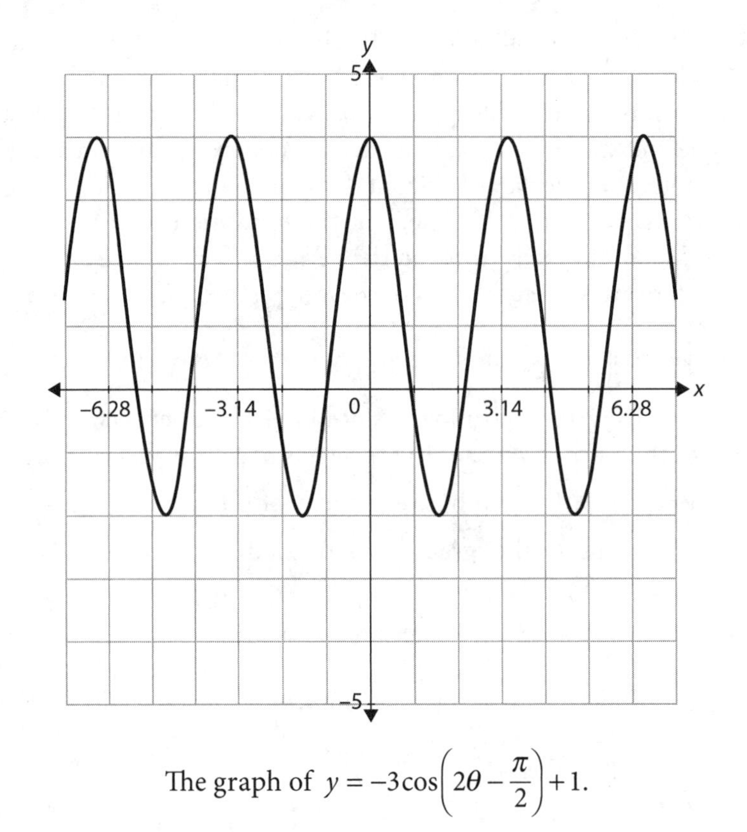

The graph of $y = -3\cos\left(2\theta - \frac{\pi}{2}\right) + 1$.

Step 4. Explore Inverse Trigonometric Functions

In order for a function to have an inverse function, the function must be one-to-one. This means that not only does every x-value have only one y-value, but every y-value comes from only one x-value. The periodic nature of the trig functions means that the same y-value is produced by many different x-values, and that in turn means that they would not have an inverse that is a function.

Each function can be restricted so that it is one-to-one, however. For each of the parent functions, choose first-quadrant values, where the function is positive, and the nearest continuous quadrant in which the function is negative. The sine function will be one-to-one if restricted to the interval $\left[-\frac{\pi}{2},\frac{\pi}{2}\right]$. The cosine function is one-to-one on the interval $[0,\pi]$, and the tangent function on the interval $\left(-\frac{\pi}{2},\frac{\pi}{2}\right)$. Each of these restricted functions has an inverse, denoted as $\sin^{-1} x$, $\cos^{-1} x$, and $\tan^{-1} x$.

INVERSE FUNCTION	***DOMAIN***	***RANGE***
$\sin^{-1} x$	$[-1,1]$	$\left[-\frac{\pi}{2},\frac{\pi}{2}\right]$
$\cos^{-1} x$	$[-1,1]$	$[0,\pi]$
$\tan^{-1} x$	$(-\infty,\infty)$	$\left(-\frac{\pi}{2},\frac{\pi}{2}\right)$

Evaluating Inverse Functions

Because the inverse functions have limited ranges, you need to think carefully when evaluating. It's tempting, for example, to say that $\sin^{-1}\left(\sin\left(\frac{5\pi}{4}\right)\right)$ would equal $\frac{5\pi}{4}$ because the sin function and the $\sin^{-1}$ counteract each other. Unfortunately, that's not quite true. Working from the inside out, you can say that $\sin^{-1}\left(\sin\left(\frac{5\pi}{4}\right)\right)=\sin^{-1}\left(-\frac{\sqrt{2}}{2}\right)$, but $\sin^{-1}$ will never return $\frac{5\pi}{4}$. Instead, $\sin^{-1}\left(-\frac{\sqrt{2}}{2}\right)=-\frac{\pi}{4}$. Take things a step at a time, and be aware of domains and ranges when evaluating.

Exercise 12.4

Evaluate each expression. Be aware of the domains and ranges of the inverse functions.

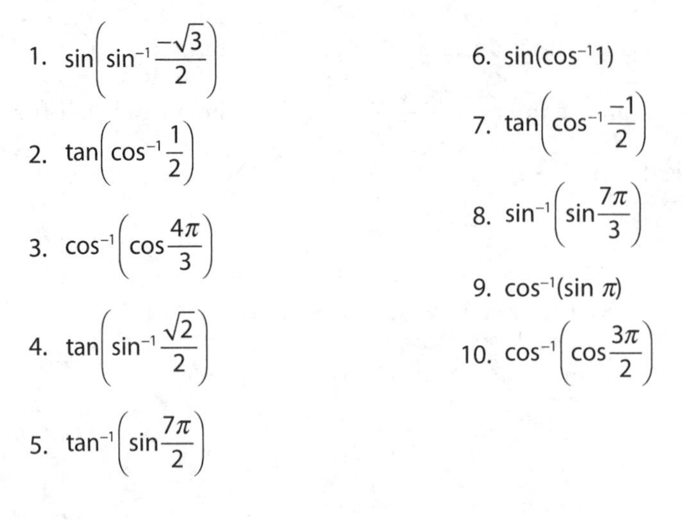

1. $\sin\left(\sin^{-1}\frac{-\sqrt{3}}{2}\right)$
2. $\tan\left(\cos^{-1}\frac{1}{2}\right)$
3. $\cos^{-1}\left(\cos\frac{4\pi}{3}\right)$
4. $\tan\left(\sin^{-1}\frac{\sqrt{2}}{2}\right)$
5. $\tan^{-1}\left(\sin\frac{7\pi}{2}\right)$
6. $\sin(\cos^{-1}1)$
7. $\tan\left(\cos^{-1}\frac{-1}{2}\right)$
8. $\sin^{-1}\left(\sin\frac{7\pi}{3}\right)$
9. $\cos^{-1}(\sin\pi)$
10. $\cos^{-1}\left(\cos\frac{3\pi}{2}\right)$

Step 5. Verify Identities

An identity is an equation that is true for all values of the variable involved. Because an identity is a statement that is always true, the two sides of the equation are interchangeable, and one can be substituted for the other whenever convenient. You'll memorize some key identities and use them to substitute into other equations, either to demonstrate that those equations are also identities or to simplify an equation so that it can be solved.

Fundamental Identities

There are many trigonometric identities, but a few are particularly important for simplifying trigonometric equations. There are four groups of fundamental identities: reciprocal, quotient, cofunction, and Pythagorean identities.

Reciprocal Identities. The first group is the reciprocal identities, based on the definitions of the trig functions.

$$\sin\theta = \frac{1}{\csc\theta} \quad \cos\theta = \frac{1}{\sec\theta} \quad \tan\theta = \frac{1}{\cot\theta}$$

$$\csc\theta = \frac{1}{\sin\theta} \quad \sec\theta = \frac{1}{\cos\theta} \quad \cot\theta = \frac{1}{\tan\theta}$$

Quotient Identities. Some quotient identities are commonly used.

$$\tan\theta = \frac{\sin\theta}{\cos\theta} \quad \cot\theta = \frac{\cos\theta}{\sin\theta}$$

These are the less frequently used quotient identities.

$$\sin\theta = \frac{\tan\theta}{\sec\theta} \quad \cos\theta = \frac{\cot\theta}{\csc\theta}$$

$$\csc\theta = \frac{\cot\theta}{\cos\theta} \quad \sec\theta = \frac{\tan\theta}{\sin\theta}$$

Cofunction Identities. The cofunction identities get their names because they talk about functions of complementary angles.

$$\sin\theta = \cos\left(\frac{\pi}{2} - \theta\right) \quad \cos\theta = \sin\left(\frac{\pi}{2} - \theta\right)$$

$$\sec\theta = \csc\left(\frac{\pi}{2} - \theta\right) \quad \csc\theta = \sec\left(\frac{\pi}{2} - \theta\right)$$

$$\tan\theta = \cot\left(\frac{\pi}{2} - \theta\right) \quad \cot\theta = \tan\left(\frac{\pi}{2} - \theta\right)$$

Pythagorean Identities. The Pythagorean identities take their names from the Pythagorean theorem by which they're derived.

$$\sin^2\theta + \cos^2\theta = 1$$

$$1 + \cot^2\theta = \csc^2\theta$$

Divide $\sin^2\theta + \cos^2\theta = 1$ by $\sin^2\theta$.

$$\tan^2\theta + 1 = \sec^2\theta$$

Divide $\sin^2\theta + \cos^2\theta = 1$ by $\cos^2\theta$.

Probably the most commonly used identity, $\sin^2 \theta + \cos^2 \theta = 1$, may be more helpful to you when rearranged as $\sin^2 \theta = 1 - \cos^2 \theta$ or $\cos^2 \theta = 1 - \sin^2 \theta$.

Sum and Difference Identities

A brief examination of common values will tell you that the trig functions are not additive. For example, $\sin\frac{\pi}{6} + \sin\frac{\pi}{3} \neq \sin\left(\frac{\pi}{6} + \frac{\pi}{3}\right)$. It is helpful, therefore, to have identities for the trig functions of sums and differences.

$$\sin(\alpha + \beta) = \sin\alpha \cos\beta + \cos\alpha \sin\beta$$

$$\cos(\alpha + \beta) = \cos\alpha \cos\beta - \sin\alpha \sin\beta$$

$$\tan(\alpha + \beta) = \frac{\tan\alpha + \tan\beta}{1 - \tan\alpha \tan\beta}$$

$$\sin(\alpha - \beta) = \sin\alpha \cos\beta - \cos\alpha \sin\beta$$

$$\cos(\alpha - \beta) = \cos\alpha \cos\beta + \sin\alpha \sin\beta$$

$$\tan(\alpha - \beta) = \frac{\tan\alpha - \tan\beta}{1 + \tan\alpha \tan\beta}$$

Double-Angle and Half-Angle Identities

If the sum identities are applied to two angles of equal measure, the results are called the double-angle identities. From the identity for the cosine of a double angle, you can derive the half-angle identities, and those can be useful both in identities and equations and in finding the exact values of the trig functions of angles that are half of the common angles.

Double-Angle Identities. If you begin with the sum identities and substitute α for β, you can derive identities for the sine, cosine, and tangent of 2α.

$$\sin(2\alpha) = 2\sin\alpha \cos\alpha$$

$$\begin{aligned}\cos(2\alpha) &= \cos^2\alpha - \sin^2\alpha \\ &= 1 - 2\sin^2\alpha \\ &= 2\cos^2\alpha - 1\end{aligned}$$

$$\tan(2\alpha) = \frac{2\tan\alpha}{1 - \tan^2\alpha}$$

Half-Angle Identities. To derive the half-angle identities, begin with the identity for the cosine of a double angle, and let $\theta = 2\alpha$. Then $\alpha = \frac{\theta}{2}$ and

$$\sin\left(\frac{\theta}{2}\right) = \pm\sqrt{\frac{1-\cos\theta}{2}}$$

$$\cos\left(\frac{\theta}{2}\right) = \pm\sqrt{\frac{1+\cos\theta}{2}}$$

$$\tan\left(\frac{\theta}{2}\right) = \pm\sqrt{\frac{1-\cos\theta}{1+\cos\theta}}$$

Whether you use the positive or the negative version will be determined by the quadrant in which $\frac{\theta}{2}$ falls.

Verifying Identities

To verify an identity, you need to demonstrate that the two sides of the equation are identical. To do this, first decide which side of the equation seems simpler. Leave this side untouched, as your goal.

On the side you've decided to change, use known identities to substitute for pieces of the expression. When the identity you're trying to prove contains more than one trig function, it's usually helpful to make substitutions that will reduce the number of functions involved. For that reason, it's often a wise idea to put as much of the equation as possible in terms of sine and cosine. Then use algebraic techniques to simplify until the more complicated side matches the goal.

$$\text{Verify: } \sin\theta + \cot\theta\cos\theta = \csc\theta$$

The right side, $\csc\theta$, seems simpler, so keep that as the goal. To reduce the number of functions involved, replace $\cot\theta$ by the quotient identity $\frac{\cos\theta}{\sin\theta}$.

$$\sin\theta + \cot\theta\cos\theta = \csc\theta$$

$$\sin\theta + \frac{\cos\theta}{\sin\theta}\cdot\cos\theta = \csc\theta$$

$$\sin\theta + \frac{\cos^2\theta}{\sin\theta} = \csc\theta$$

Find a common denominator, and add the terms on the left side.

$$\sin\theta + \frac{\cos^2\theta}{\sin\theta} = \csc\theta$$

$$\frac{\sin^2\theta}{\sin\theta} + \frac{\cos^2\theta}{\sin\theta} = \csc\theta$$

$$\frac{\sin^2\theta + \cos^2\theta}{\sin\theta} = \csc\theta$$

Use the Pythagorean identity to replace the numerator with 1, and then replace $\frac{1}{\sin\theta}$ with $\csc\theta$.

$$\frac{1}{\sin\theta} = \csc\theta$$

$$\csc\theta = \csc\theta$$

Exercise 12.5

Verify each identity.

1. $\csc x + \cot x \cdot \sec x = 2\csc x$
2. $\frac{\cos x \csc x}{\cot^2 x} = \tan x$
3. $\tan\left(\frac{\pi}{2} - \theta\right) \cdot (1 - \cos^2\theta) \cdot \sec\theta = \sin\theta$
4. $\cos\left(\frac{\pi}{2} - \theta\right) \cdot (\sin\theta + \cot\theta\cos\theta) = 1$
5. $\frac{\sin x - 1}{\csc x} \cdot (\csc x + 1) = -\cos^2 x$
6. $\csc\theta\tan\theta - \sin\theta = \frac{\tan\theta - \sin^2\theta}{\sin\theta}$
7. $\sec^2\theta(\cos^2\theta - 1) = -\tan^2\theta$

8. $\sec\theta \tan\theta - \csc^2\theta = \dfrac{\sin^3\theta - \cos^2\theta}{(\sin\theta \cos\theta)^2}$

9. $\sin(\alpha+\beta)\cdot\sin(\alpha-\beta) = \sin^2\alpha - \sin^2\beta$

10. $\csc(2\theta) = \dfrac{\sec\theta \csc\theta}{2}$

Step 6. Use Identities

The sum and difference identities can be used in simplifying identities or simplifying equations before solving, but they're also useful for finding the exact value of trig functions of angles that can be expressed as the sum or difference of angles whose values you've memorized. To find $\tan\frac{7\pi}{12}$, rewrite it as $\tan\left(\frac{\pi}{3}+\frac{\pi}{4}\right)$ and apply the identity.

$$\tan\left(\frac{7\pi}{12}\right) = \tan\left(\frac{\pi}{3}+\frac{\pi}{4}\right)$$

$$= \frac{\tan\frac{\pi}{3} + \tan\frac{\pi}{4}}{1 - \tan\frac{\pi}{3}\tan\frac{\pi}{4}}$$

Evaluate each function.

$$\tan\frac{7\pi}{12} = \frac{\sqrt{3}+1}{1-\sqrt{3}\cdot 1}$$

Simplify, and rationalize the denominator.

$$\tan\left(\frac{7\pi}{12}\right) = \frac{1+\sqrt{3}}{1-\sqrt{3}}\cdot\frac{1+\sqrt{3}}{1+\sqrt{3}}$$

$$= \frac{(1+\sqrt{3})^2}{1-3}$$

$$= -\frac{4+2\sqrt{3}}{2}$$

$$= -2-\sqrt{3}$$

Exercise 12.6

Use the sum, difference, double-angle, or half-angle identities to find the value of each expression.

1. $\cos\left(\frac{7\pi}{12}\right)$
2. $\tan\left(\frac{5\pi}{12}\right)$
3. $\sin\left(\frac{17\pi}{12}\right)$
4. $\cos\frac{\pi}{12}$
5. $\sin\frac{5\pi}{12}$

Step 7. Solve Trigonometric Equations

Solving trigonometric equations can often be made simpler by following these steps.

1. If the equation involves more than one trig function, use identities to substitute and simplify. Whenever possible, try to put the equation in terms of sine or cosine.
2. Let a single variable represent the remaining trig function. For example, replace all occurrences of $\cos\theta$ with t.
3. Solve the equation for this new placeholder variable.
4. Reinsert the trig function, and determine the value of the argument that will produce the desired value.
5. Remember that trigonometric functions are periodic. Equations commonly have multiple solutions. Be sure to give all the values of the variable that satisfy the equation. To specify all solutions, use the period of the function to summarize the repetition.

To find all solutions of the equation $3 - 3\sin\theta - 2\cos^2\theta = 0$ in the interval $[0,2\pi)$, rewrite the equation in terms of a single function. Use the Pythagorean identity to replace $\cos^2\theta$ with $1 - \sin^2\theta$.

$$3 - 3\sin\theta - 2\cos^2\theta = 0$$

$$3 - 3\sin\theta - 2(1 - \sin^2\theta) = 0$$

$$3 - 3\sin\theta - 2 + 2\sin^2\theta = 0$$

$$1 - 3\sin\theta + 2\sin^2\theta = 0$$

$$2\sin^2\theta - 3\sin\theta + 1 = 0$$

The equation has a quadratic form, and it will be easier to solve if you let $t = \sin\theta$ and $2\sin^2\theta - 3\sin\theta + 1 = 0$ become $2t^2 - 3t + 1 = 0$. Factor and solve.

$$2t^2 - 3t + 1 = 0$$
$$(2t - 1)(t - 1) = 0$$
$$2t - 1 = 0 \quad t - 1 = 0$$
$$2t = 1 \quad t = 1$$
$$t = \frac{1}{2}$$

Replace $t = \sin\theta$.

$$\sin\theta = \frac{1}{2} \quad \text{when } \theta = \frac{\pi}{6} \quad \text{or} \quad \frac{5\pi}{6}$$

and

$$\sin\theta = 1 \quad \text{when } \theta = \frac{\pi}{2}$$

So the solutions are $\theta = \frac{\pi}{6}, \frac{5\pi}{6}$, and $\frac{\pi}{2}$.

Exercise 12.7

Solve each equation over the domain [0,2π].

1. $2\tan x\cos x - \tan x = 0$
2. $2\cos^2 x + 3\cos x + 1 = 0$
3. $4\cos\theta = 3\sec\theta$
4. $2\sin^2\theta + 12\cos^2\theta = 2 - 6\cos\theta$
5. $\sin\frac{\theta}{2} = \cos\theta$

Solve each equation over the real numbers.

6. $4\sin^2\theta\tan\theta - 3\tan\theta = 0$
7. $4\cos^2\theta + 3\sin^2\theta = 5$
8. $\tan\left(\theta + \frac{\pi}{4}\right) = 6\tan\theta$
9. $2\sin^2(2\theta) = 1$
10. $2\csc\theta - \tan\theta\csc^2\theta = 0$

13

Polar and Parametric Equations

In the Cartesian, or rectangular, coordinate system, points in the plane are located by an ordered pair of coordinates. The first coordinate indicates movement left or right of the origin, and the second, motion up or down. The polar coordinate system also uses an ordered pair of coordinates to locate a point in the plane, but the movement from the origin, or pole, is indicated by an angle of rotation and a distance.

Step 1. Plot Points in the Polar Coordinate System

In the polar coordinate system, a point is located by a radius, or distance from the pole, and an angle, representing a rotation, which you can think of as an angle in standard form. The ordered pair that denotes a point is given in the form (r, θ), where r is the radius, or distance from the pole, and θ is the angle.

To plot the point $\left(3, \frac{\pi}{4}\right)$, count out to the third ring, and then move around the circle to the spoke that represents $\frac{\pi}{4}$. You may find it easier to first locate the $\frac{\pi}{4}$ spoke and then follow it out to the third ring. The ordered pair $\left(8, -\frac{\pi}{2}\right)$ tells you to plot a point on the eighth circle but to move $\frac{\pi}{2}$ units clockwise, or in the negative direction. To plot $\left(-4, \frac{7\pi}{6}\right)$, imagine you are standing at the pole. Turn until you are looking down the spoke that represents $\frac{7\pi}{6}$, and then move backward four rings. If you think this

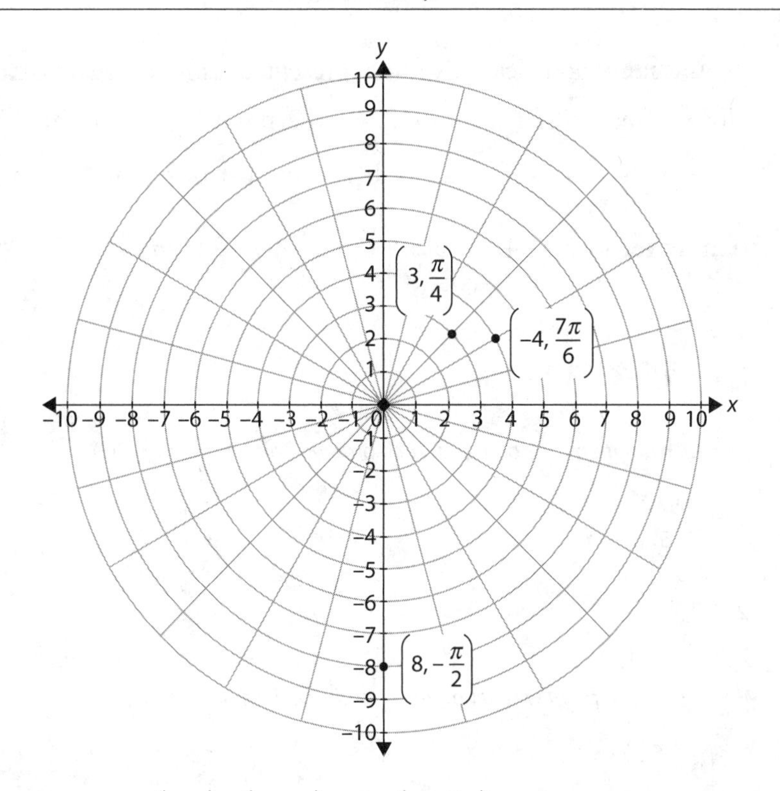

The points $\left(3, \frac{\pi}{4}\right)$, $\left(8, -\frac{\pi}{2}\right)$, and $\left(-4, \frac{7\pi}{6}\right)$ plotted on a polar grid

seems to take you to the point $\left(4, \frac{\pi}{6}\right)$, you're absolutely correct. In the polar coordinate system, it's quite possible to name the same point with different coordinates.

One of the key features of polar coordinates is that a single point may be represented by more than one set of coordinates. In fact, each point in the plane has infinitely many sets of coordinates. Because rotation can be counterclockwise or clockwise, a point such as $\left(4, \frac{\pi}{3}\right)$ can also be represented as $\left(4, -\frac{5\pi}{3}\right)$. Because it's possible to have angles of greater than 2π, it could also be named as $\left(4, \frac{7\pi}{3}\right)$ or $\left(4, -\frac{11\pi}{3}\right)$ or $\left(4, \frac{37\pi}{3}\right)$ or any expression of the form $\left(4, \frac{\pi}{3} \pm 2n\pi\right)$ or $\left(4, -\frac{5\pi}{3} \pm 2n\pi\right)$.

The same point can also be represented using a negative radius with an angle offset by half a rotation. So the point $\left(4,\frac{\pi}{3}\right)$ can also be named as $\left(-4,\frac{4\pi}{3}\right)$ or $\left(-4,-\frac{2\pi}{3}\right)$. That adds to the list of possibilities all representations of the form $\left(-4,\frac{4\pi}{3}\pm 2n\pi\right)$ and $\left(-4,-\frac{2\pi}{3}\pm 2n\pi\right)$.

Exercise 13.1

Plot each point on the polar grid. Each ordered pair has the form (r,θ).

1. $\left(2,\frac{5\pi}{6}\right)$
2. $\left(9,-\frac{5\pi}{4}\right)$
3. $\left(-6,\frac{2\pi}{3}\right)$
4. $\left(-5,-\frac{5\pi}{6}\right)$

Represent each point with a negative r and a positive θ.

5. $\left(9,-\frac{5\pi}{4}\right)$
6. $\left(4,-\frac{\pi}{3}\right)$

Represent each point with a positive r and a positive θ.

7. $\left(-5,-\frac{5\pi}{6}\right)$
8. $\left(-6,\frac{2\pi}{3}\right)$

Represent each point with a positive r and a negative θ.

9. $\left(2,\frac{5\pi}{6}\right)$
10. $\left(5,\frac{\pi}{3}\right)$

Step 2. Convert Coordinates Between Systems

Follow these steps to convert to polar coordinates, given a point in rectangular coordinates (x,y):

1. Note in which quadrant the point lies.
2. Use the Pythagorean theorem to calculate $r=\sqrt{x^2+y^2}$.
3. Find $\tan^{-1}\left(\frac{y}{x}\right)$.

4. If the point lies in quadrant I or quadrant IV, $\theta = \tan^{-1}\left(\frac{y}{x}\right)$.

5. If the point falls in quadrant II or quadrant III, $\theta = \pi + \tan^{-1}\left(\frac{y}{x}\right)$.

To convert the rectangular point (4,−4) to polar coordinates, first notice that it is a fourth-quadrant point, then find $r = \sqrt{4^2 + (-4)^2} = \sqrt{32} = 4\sqrt{2}$, and calculate $\tan^{-1}\left(\frac{-4}{4}\right) = \tan^{-1}(-1) = -\frac{\pi}{4}$. Because the point is in quadrant IV, you can use the value of $\tan^{-1}\left(\frac{y}{x}\right)$ as θ, so the point can be named as $\left(4\sqrt{2}, -\frac{\pi}{4}\right)$.

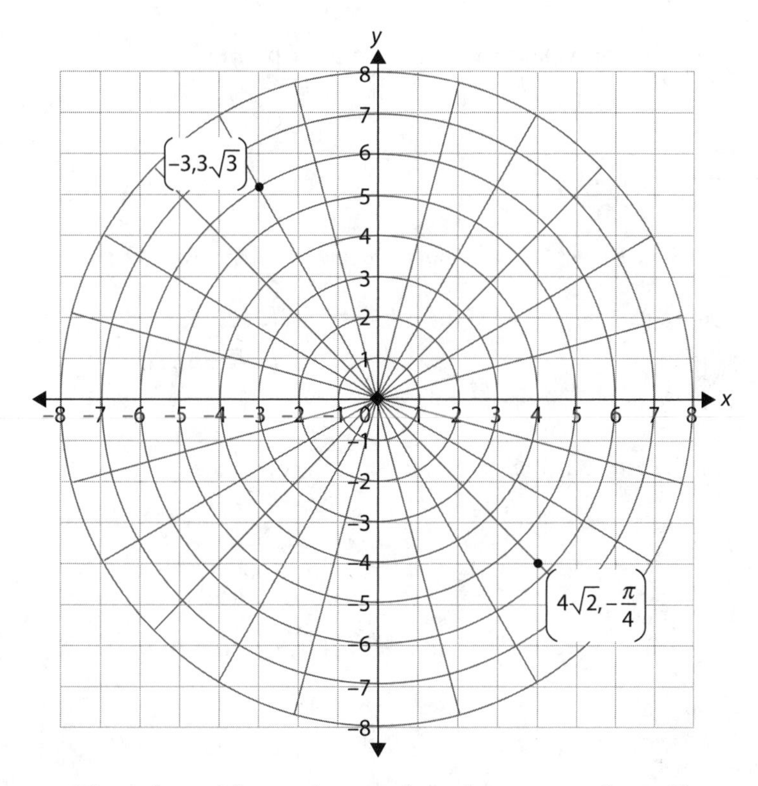

The polar grid superimposed on the rectangular grid

To convert the rectangular point $(-3, 3\sqrt{3})$ to polar form, note its position in quadrant II. Find $r = \sqrt{(-3)^2 + (3\sqrt{3})^2} = \sqrt{9 + 27} = 6$. Then calculate the value of $\tan^{-1}\left(\frac{3\sqrt{3}}{-3}\right) = \tan^{-1}(-\sqrt{3}) = -\frac{\pi}{3}$, but remember that the point is in quadrant II. Adjust by adding π to the value $\tan^{-1}\left(\frac{y}{x}\right)$ gave, and find that θ is $\frac{2\pi}{3}$, so the point is $\left(6,\frac{2\pi}{3}\right)$. For points that fall on the y-axis, use $\theta=\frac{\pi}{2}$ or $\theta=-\frac{\pi}{2}$. The rectangular point $(0,5)$ is the polar point $\left(5,\frac{\pi}{2}\right)$, and the rectangular point $(0,-3)$ is $\left(3,-\frac{\pi}{2}\right)$.

To convert the polar point (r,θ) to rectangular coordinates, find $x=r\cos\theta$ and $y=r\sin\theta$.

The polar point $\left(8,\frac{3\pi}{4}\right)$ has an x-coordinate of $8\cos\frac{3\pi}{4}=8\left(-\frac{\sqrt{2}}{2}\right)=-4\sqrt{2}$ and a y-coordinate that is equal to $8\sin\frac{3\pi}{4}=8\left(\frac{\sqrt{2}}{2}\right)=4\sqrt{2}$. The polar point $\left(8,\frac{3\pi}{4}\right)$ is equivalent to the rectangular point $(-4\sqrt{2},4\sqrt{2})$.

Exercise 13.2

Convert each point from rectangular to polar coordinates.

1. $(-4,-4\sqrt{3})$
2. $(0,-1)$
3. $(4,-4)$
4. $(6\sqrt{3},-6)$
5. $(9\sqrt{6},-9\sqrt{2})$

Convert each point from polar to rectangular coordinates.

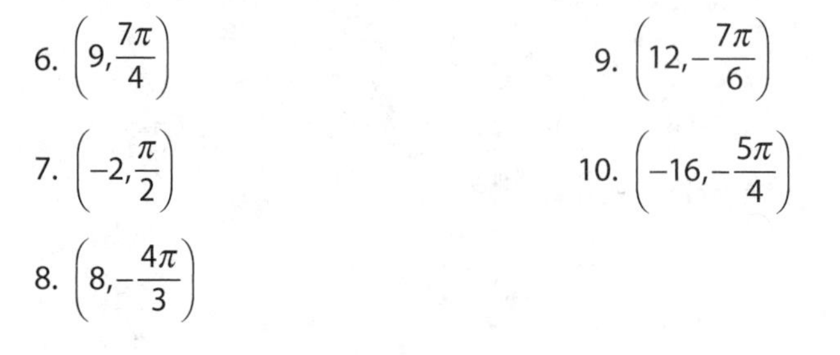

6. $\left(9,\frac{7\pi}{4}\right)$
7. $\left(-2,\frac{\pi}{2}\right)$
8. $\left(8,-\frac{4\pi}{3}\right)$
9. $\left(12,-\frac{7\pi}{6}\right)$
10. $\left(-16,-\frac{5\pi}{4}\right)$

Step 3. Graph Polar Equations

Equations in polar form generally express r in terms of θ and often define distinctive curves that would be difficult to represent in rectangular coordinates. To sketch a graph, build a table of values and plot points using values of θ in order and connecting as you go along. Because values of r may be negative, the graph may trace in unexpected ways. It can even trace over itself, because the periodic nature of the trig functions can give multiple representations of the same point. Remember that most polar graphs are curves, so avoid straight line connections.

Lines, Circles, and Spirals

It is possible to find a polar equation that has a linear graph. A line passing through the polar has a constant value of θ, and vertical and horizontal lines have fairly simple equations. Other lines have more complicated equations.

Line passing through the pole $\theta = c$

Vertical line $r = \dfrac{a}{\cos\theta}$

Horizontal line $r = \dfrac{a}{\sin\theta}$

General form of a line $r = \dfrac{b\cos\theta}{\sin(\theta - \phi)}$, where b is the y-intercept of the line in rectangular form, and ϕ is the arctan of the slope of the line

The line $y = \sqrt{3}x - 2$ becomes

$$r = \frac{-2\cos\frac{\pi}{3}}{\sin\left(\theta - \frac{\pi}{3}\right)}$$

$$r = \frac{-2\left(\frac{1}{2}\right)}{\sin\theta\cos\frac{\pi}{3} - \cos\theta\sin\frac{\pi}{3}}$$

$$r = \frac{-1}{\frac{1}{2}\sin\theta - \frac{\sqrt{3}}{2}\cos\theta}$$

$$r = \frac{-2}{\sin\theta - \sqrt{3}\cos\theta}$$

Creating a polar equation that describes a circle centered at the pole only requires specifying the radius. The equations of circles tangent to the pole and symmetric about one of the axes are simple as well. When the center starts to wander, the equation becomes more complex.

Circle of radius c centered at the pole — $r = c$

Circle of radius a tangent to the pole — $r = 2a\sin\theta$ symmetric about the vertical axis

Circle of radius a tangent to the pole — $r = 2a\cos\theta$ symmetric about the horizontal axis

General equation of a circle of radius c — $r^2 - 2rd\cos(\theta - \phi) + d^2 = c^2$, where d is the distance from the pole to the center, (h,k), and $\phi = \tan^{-1}\left(\frac{k}{h}\right)$

The circle centered at $(2\sqrt{3},-2)$ with radius 3 has the rectangular equation $(x - 2\sqrt{3})^2 + (y+2)^2 = 9$. The distance of the center from the origin is $d = \sqrt{(2\sqrt{3})^2 + (-2)^2} = 4$, and $\phi = \tan^{-1}\left(\frac{-2}{2\sqrt{3}}\right) = -\frac{\pi}{6}$. In polar form, this would become $r^2 - 2r \cdot 4\cos\left(\theta - -\frac{\pi}{6}\right) + 4^2 = 3^2$ or $r^2 - 8r\cos\left(\theta + \frac{\pi}{6}\right) + 7 = 0$.

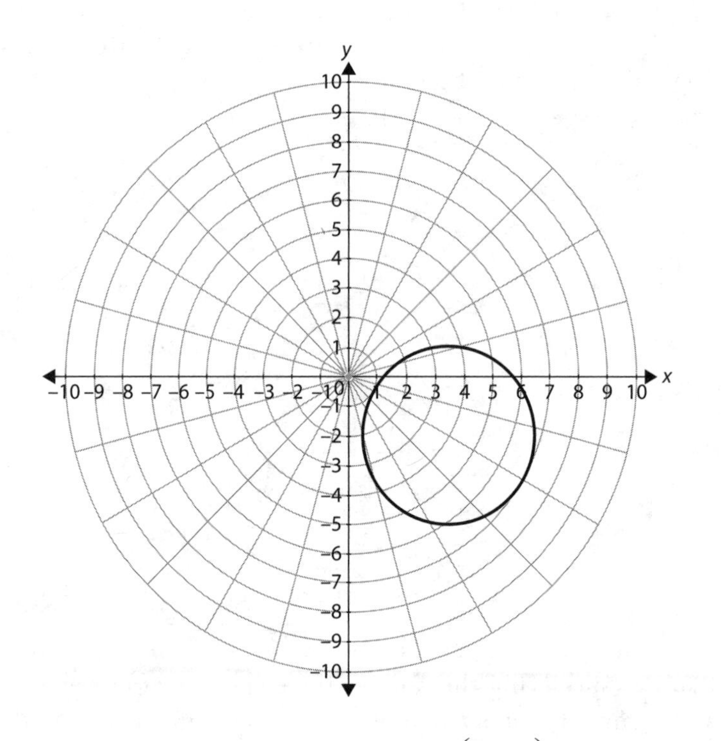

The graph of the circle $r^2 - 8r\cos\left(\theta + \frac{\pi}{6}\right) + 7 = 0$

When the value of r depends directly on the value of θ, rather than a trig frunction of θ, the resulting graph spirals.

Archimedean spiral $r = a\theta$
Logarithmic spiral $r = e^{a\theta}$

Limaçons, Cardioids, and Roses

Classic polar graphs include the limaçons, named for snails, the heart-shaped cardioids, and the polar rose.

Limaçon $r = a \pm b \cos\theta$ symmetric about the horizontal axis

If $a < b$, the limaçon will have an inner loop, but if $a > b$, there will be no inner loop.

Limaçon $r = a \pm b \sin\theta$ symmetric about the vertical axis
Cardioid $r = a \pm a \cos\theta$ symmetric about the horizontal axis
Cardioid $r = a \pm a \sin\theta$ symmetric about the vertical axis
Polar rose $r = a \cos b\theta$ symmetric about the horizontal axis

The value of a gives the length of the petal, while b controls the number of petals. If b is an odd number, there will be b petals, but if b is even, there will be $2b$ petals.

Polar rose $r = a \sin b\theta$ symmetric about the vertical axis

The equation $r = 2 - 3\sin\theta$ yields a limaçon with an inner loop, symmetric about the vertical axis.

The equation $r = 2 + 2 \cos\theta$ is a cardioid symmetric about the horizontal axis.

The equation $r = 3\sin(4\theta)$ is a rose with eight petals, each three units long. (See figure on page 205)

Conics

Conic sections with focus at the pole have equations of the form $r = \frac{ep}{1 \pm e\cos\theta}$ or $r = \frac{ep}{1 \pm e\sin\theta}$, where p is the distance from the pole to the directrix and e is the eccentricity of the conic. The eccentricity of an ellipse is $0 < e < 1$, the eccentricity of a hyperbola is $e > 1$, and the parabola has $e = 1$. The equation $r = \frac{3}{2 + \cos\theta} = \frac{.5(3)}{1 + .5\cos\theta}$ describes an ellipse, while $r = \frac{9}{2 + 3\cos\theta} = \frac{1.5(3)}{1 + 1.5\cos\theta}$ is a hyperbola, and $r = \frac{3}{1 + \cos\theta}$ is a parabola.

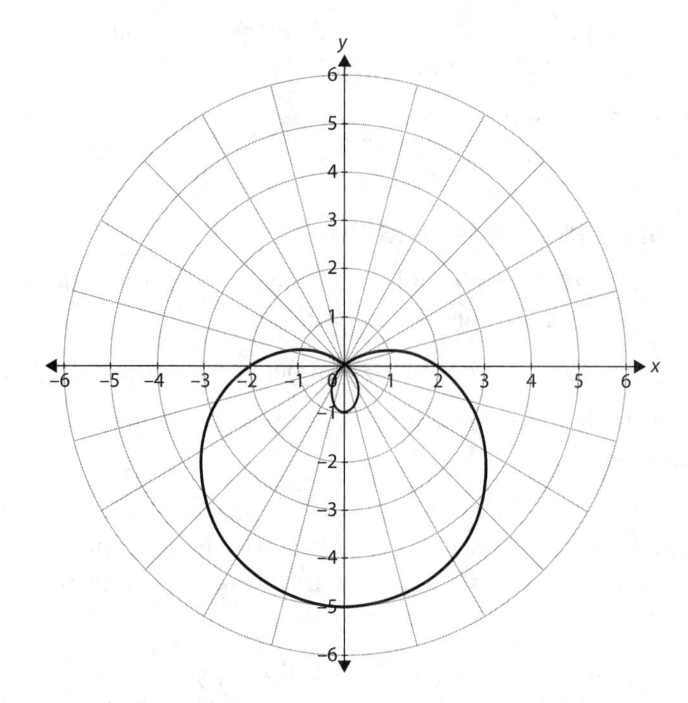

The limaçon $r = 2 - 3\sin\theta$ with its inner loop

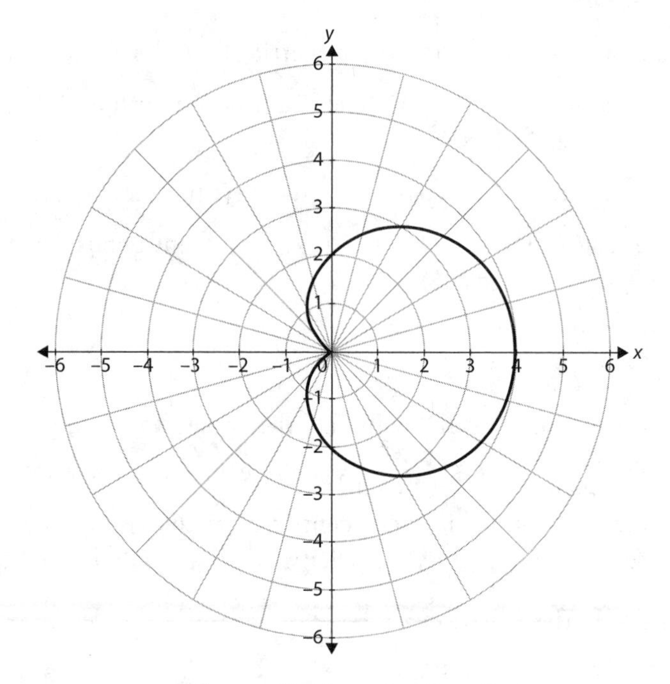

The cardioid $r = 2 + 2\cos\theta$

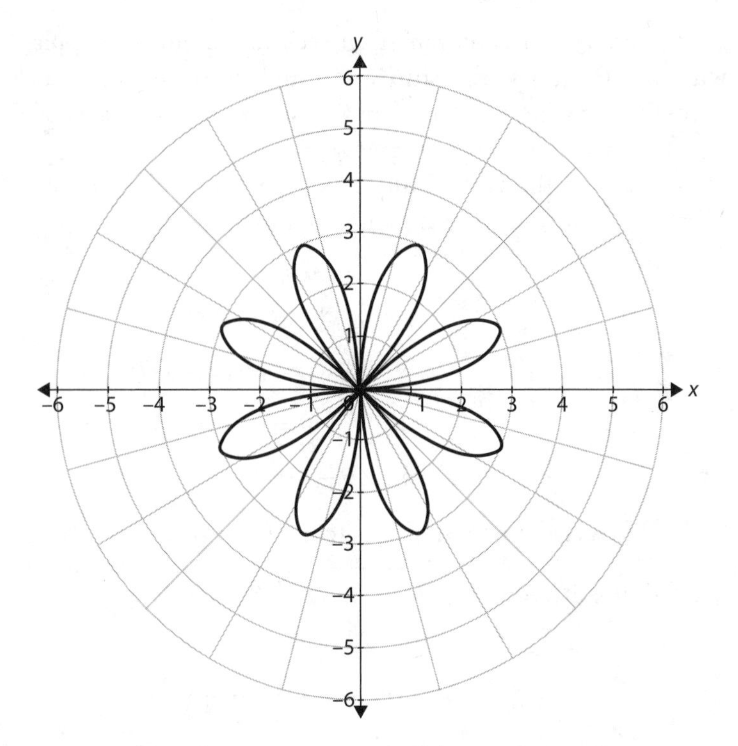

The polar rose $r = 3\sin(4\theta)$ has eight petals

Exercise 13.3

Sketch a graph of each equation.

1. $r = 3 + 5\cos\theta$
2. $r = 4\cos 3\theta$
3. $r = 5\sin 4\theta$
4. $r = 4 - 2\sin\theta$
5. $r = \dfrac{4}{2 - \cos\theta}$
6. $r = 2\sin\theta$
7. $r = 4\cos 2\theta$
8. $r = 2 - 5\sin\theta$
9. $r = 4 + \cos\theta$
10. $r = 3\sin 2\theta$

Step 4. Convert Equations Between Polar and Rectangular Forms

The same conversions that you used to change points from rectangular to polar coordinates or polar to rectangular coordinates can be used to change equations from one form to another.

To convert an equation from rectangular form to polar form, replace x with $r\cos\theta$ and y with $r\sin\theta$, and simplify. The equation $x^2 + y^2 = 9$ becomes $(r\cos\theta)^2 + (r\sin\theta)^2 = 9$ or $r^2(\cos^2\theta + \sin^2\theta) = 9$. Because $\cos^2\theta + \sin^2\theta = 1$, the equation becomes $r^2 = 9$ or simply $r = 3$.

To convert the rectangular equation $x = y^2 - 4$ to polar form, make the substitutions of $r\cos\theta$ and $r\sin\theta$ for x and y.

$$x = y^2 - 4$$

$$r\cos\theta = (r\sin\theta)^2 - 4$$

$$r\cos\theta = r^2(1 - \cos^2\theta) - 4$$

$$r\cos\theta = r^2 - r^2\cos^2\theta - 4$$

$$r^2 - r^2\cos^2\theta - r\cos\theta - 4 = 0$$

To convert the equation $r = 4\cos\theta$ from polar form to rectangular form, use the substitutions $r = \sqrt{x^2 + y^2}$ and $\theta = \tan^{-1}\left(\frac{y}{x}\right)$.

$$\sqrt{x^2 + y^2} = 4\cos\left(\tan^{-1}\left(\frac{y}{x}\right)\right)$$

$$\sqrt{x^2 + y^2} = \frac{4x}{\sqrt{x^2 + y^2}}$$

$$x^2 + y^2 = 4x$$

$$x^2 - 4x + y^2 = 0$$

Exercise 13.4

Convert to polar form.

1. $x = 7$
2. $(x - 1)^2 + y^2 = 25$
3. $y = x^2 - 4$
4. $4x^2 + 9y^2 = 36$
5. $16y^2 - 4x^2 = 64$

Convert to rectangular form.

6. $r = \dfrac{3}{\sin\theta}$
7. $\theta = \dfrac{\pi}{3}$
8. $r = 2 - 3\cos\theta$
9. $r = 1 + \tan\theta$
10. $r = \dfrac{5}{3 + 2\sin\theta}$

Step 5. Convert Equations Between Parametric and Function Forms

Rather than expressing the relationship between variables x and y directly, with y as a function of x, a parametric equation gives both x and y as functions of a parameter, t. The parameter t is often time, and one equation talks about the horizontal motion of an object over time and the other about the vertical motion over time. In addition to adding a consideration for time and speed to the equation, parametric equations also introduce a directionality not present in standard function notation. As the parameter t increases, x and y change, but plotting the points (x,y) may take you in any direction, even looping back over itself.

If a parametric function defines $x(t) = t + 2$ and $y(t) = t^2 - 1$, for $t \geq 0$, the set of points $(x(t), y(t))$ describes the path of an object over time. The parametric equation introduces a notion of directionality not present in equations in function form. As t increases, the object traces out a path, in this example moving to the right and upward. In other cases, the motion may change direction or double back on itself. As you plot the graph, you'll want to denote the direction with arrows.

Understanding the shape of the graph may be easier if you express the relationship in function form, defining y in terms of x. To do this, you'll need to eliminate the parameter t. To eliminate the parameter, solve $x(t)$ for t. Substitute the expression for t derived from $x(t)$ for t in $y(t)$ and simplify. A parametric equation is defined by

$$x(t) = 2t - 5$$

$$y(t) = t^2 + 3t - 1$$

To eliminate the parameter, solve $x = 2t - 5$ for t.

$$\begin{aligned} x &= 2t - 5 \\ x + 5 &= 2t \\ \frac{x+5}{2} &= t \end{aligned}$$

Substitute $\frac{x+5}{2} = t$ for t in the $y(t)$ equation.

$$
\begin{aligned}
y &= t^2 + 3t - 1 \\
&= \left(\frac{x+5}{2}\right)^2 + 3\left(\frac{x+5}{2}\right) - 1 \\
&= \frac{x^2 + 10x + 25}{4} + \frac{3x + 15}{2} - 1 \\
&= \frac{x^2 + 10x + 25}{4} + \frac{6x + 30}{4} + \frac{4}{4} \\
&= \frac{1}{4}(x^2 + 16x + 59)
\end{aligned}
$$

Exercise 13.5

Express each relation as a function y in terms of x.

1. $x = t + 1$
   $y = t^2$
2. $x = t^2$
   $y = t^2 - 4$
3. $x = \sqrt{t}$
   $y = t + 4$
4. $x = \frac{1}{t}$
   $y = t - 1$
5. $x = 3t$
   $y = \cos t$
6. $x = \pi t$
   $y = \sin t$
7. $x = t - \frac{\pi}{4}$
   $y = 3\cos t$
8. $x = \sin t$
   $y = \cos t$
9. $x = \cos t$
   $y = \sec t$
10. $x = t^2 - 4$
    $y = t^4$

Step 6. Graph Parametric Equations

To graph parametric equations, make a table of values with columns for the parameter, t, and for x and y. Choose appropriate values for t, and then evaluate x and y for each value of t. Plot points, moving in order from the smallest to the largest value of t. Connect points as you go along, using a smooth curve, and include arrows on the curve to indicate direction of movement.

Consider the curve defined by the parametic equation

$$x(t) = \frac{t}{2}$$

$$y(t) = 8t - t^2$$

Build a table of values, choosing values of t and evaluating x and y.

t	0	1	2	3	4	5	6	7	8
x	0	0.5	1	1.5	2	2.5	3	3.5	4
y	0	7	12	15	16	15	12	7	0
Point	(0,0)	(0.5,7)	(1,12)	(1.5,15)	(2,16)	(2.5,15)	(3,12)	(3.5,7)	(4,0)

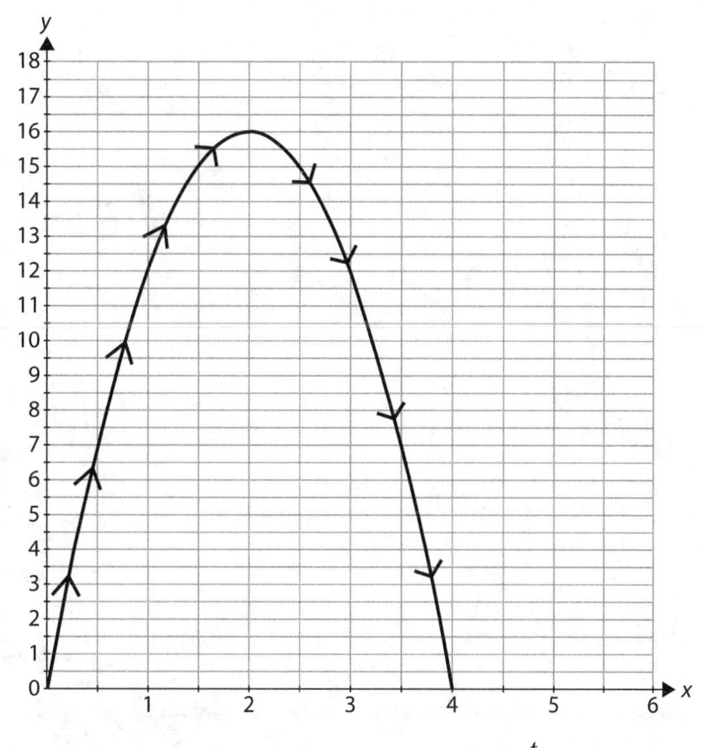

The graph of the parametric equation $x(t) = \frac{t}{2}$, $y(t) = 8t - t^2$, with arrows showing direction

Exercise 13.6

Sketch a graph of each parametric equation.

1. $x = 3 - 2t$
   $y = 4 + t$
2. $x = t^2$
   $y = t + 1$
3. $x = t - 1$
   $y = 1 - t^2$
4. $x = t^2 - t$
   $y = t^2 + t$
5. $x = \dfrac{t^2}{2}$
   $y = \sqrt{t - 1}$
6. $x = \cos t$
   $y = \sin t$
7. $x = 2\cos t$
   $y = \sin^2 t$
8. $x = 4 - \sin t$
   $y = \sin^2 t - 1$
9. $x = 5\cos t$
   $y = 3\sin t$
10. $x = t\cos(\pi t)$
    $y = \tan(\pi t)$

14

Transformations

The word *transformation* refers to any one of several operations on a graph or figure in the plane, each of which maps every point of the original figure, the pre-image, to a point of a new figure, called the image. Transformations appear in geometry, acting on points, segments, and polygons, but also in algebra, acting on the graphs of relations and functions. In fact, a transformation can be thought of as a function whose domain is a set of points. The transformation maps each point in its domain to a point in the range, and transformations can be composed to create new transformations.

Step 1. Understand the Geometry of Reflections

Reflection is a transform that maps every point of the pre-image to a point on the opposite side of a line, called the reflecting line, according to two rules: the image point and the pre-image point both line on a line perpendicular to the reflecting line, and the distance from the pre-image to the reflecting line is equal to the distance from the reflecting line to the image point.

Rigid Transformations Create Congruent Figures

The fundamental transformation, reflection, is called a rigid transformation that preserves distance and angle measure. Any line segment in the image is congruent to the corresponding segment from the pre-image, and any angle from the pre-image is sent to a corresponding angle in the image which has the same measure. The reflection of the graph or image across a reflecting line has the effect of moving the image to a different location and the reflection changes the orientation of the figure. If the pre-image is a pentagon whose vertex labels, read in clockwise order, spell out FRESH, the image after a reflection will have vertices labeled F′, R′, E′, S′, and H′, but you'll have to read counterclockwise to get F′R′E′S′H′.

Reflection Symmetry

Reflection is probably not a new idea for you. You may have talked about reflection symmetry in a geometry course or when investigating the graphs of functions. Generally, the reflecting lines for those discussions were the x- or y-axis or a simple line like $y = x$ or $y = -x$. In fact, any line can serve as a reflecting line.

Using Coordinates

With the traditional reflecting lines, the effect of reflection is simple to state.

REFLECTING LINE	PRE-IMAGE POINT	IMAGE POINT
y-axis ($x = 0$)	(a, b)	$(a, -b)$
x-axis ($y = 0$)	(a, b)	$(-a, b)$
$y = x$	(a, b)	(b, a)
$y = -x$	(a, b)	$(-b, -a)$

ΔABC has vertices A(−4, 5), B(0, −1), and C(2, 3). When the triangle is reflected over the y-axis, the image triangle has vertices A′(4, 5), B′(0, −1), and C′(−2, 3).

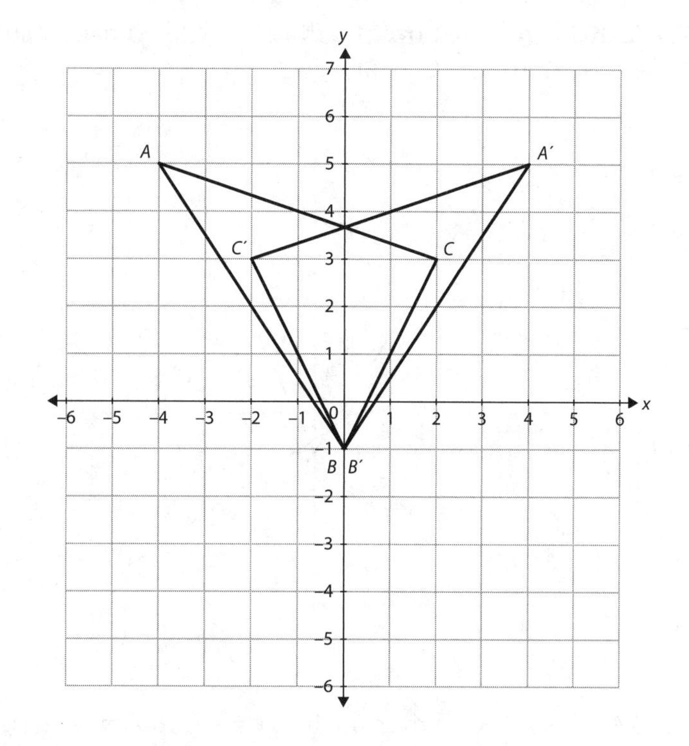

If ΔABC is reflected over the x-axis, the image triangle has vertices A′(−4, −5), B′(0, 1), and C′(2, −3).

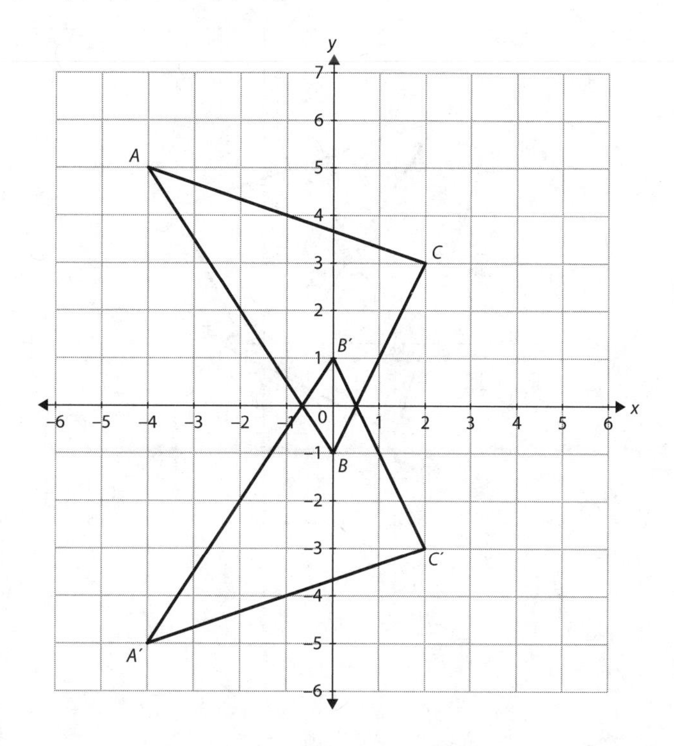

If ΔABC is reflected over the line $y = x$, the image triangle has vertices A′(5, −4), B′(−1, 0), and C′(3, 2).

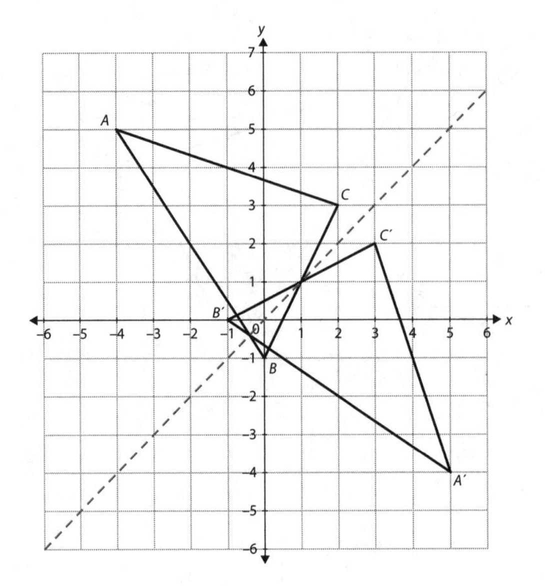

If ΔABC is reflected over the line $y = -x$, the image triangle has vertices A′(−5, 4), B′(1, 0), and C′(3, −2).

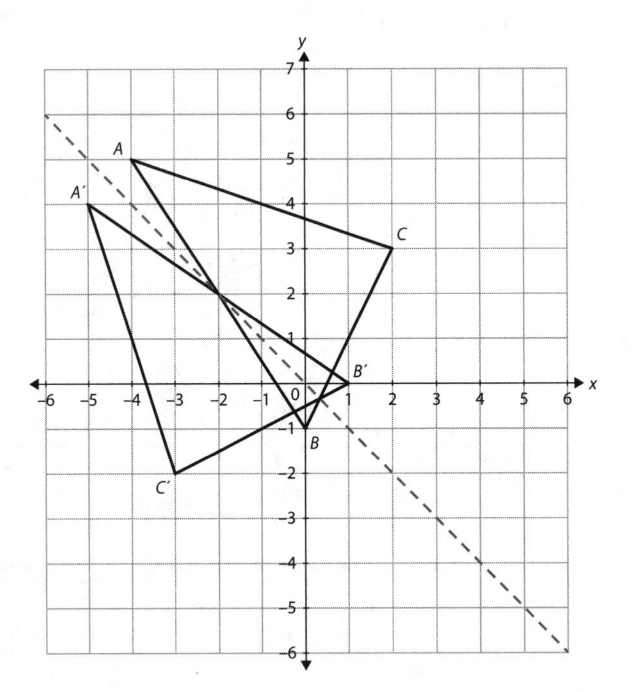

Changing the Line

In the last two cases, if the reflecting line is changed so that $m \neq \pm 1$, that change in slope will affect the reflection image. When a point is reflected over the line $y = x$, its image will line on a line perpendicular to $y = x$ that passes through the pre-image point. If you find that perpendicular line and find the point at which it intersects the reflecting line, that point of intersection is the midpoint of the segment connecting pre-image (a, b) to image (c, d). You can use the midpoint formula to find the coordinates of the image point. In the table below, you'll see the steps necessary to find that image point algebraically. In the example below the table, we'll apply the same steps to a reflecting line with a slope other than 1 or −1.

Pre-image Point	(a, b)
Reflecting Line	$y = x$
Perpendicular Line	$y = -1(x - a) + b$
Find Point of Intersection Solve the system $\begin{cases} y = x \\ y = -1(x-a)+b \end{cases}$	$y = -1(x - a) + b$ $x = -1(x - a) + b$ $x = -x + a + b$ $2x = a + b$ $\boxed{x = \frac{a+b}{2}}$ $\boxed{y = \frac{a+b}{2}}$
Point of Intersection is midpoint of segment connecting pre-image and image $M = \left(\frac{a+c}{2}, \frac{b+d}{2}\right)$	Intersection point = Midpoint $x = \frac{a+b}{2} = \frac{a+c}{2}$ $c = b$ Intersection point = Midpoint $y = \frac{a+b}{2} = \frac{b+d}{2}$ $d = a$
Image Point	(b, a)

While there's no reason to execute all those steps when the reflecting line is $y = x$, it provides a method for handling reflections over lines with different slopes.

Let's suppose you want to reflect the point (6, 2) over the line $y = 2x$.

Reflecting Line: $y = 2x$

Perpendicular Line: $y = -\frac{1}{2}(x-6)+2$

Point of intersection is midpoint of segment connecting pre-image and image

Point of Intersection: Solve the system $\begin{cases} y = 2x \\ y = -\frac{1}{2}(x-6)+2 \end{cases}$

$$2x = -\tfrac{1}{2}(x-6)+2$$

$$2x = -\tfrac{1}{2}x+5$$

$$2\tfrac{1}{2}x = 5$$

$$\boxed{x = 2}$$

$$\boxed{y = 2x = 4}$$

The point of intersection is (2, 4), and that point is the midpoint of the segment connecting pre-image (6, 2) with image (c, d). You can use the midpoint formula $M = \left(\frac{6+c}{2}, \frac{2+d}{2}\right)$ and set $\frac{6+c}{2} = 2$ and $\frac{2+d}{2} = 4$. Solve to find $c = -2$ and $d = 6$. The image point is (–2, 6).

Invisible Reflections

Some reflections may seem to have no effect. If you reflect the graph of $y = x^2$ across the y-axis, you change the sign of the x-coordinate, and the change is not apparent because the parabola is symmetric about the y-axis and the squaring obscures the sign of the original x-coordinate. If you reflect that same graph across the x-axis, only the vertex is unchanged. The image produced is the graph of $y = -x^2$ because the signs of the y-coordinates are changed. Reflecting the parabola that is the graph of the function $y = x^2$ over the line $y = x$ switches x and y and produces the parabola that is the graph of $x = y^2$, which is not a function.

Exercise 14.1

In questions 1 through 7, use properties of reflection to find the coordinates of the image point.

1. Find the image of the point (−3, 7) after a reflection over the x-axis.
2. Find the image of the point (5, −2) after a reflection over the line $y = x$.
3. Find the image of the point (−6, 5) after a reflection over the line $y = -x$.
4. If the triangle with vertices A(−4, −1), B(−1, 6), and C(3, 1) is reflected over the line $y = -x$, find the vertices of the image triangle $\Delta A'B'C'$.
5. What is the image of the point (2, −2) under a reflection over the line $y = 4x$?
6. What is the image of the point (0, −4) under a reflection over the line $y = -2x$?
7. What is the image of the point (−5, 3) under a reflection over the line $y = \frac{1}{3}x$?

In questions 8 through 10, sketch the graph and use the definition of reflection to find its image. Use properties of conics to find the equation of the image.

8. Sketch the graph of $y = x^2$ and its image under a reflection across the line $x = -2$. What is the equation of the image?
9. Sketch the graph of $x^2 + y^2 = 4$ and its image under a reflection across the line $y = 3$. What is the equation of the image?
10. Sketch the graph of $\frac{x^2}{4} + \frac{y^2}{9} = 1$ and its image under a reflection over the line $y = 2$. Then sketch the graph that results when that image is reflected over the line $x = 3$. What is the equation of this final image?

Step 2. Translate by Reflecting

In working with functions, you frequently recognized that the equation revealed that the graph was a translation, or slide, of a parent function. A translation can shift a graph or other image right, left, up, or down, or some combination of those. Translating a circle of radius 5 centered at the origin to the right four units and down three units can be seen as the composition of two translations: one that moves the circle to the right and one that moves it down, although most of us would think of it as a single operation.

Translation as a Composition

If we break down the operation of translation to its roots, every translation is a composition of two reflections. Shifting the circle four units to the right requires reflecting it over a vertical line and then over a second vertical line parallel to the first. Reflection reverses orientation, but in a translation, the two reflections in sequence restore the original orientation. Reflecting over two parallel vertical lines translates left or right. Reflecting over two parallel horizontal lines moves the image up or down. Reflecting over two parallel oblique lines with slope m will shift the image along a line perpendicular to the parallels.

Magnitude

The distance the pre-image moves depends upon the distance between the two reflecting lines and the order of reflections.

If the point (–7, 2) is reflected over the vertical line $x = 3$, it is sent to the point (13, 2). The distance from (–7, 2) to (3, 2) is 10 units, so the image point (13, 2) is 10 units to the other side of the line. If the image point (13, 2) is then reflected across the vertical line $x = 5$, (13, 2) is 8 units from (5, 2) and so maps onto a point that is 8 units on the other side of (5, 2) landing at (–3, 2).

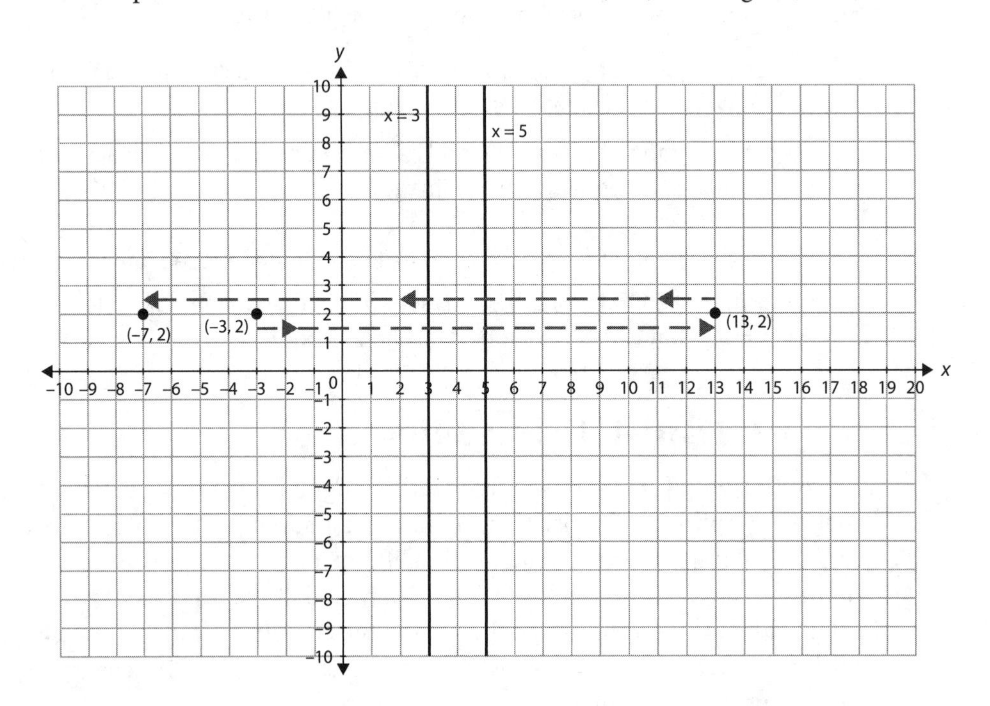

Order of Reflections

The two reflecting lines, $x = 3$ and $x = 5$, are 2 units apart and the distance between the original pre-image (–7, 2) and the final image (–3, 2) is 4 units to the left, twice the distance between the reflecting lines.

If (−7, 2) is first reflected over $x = 5$, it is mapped onto (17, 2). If (17, 2) is reflected over $x = 3$, it is sent to (−11, 2). The distance the point is translated is again 4 units, but in the opposite direction.

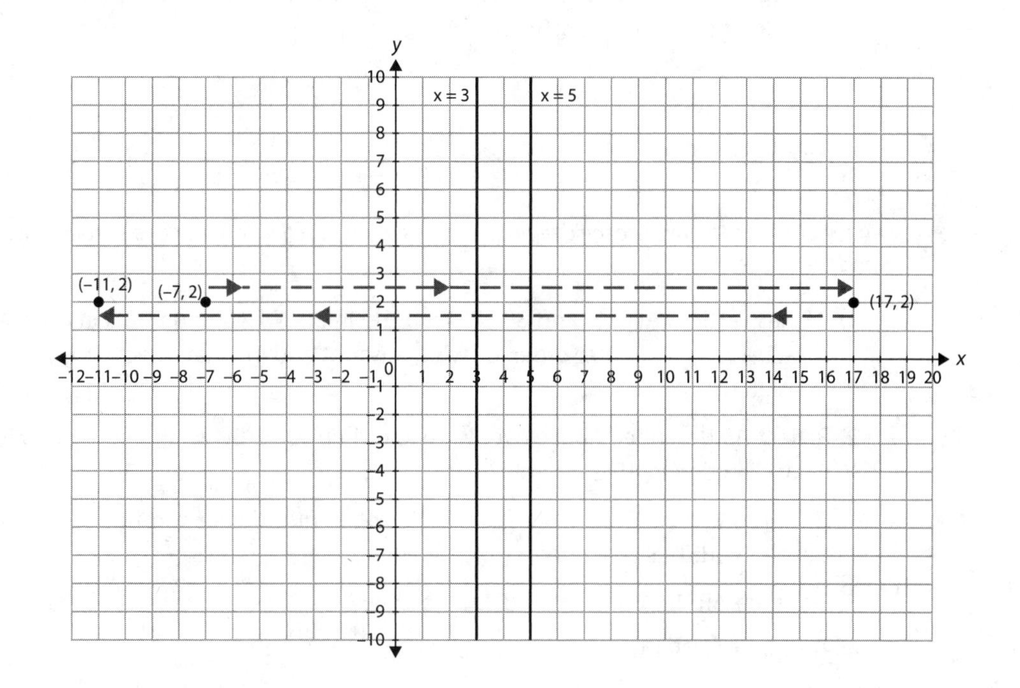

Vertical Reflecting Lines

A translation that is created by reflecting over two vertical lines $x = a$ and $x = b$, where $a < b$, will shift the pre-image horizontally for a distance of $2(b - a)$ units. If the first reflection is over $x = a$, then the movement will be to the right. If the line $x = b$ is the first reflecting line, the movement is to the left.

Horizontal Reflection Lines

A translation that is created by reflecting over two horizontal lines $y = a$ and $y = b$, where $a < b$, will shift the pre-image vertically for a distance of $2(b - a)$ units. If the first reflection is over $y = a$, then the movement will be upward. If the line $y = b$ is the first reflecting line, the movement is downward.

Rigid Transformation

Because translation is the composition of two reflections, and reflection preserves both length and angle measure, translation also preserves length and angle measure. Translation is a rigid transformation and the image under a translation is congruent to the pre-image.

Exercise 14.2

In the questions below, use the definition of translation as reflection over two parallel lines to find the image.

1. $\overline{AB}$ has vertices A(−1,5) and B(2, 7). Find the image of $\overline{AB}$ under reflection over $x = -2$ and $x = 4$, in that order. By what amount and in what direction is $\overline{AB}$ translated?
2. Reflect $\overline{AB}$ described above over $y = 0$ and then $y = 3$. By what amount and in what direction is $\overline{AB}$ translated?

   ΔRST has vertices R(3, 5), S(6,1), and T(5, −4). In questions 3 through 5, translate ΔRST as described.
3. Translate ΔRST 6 units left and 4 units down.
4. Translate ΔRST 1 unit right and 3 units up.
5. Translate ΔRST 5 units left and 5 units down. What is the straight-line distance ΔRST has moved?
6. If you wished to shift ΔRST to the upper right at an angle of 60° from horizontal so that it traveled a straight-line distance of 10 units, what two translations might you apply?
7. If ΔABC is reflected over $x = 7$ and then over $x = -2$, how far and in what direction is ΔABC translated?
8. Line segment $\overline{XY}$ is reflected over $y = 4$ and then over $y = -3$. The final image $\overline{X'Y'}$ has endpoints X′(−2, −15) and Y′(4, −12). What are the endpoints of $\overline{XY}$?
9. If the graph of $y = x^2 - 4$ is reflected over the x-axis and then over $y = -2$, what is the equation of the image?
10. If the image from the previous question is then reflected over $x = -3$ and then the y-axis, what is the equation of the resulting graph?

Step 3. Rotate by Reflecting

When we think about rotating a figure in the plane, we generally think about movement around a point, usually the origin. That point, the center of rotation, is the intersection point of two reflecting lines. The composition of two reflections over intersecting reflecting lines is the transformation we label as rotation. The point at which the two lines intersect is the center of rotation. The size of the rotation is twice the measure of the smaller angle formed by the intersecting lines.

ΔABC has vertices A(2, 1), B(7, 1), and C(4, 3). Let's choose two reflecting lines that will intersect at the origin: $y = x$ and the y-axis $x = 0$. If we reflect first over the line $y = x$, the intermediate image ΔA′B′C′ will have vertices A′(1, 2), B′(1, 7), and C′(3, 4). Reflecting that intermediate image over the y-axis gives the final triangle ΔA″B″C″ with vertices A″(−1, 2), B″(−1, 7), and C″(−3, 4).

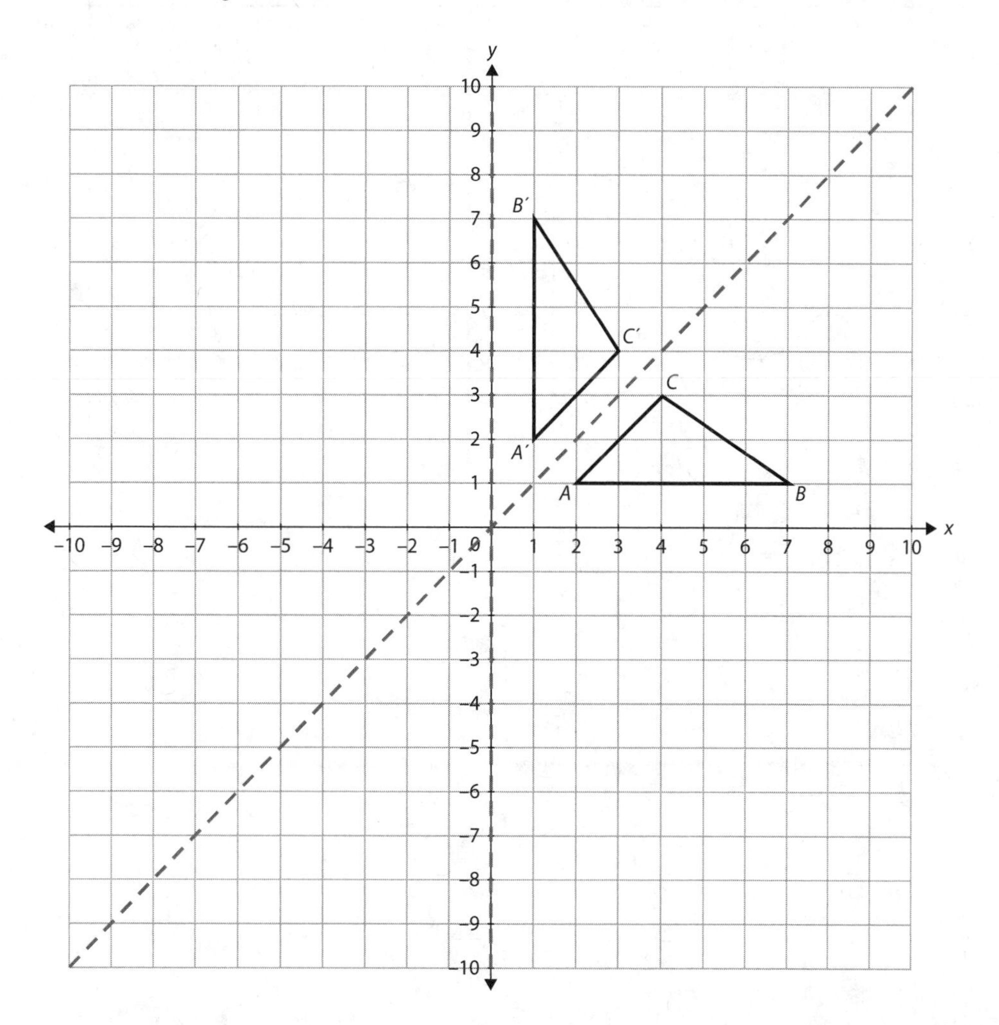

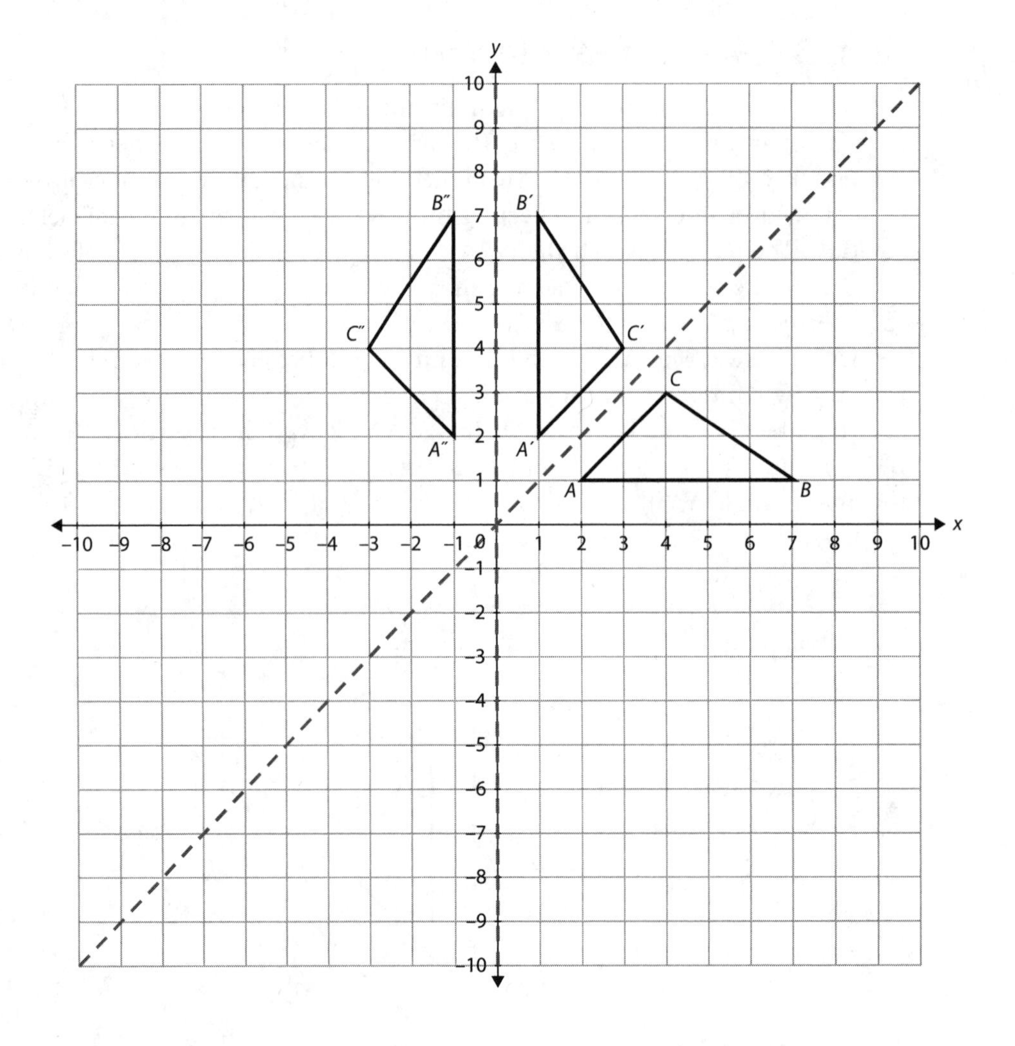

y
x
B″
B′
C″
C′
C
A″
A′
A
B

Magnitude of Rotation

The line $y = x$ and the y-axis intersect to form two pairs of vertical angles. One pair measures 45° each and the other pair 135° each. The triangle has been rotated $2 \times 45° = 90°$ counterclockwise.

Order of Reflections

If the order of the reflections is changed, the intermediate image from reflecting over the y-axis has vertices (−2, 1), (−7, 1), and (−4, 3), and the final image has vertices (1, −2), (1, −7), and (3, −4). This final image is a rotation of the original 90° clockwise.

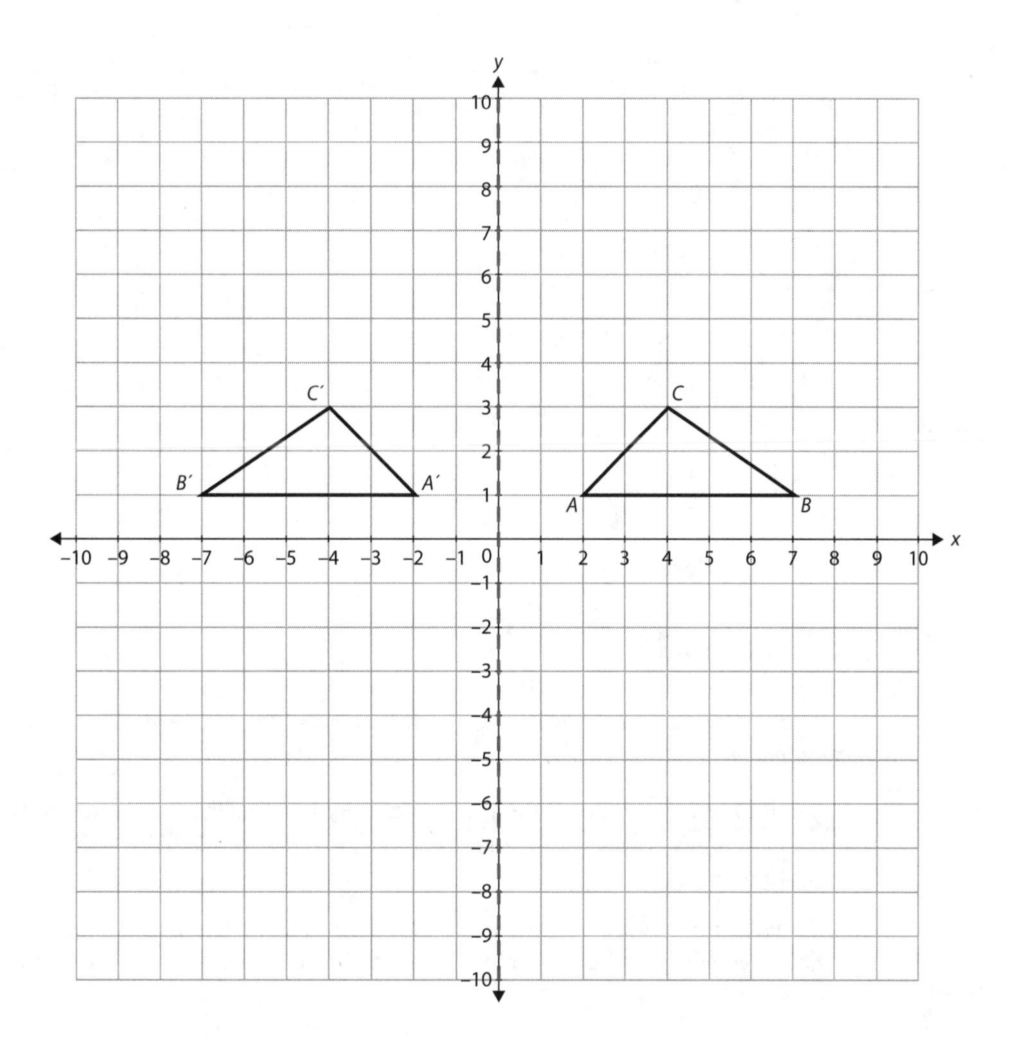

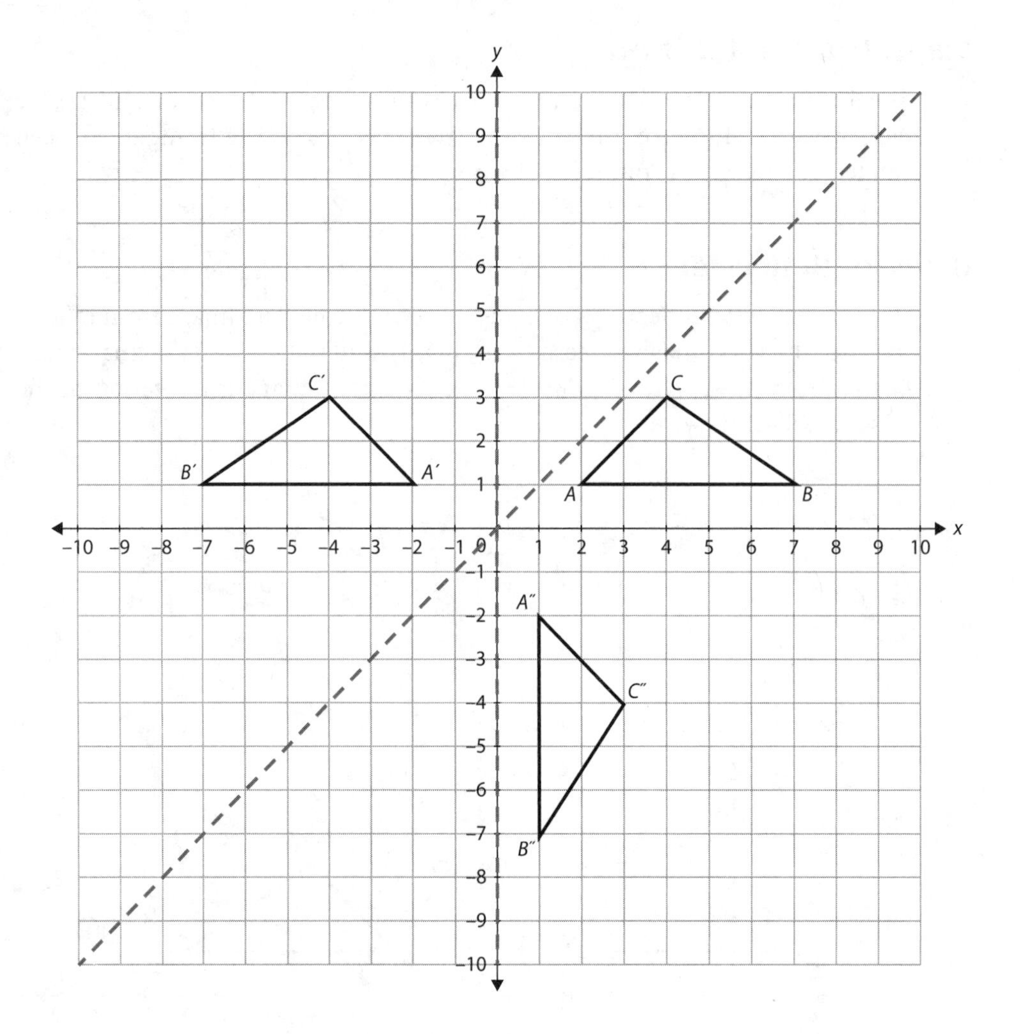

Exercise 14.3

In questions 1 through 3, use the definition of rotation as reflections over intersecting lines to find the image.

1. If point A is reflected over the y-axis and then over the line $y = x$, through how many degrees does it rotate about the origin? Is the rotation clockwise or counterclockwise?
2. If point B is reflected over the y-axis and then over the line $y = -x$, through how many degrees does it rotate about the origin? Is the rotation clockwise or counterclockwise?

3. If point C is reflected over the line $y = x$ and then over the x-axis, through how many degrees does it rotate about the origin? Is the rotation clockwise or counterclockwise?

For questions 4 through 6, ΔXYZ has vertices X(2, 4), Y(7, 3), and Z(5, –1). Use this triangle for questions 4 through 6. Find the image after rotation by measuring with a protractor.

4. Find the image of ΔXYZ under a clockwise rotation about the origin of 90°.
5. Find the image of ΔXYZ under a counterclockwise rotation about the origin of 180°.
6. If ΔXYZ is itself the image of ΔPQR, when ΔPQR has been rotated about the origin in a clockwise direction 270°, what are the vertices of the original ΔPQR?

In exercises 7 through 10, find the image under the rotation by any convenient method. Use coordinate geometry to find the equation of the image.

7. If the graph of $y = 2x + 3$ is reflected over $y = x$ and then over $y = -x$, what is the equation of the resulting graph?
8. If the graph of $y = |x - 3|$ is reflected over the y-axis and then over the x-axis, what is the equation of the graph that results?
9. If the graph of $\frac{x^2}{16} + \frac{y^2}{9} = 1$ is rotated 90° counterclockwise about the origin, what is the equation of the resulting ellipse?
10. If the graph of $y = x^2 - 4$ is rotated 90° clockwise about the origin, what is the equation of the image?

Step 4. Dilate Geometrically

Dilation differs from reflection, translation, and rotation in that it not only moves the figure to a new location, but also changes its size. Rigid transformations produce images that are congruent to the original; dilation does not, but it produces an image similar to the pre-image. The image may be larger or smaller than the pre-image.

Center and Scale Factor

Every dilation has a center, which may be a point of the pre-image, or in the interior of the pre-image, or outside of it. In addition, the dilation has a scale factor that indicates the ratio of the image to the pre-image. If the scale factor is greater than one, the dilation is an enlargement; if the scale factor is less than one, the dilation is a reduction.

If a segment is drawn from the center of the dilation to each vertex of the pre-image, and the length of that segment is multiplied by the scale factor, the result is the distance of the image point from the center. If the scale factor is greater than one, the image is farther away from the center. If the scale factor is less than one, the image is closer to the center than the pre-image.

ΔABC has vertices A(−4, 3), B(2, 2), and C(0, −4). If we apply a dilation centered at the origin with a scale factor of 2, the segment connecting the origin to A, which has a length of 5 units, is extended to have a length of 10, giving the point A′ (−8, 6). The segment from the origin to B is doubled, making B′(4, 4) and C′ falls twice as far down the *y*-axis, so C′(0, −8).

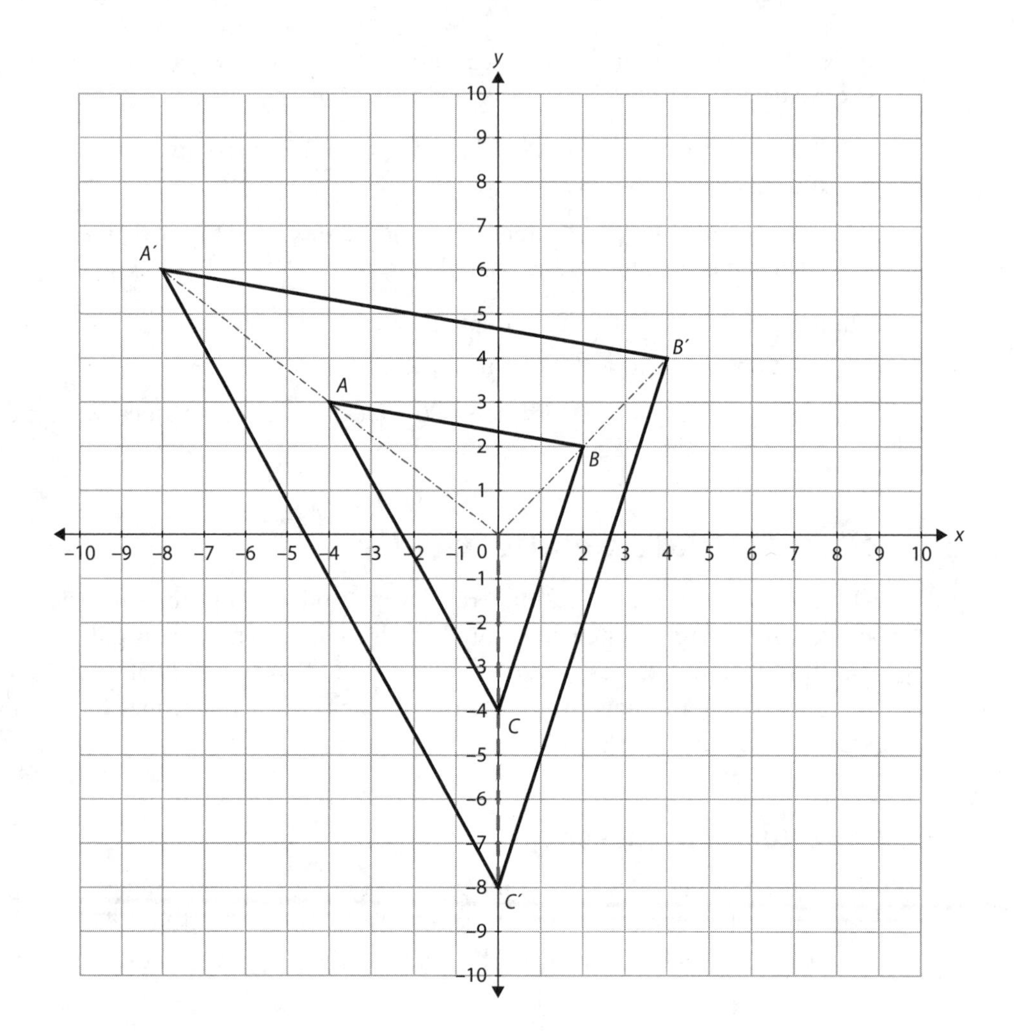

If we use the same triangle as the pre-image for a dilation centered at the origin but with a scale factor of 0.4, the image is ΔA′B′C′ with A′(−1.6, 1.2), B′(0.8, 0.8), and C′(0, −1.6).

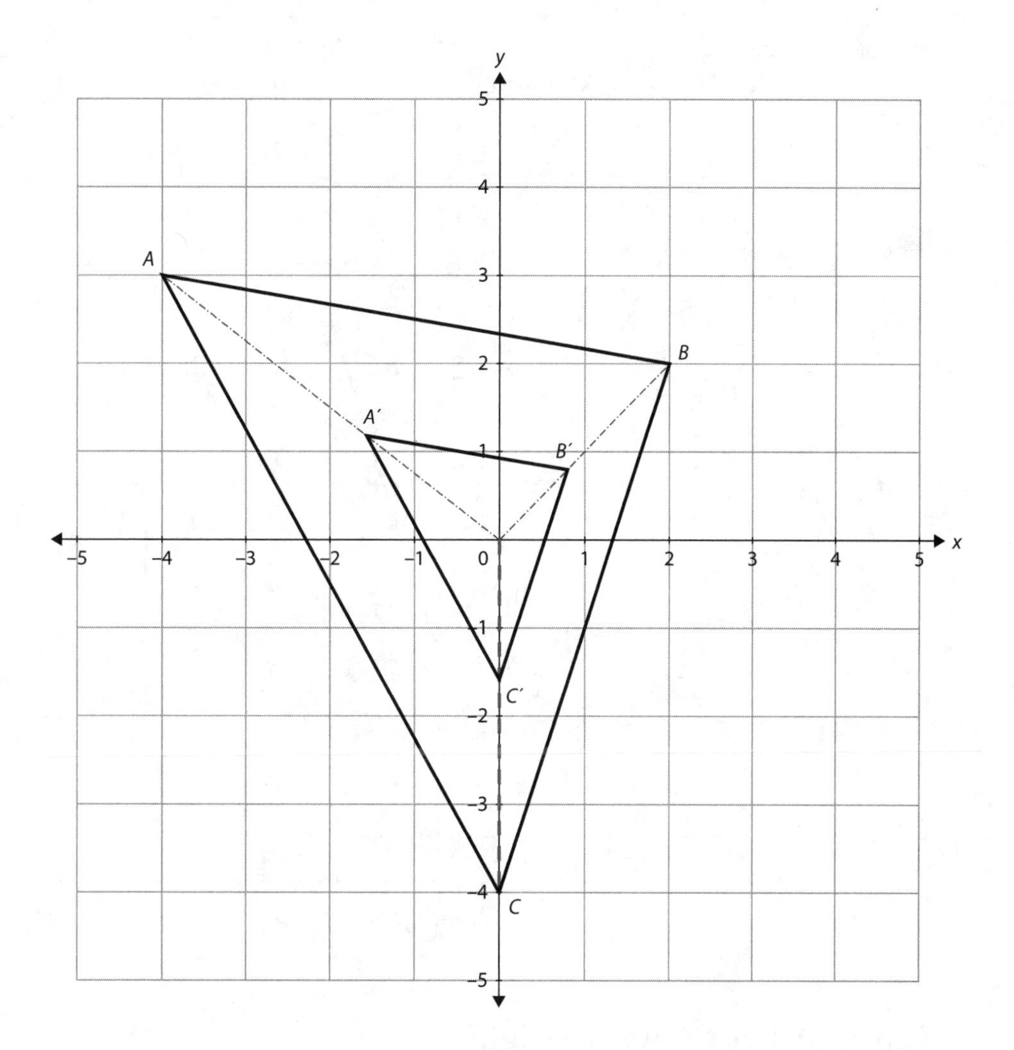

If we use a dilation with a scale factor of 2, but centered at C, the image of C is C. That point does not change, but A′ is (−8, 10) and B′ is (4, 8).

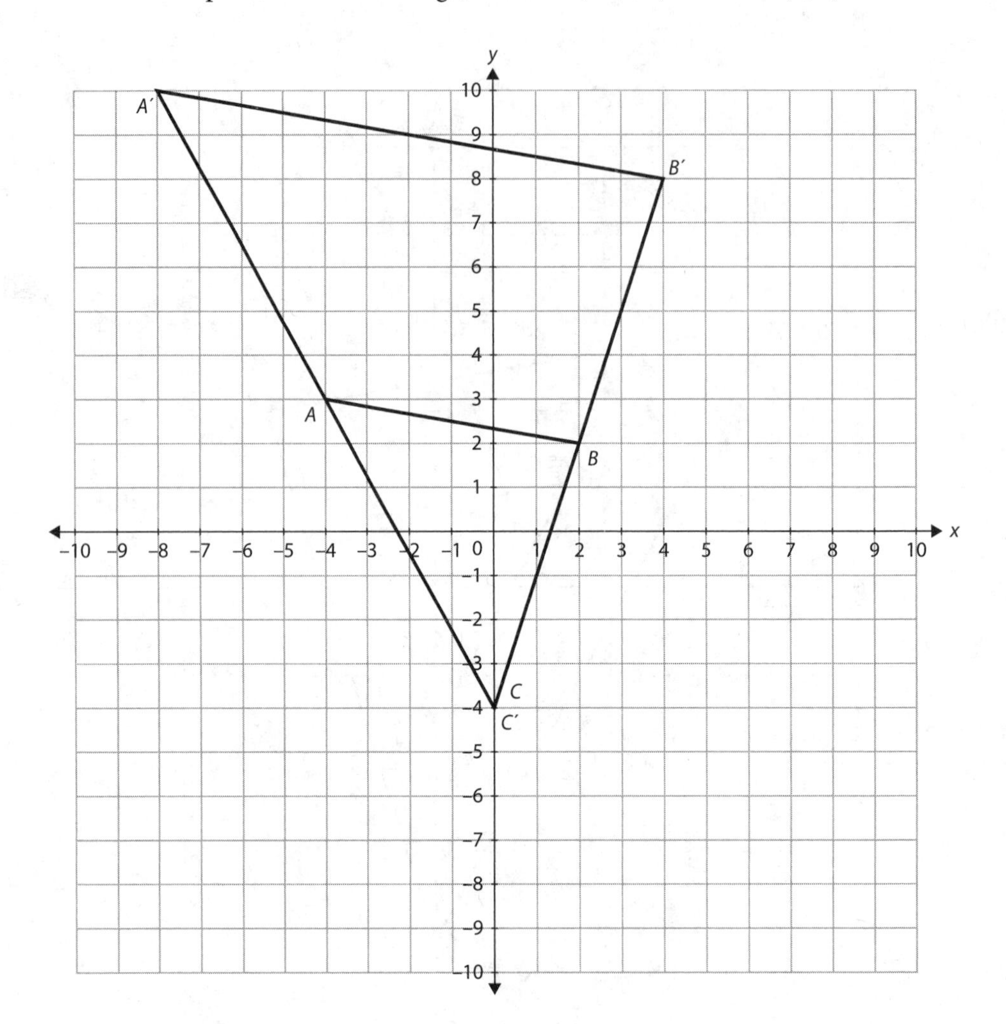

Center Other than the Origin

The simple multiplication of the coordinates by scale factor seems to have disappeared here, but another transformation may be helpful. When the center is not at the origin:

1. Translate the center of the dilation to the origin and apply the same translation to the pre-image.
2. Multiply the coordinates of the translated pre-image by the scale factor.

3. Translate the center of the dilation back to its original location and apply the same translation to the scaled coordinates.

PRE-IMAGE	TRANSLATE TO ORIGIN $T(X, Y) \rightarrow (X, Y + 4)$	MULTIPLY BY SCALE FACTOR	TRANSLATE BACK $T(X, Y) \rightarrow (X, Y - 4)$
A(–4, 3)	(–4, 7)	(–8, 14)	(–8, 10)
B(2, 2)	(2, 6)	(4, 12)	(4, 8)
C(0, –4)	(0, 0)	(0, 0)	(0, –4)

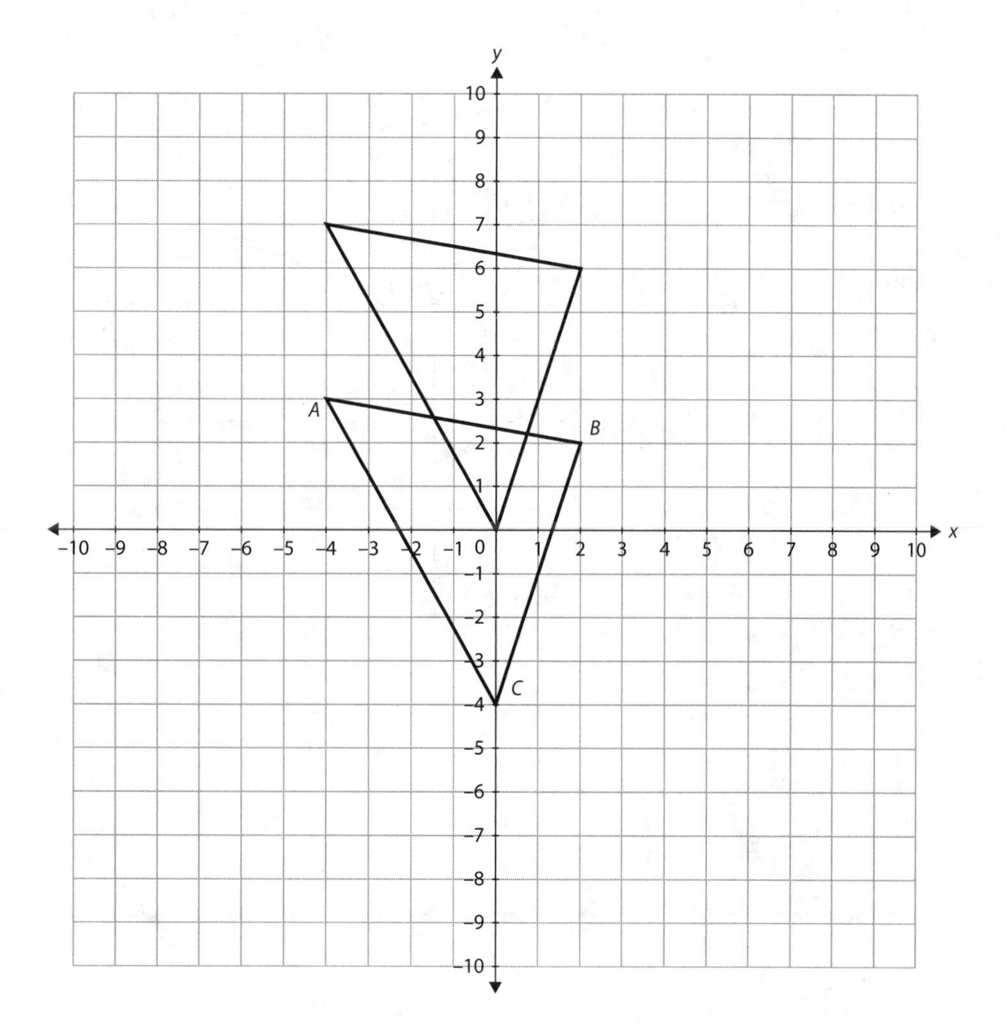

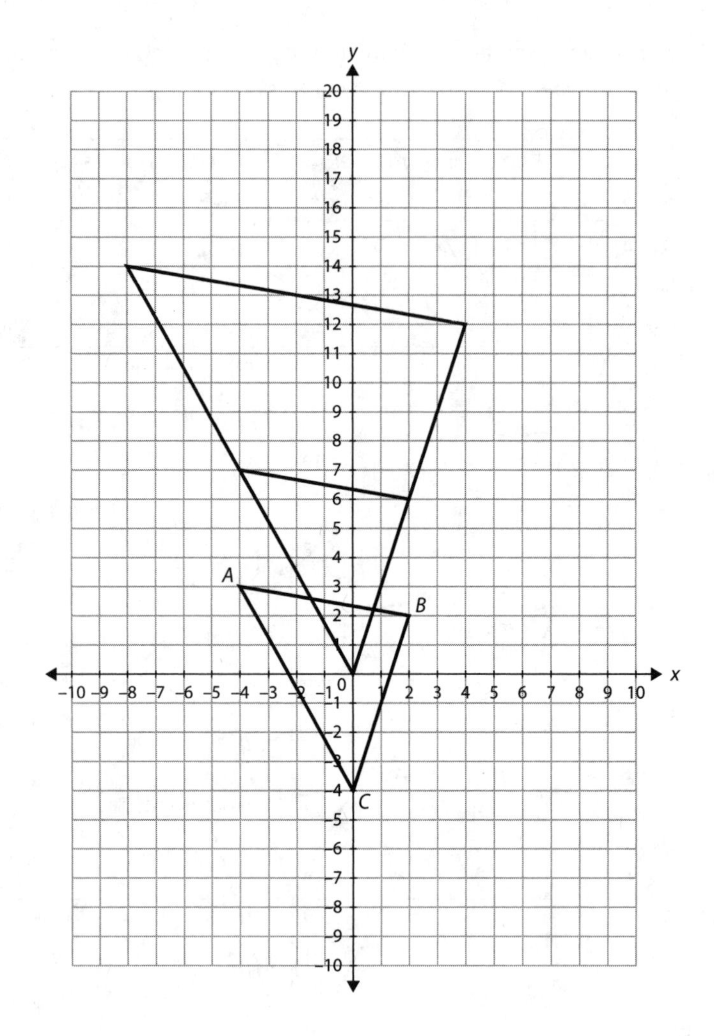
y
x
A
B
C

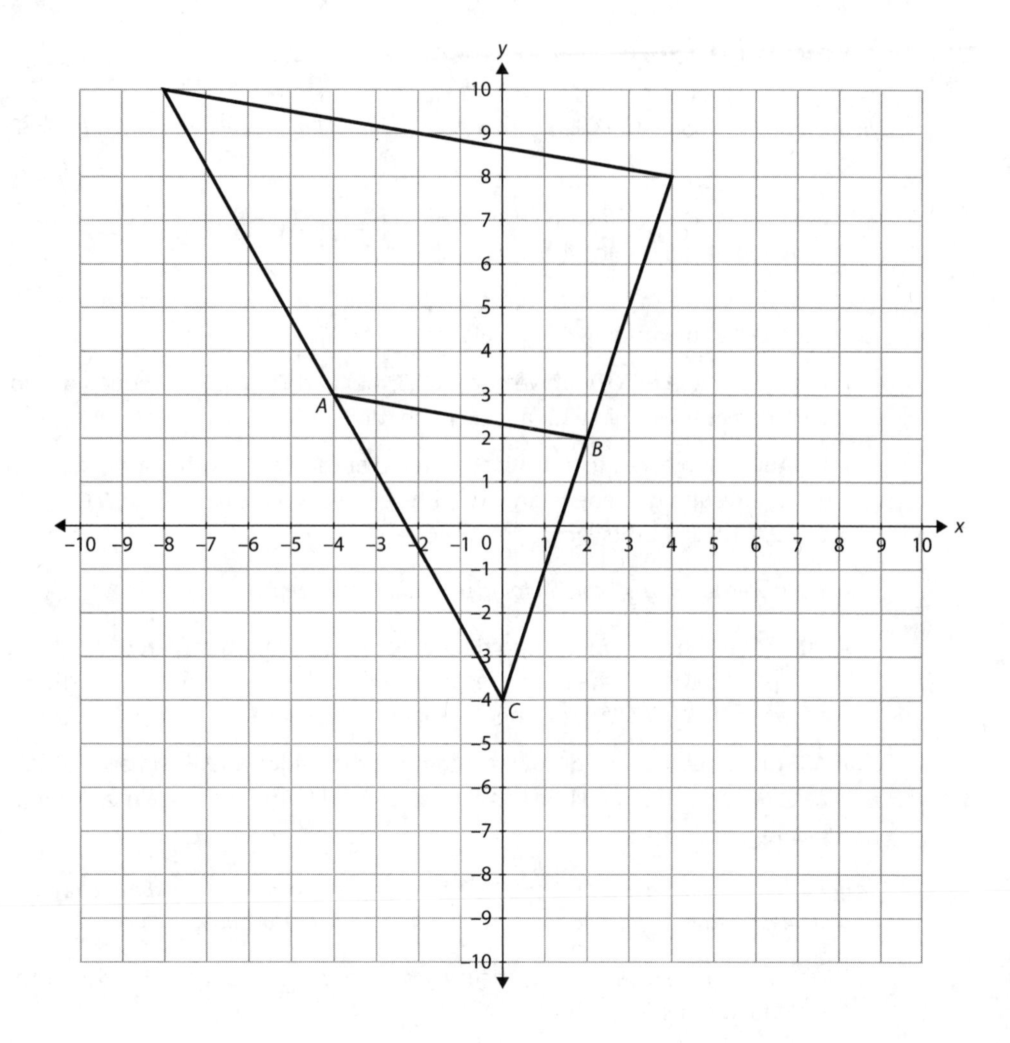

y
x
A
B
C
−10
−9
−8
−7
−6
−5
−4
−3
−2
−1
0
1
2
3
4
5
6
7
8
9
10

Exercise 14.4

In questions 1 through 4, draw rays from the center of dilation and locate the vertices of the image.

1. Find the image of $\overline{AB}$ with vertices A(−2, 5) and B(0, −3) under a dilation centered at the origin with a scale factor of 3.
2. Find the image of $\overline{CD}$ with vertices C(2, 4) and D(6, −4) under a dilation centered at the origin with a scale factor of $\frac{1}{2}$.
3. Find the image of $\overline{XY}$ with vertices X(0.3, 1.2) and Y(−3.6, 2.2) under a dilation centered at the origin with a scale factor of 5.
4. If ΔABC is subjected to a dilation centered at the origin with a scale factor of 1.2, the resulting image is ΔA′B′C′ with vertices A′(−6, 10.8), B′(4.8, 9.6), and C′(7.2, −12). Find the vertices of ΔABC.

Use your understanding of scale factor to solve questions 5 and 6.

5. WXYZ is a rectangle in the coordinate plane. If a dilation centered at the origin is applied to WXYZ, the image rectangle W′X′Y′Z′ has an area one-fourth that of rectangle WXYZ. What is the scale factor of the dilation?
6. ΔRST is subjected to a dilation centered at the origin, and the image ΔR′S′T′ has an area that is 17.64 times the area of ΔRST. What is the scale factor of the dilation?

In questions 7 and 8, translate the given information to place the center at the origin. Multiply coordinates by scale factor, and translate back to the original location.

7. Find the image of $\overline{AB}$ with vertices A(−4, 5) and B(2, 1) under a dilation with a scale factor of 1.5 centered at (2, 1).
8. Find the image of ΔRST with vertices R(−2, 6), S(3, 5), and T(1, −1) under a dilation with a scale factor of 2 centered at (5, 5).

In questions 9 and 10, choose a few key points to help you find the image under the dilation. Use your experience with conics to find the equation of the image.

9. Find the image of the graph of $x^2 + y^2 = 1$ under a dilation with a scale factor of 3 centered at (0, 3). What is the equation of the image?
10. Find the image of the graph of $\frac{x^2}{9} + \frac{y^2}{4} = 1$ under a dilation with a scale factor of 2 centered at (−2, 0). What is the equation of the image?

Step 5. Use Matrices for Transformations

The calculations involved in finding an image point are not difficult, but reflecting, rotating, or dilating a figure with many vertices can become tedious. Using matrices to apply the same transformation to many points at the same time can speed the process.

Reflection

Common reflections can be accomplished by multiplying a square matrix associated with the reflection times a matrix of coordinates. We'll use a 2×3 matrix representing ΔABC with vertices $A(-4, 3)$, $B(2, 2)$, and $C(0, -4)$ for the examples.

TYPE OF REFLECTION	***ASSOCIATED MATRIX***	***EXAMPLE***
Reflection over the y-axis	$\begin{bmatrix} -1 & 0 \\ 0 & 1 \end{bmatrix}$	$\begin{bmatrix} -1 & 0 \\ 0 & 1 \end{bmatrix}\begin{bmatrix} -4 & 2 & 0 \\ 3 & 2 & -4 \end{bmatrix} = \begin{bmatrix} 4 & -2 & 0 \\ 3 & 2 & -4 \end{bmatrix}$

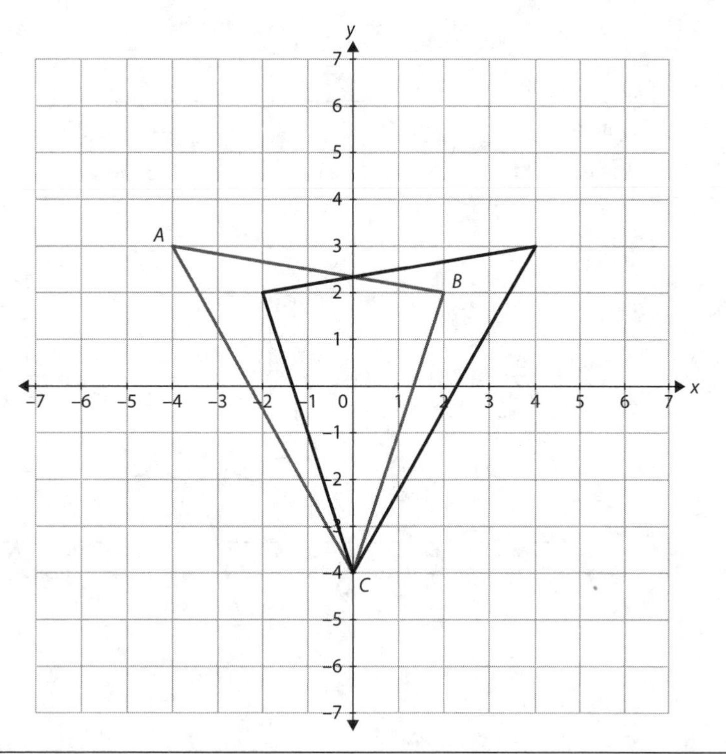

Reflection over the x-axis	$\begin{bmatrix} 1 & 0 \\ 0 & -1 \end{bmatrix}$	$\begin{bmatrix} 1 & 0 \\ 0 & -1 \end{bmatrix}\begin{bmatrix} -4 & 2 & 0 \\ 3 & 2 & -4 \end{bmatrix} = \begin{bmatrix} -4 & 2 & 0 \\ -3 & -2 & 4 \end{bmatrix}$

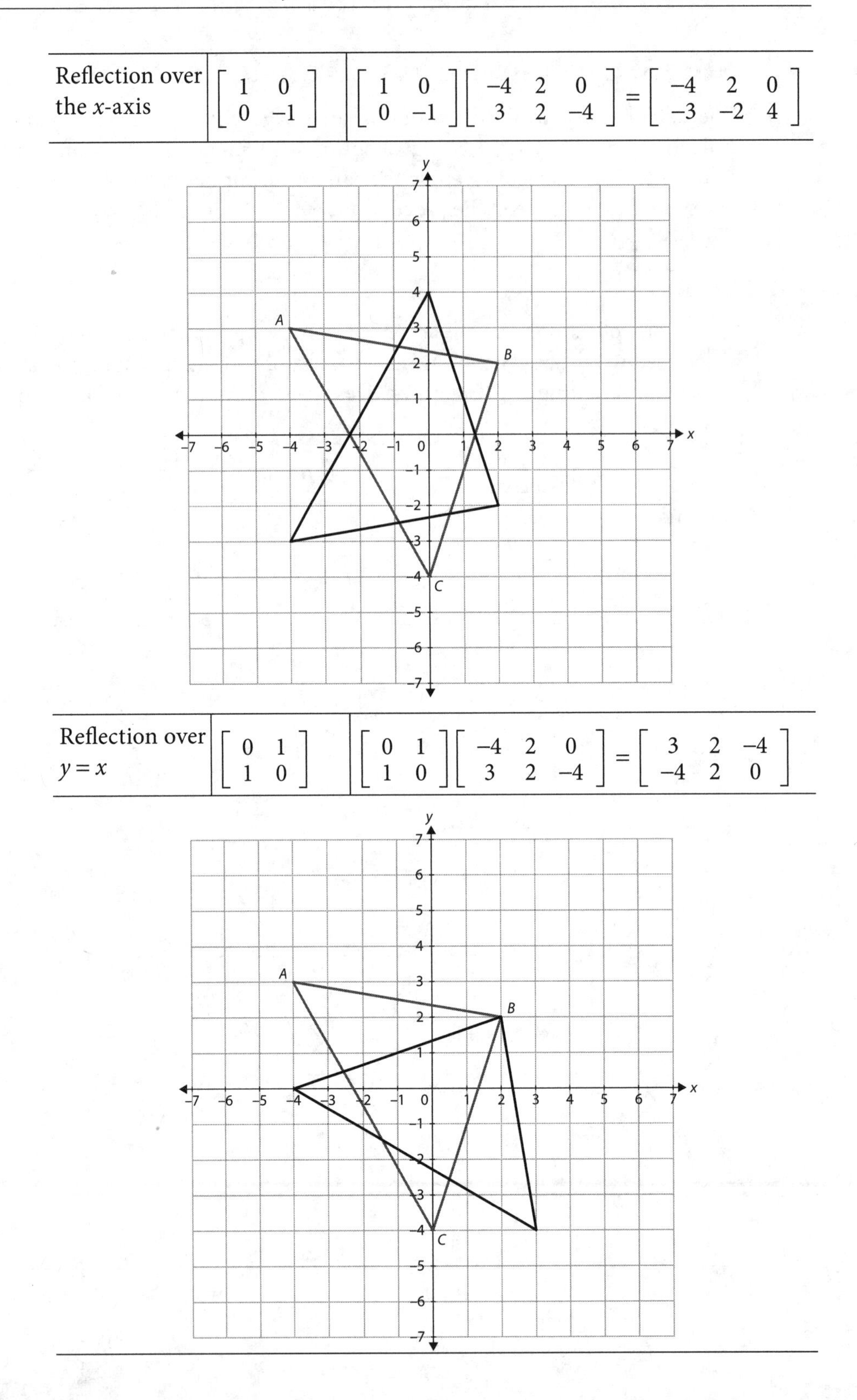

Reflection over $y = x$	$\begin{bmatrix} 0 & 1 \\ 1 & 0 \end{bmatrix}$	$\begin{bmatrix} 0 & 1 \\ 1 & 0 \end{bmatrix}\begin{bmatrix} -4 & 2 & 0 \\ 3 & 2 & -4 \end{bmatrix} = \begin{bmatrix} 3 & 2 & -4 \\ -4 & 2 & 0 \end{bmatrix}$

Reflection over $y = -x$	$\begin{bmatrix} 0 & -1 \\ -1 & 0 \end{bmatrix}$	$\begin{bmatrix} 0 & -1 \\ -1 & 0 \end{bmatrix}\begin{bmatrix} -4 & 2 & 0 \\ 3 & 2 & -4 \end{bmatrix} = \begin{bmatrix} -3 & -2 & 4 \\ 4 & -2 & 0 \end{bmatrix}$

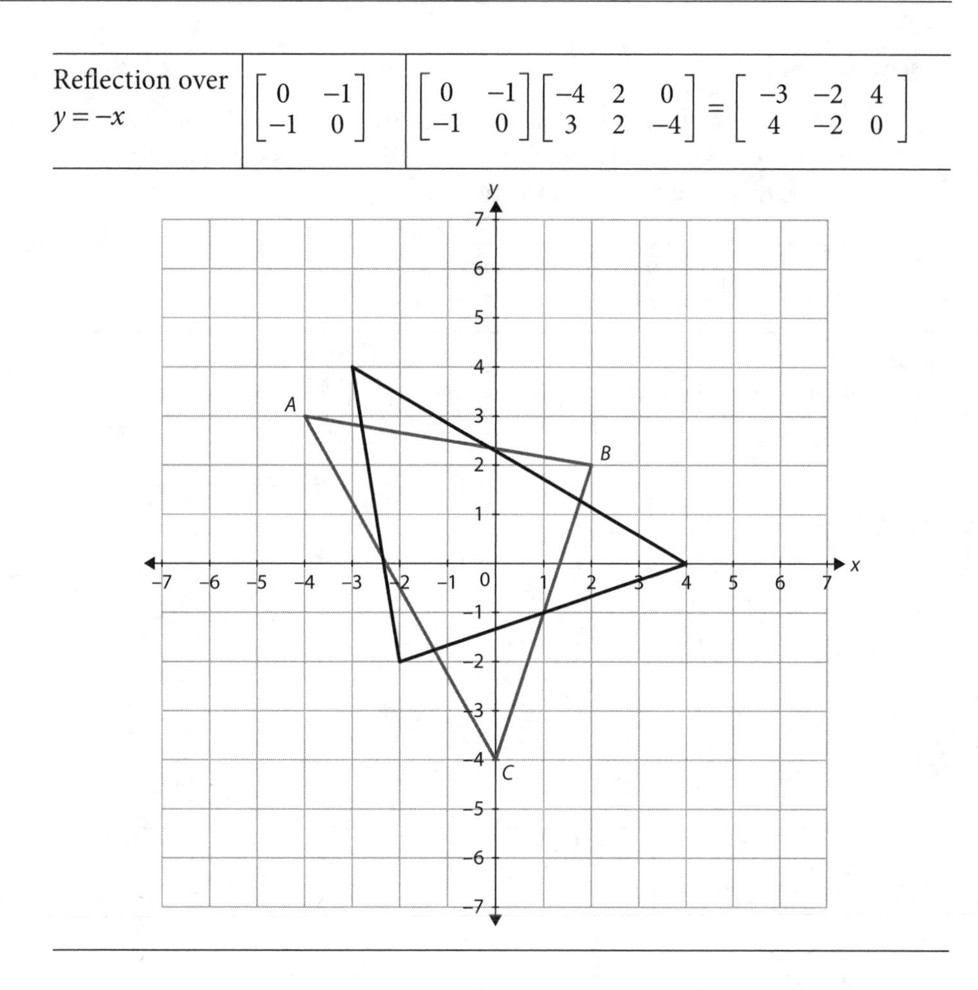

Matrices are not particularly helpful for translations. The basic relationship would be addition: $\begin{bmatrix} x \\ y \end{bmatrix} + \begin{bmatrix} h \\ k \end{bmatrix} = \begin{bmatrix} x+h \\ y+k \end{bmatrix}$. Because matrices to be added must have the same dimensions, translation of a triangle 6 units right and 3 units down would require $\begin{bmatrix} -4 & 2 & 0 \\ 3 & 2 & -4 \end{bmatrix} + \begin{bmatrix} 6 & 6 & 6 \\ -3 & -3 & -3 \end{bmatrix}$, not much of a time saver.

Rotation

A rotation of 90° counterclockwise is equivalent to a rotation of 270° clockwise, and in general, a rotation of d° in one direction is equivalent to a rotation of $(360 - d)$° in the opposite direction.

ROTATION	***EQUIVALENT ROTATION***	***ASSOCIATED MATRIX***	***EXAMPLE***
Rotate 90° counterclockwise	Rotate 270° clockwise	$\begin{bmatrix} 0 & -1 \\ 1 & 0 \end{bmatrix}$	$\begin{bmatrix} 0 & -1 \\ 1 & 0 \end{bmatrix}\begin{bmatrix} -4 & 2 & 0 \\ 3 & 2 & -4 \end{bmatrix} = \begin{bmatrix} -3 & -2 & 4 \\ 4 & -2 & 0 \end{bmatrix}$

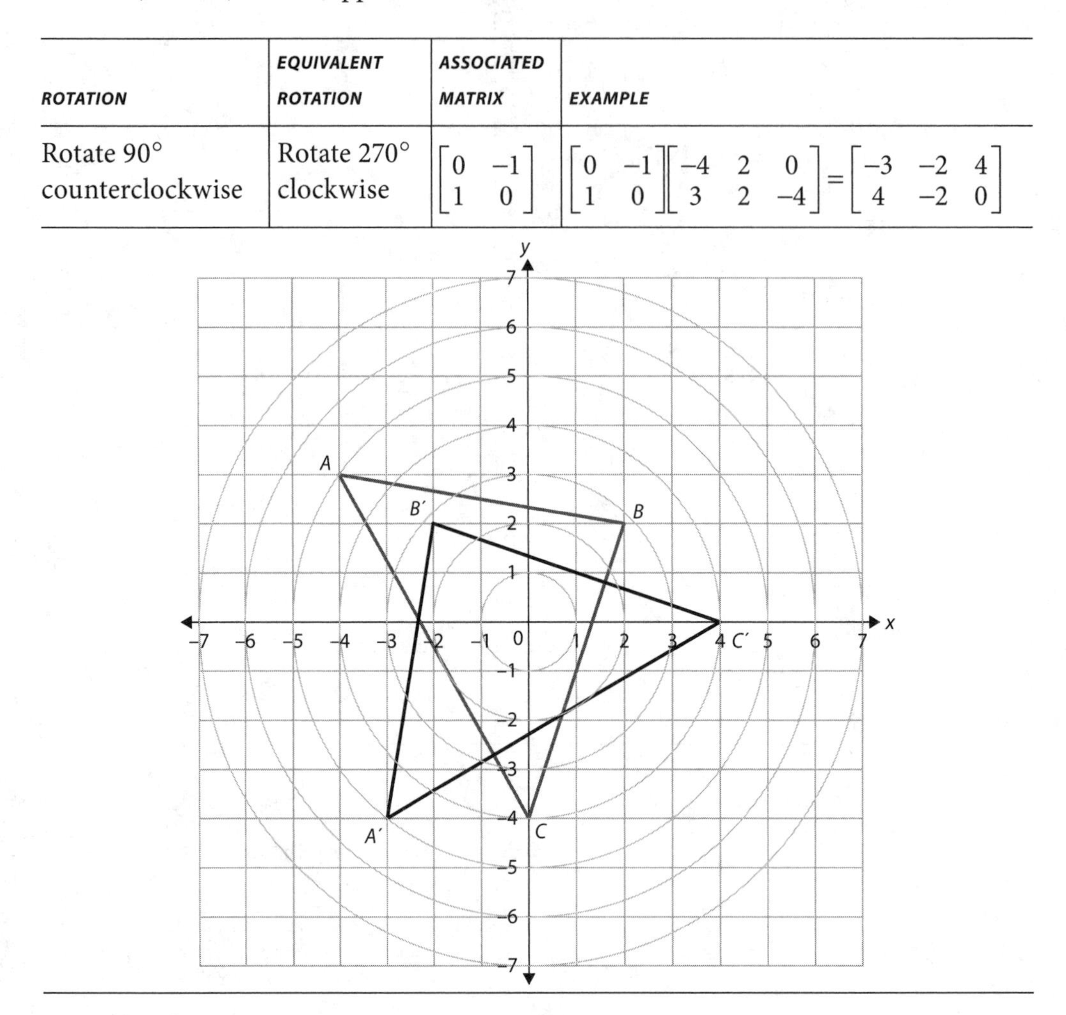

Rotate 180° counterclockwise	Rotate 180° clockwise	$\begin{bmatrix} -1 & 0 \\ 0 & -1 \end{bmatrix}$	$\begin{bmatrix} -1 & 0 \\ 0 & -1 \end{bmatrix}\begin{bmatrix} -4 & 2 & 0 \\ 3 & 2 & -4 \end{bmatrix} = \begin{bmatrix} 4 & -2 & 0 \\ -3 & -2 & 4 \end{bmatrix}$

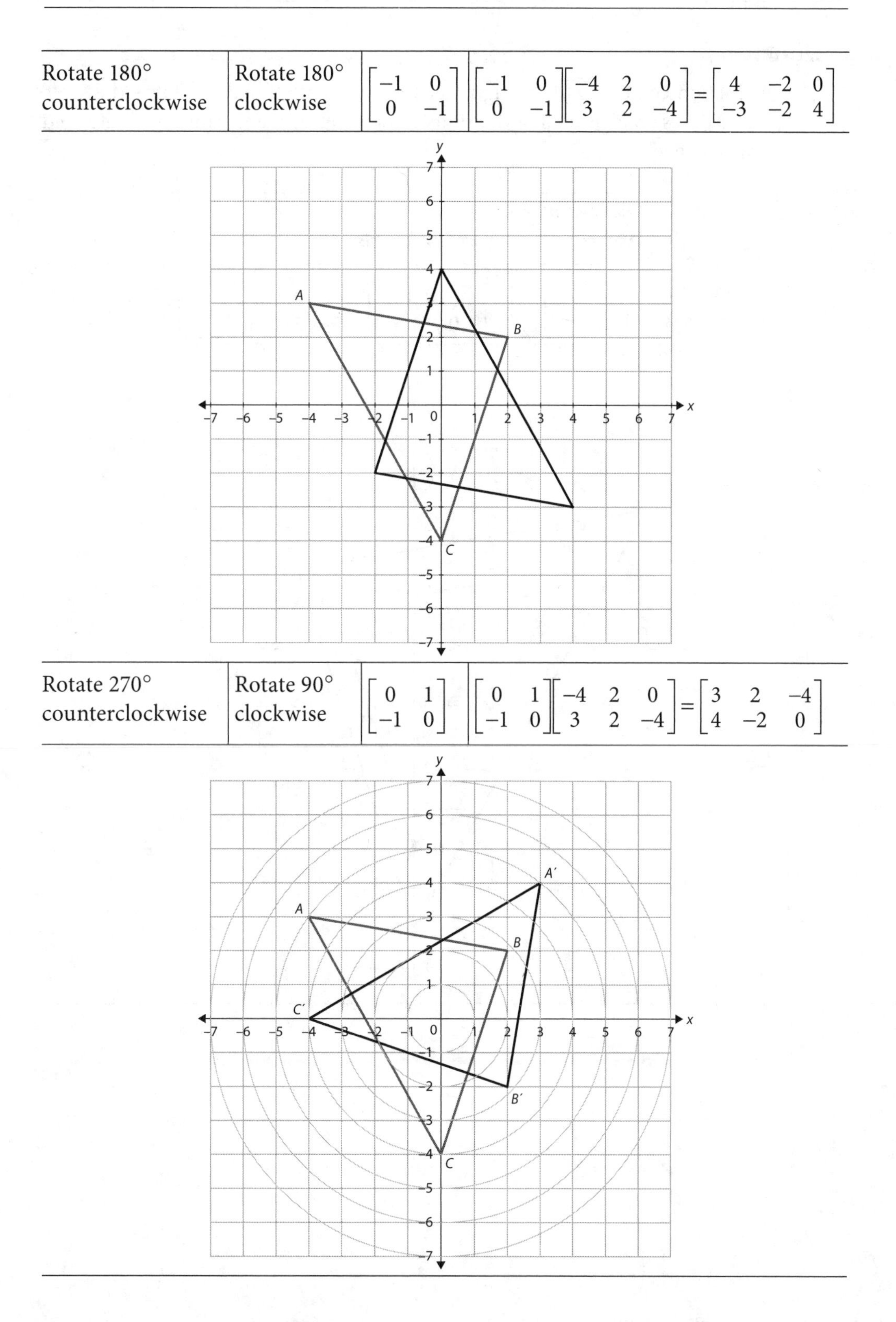

Rotate 270° counterclockwise	Rotate 90° clockwise	$\begin{bmatrix} 0 & 1 \\ -1 & 0 \end{bmatrix}$	$\begin{bmatrix} 0 & 1 \\ -1 & 0 \end{bmatrix}\begin{bmatrix} -4 & 2 & 0 \\ 3 & 2 & -4 \end{bmatrix} = \begin{bmatrix} 3 & 2 & -4 \\ 4 & -2 & 0 \end{bmatrix}$

Dilation

A dilation centered at the origin is associated with a matrix that is the scale factor times an identity. For dilations centered elsewhere, translate, dilate, and translate back.

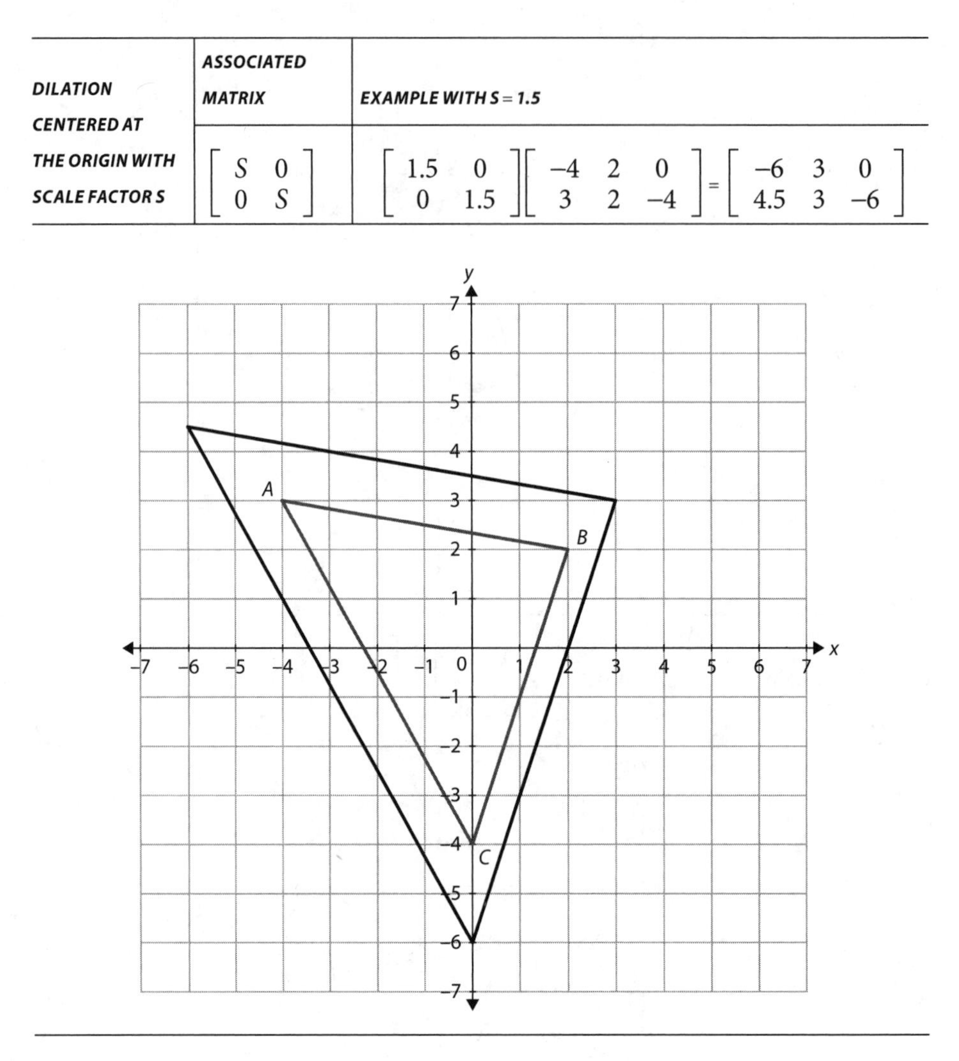

DILATION CENTERED AT THE ORIGIN WITH SCALE FACTOR S	*ASSOCIATED MATRIX*	*EXAMPLE WITH S = 1.5*
	$\begin{bmatrix} S & 0 \\ 0 & S \end{bmatrix}$	$\begin{bmatrix} 1.5 & 0 \\ 0 & 1.5 \end{bmatrix}\begin{bmatrix} -4 & 2 & 0 \\ 3 & 2 & -4 \end{bmatrix} = \begin{bmatrix} -6 & 3 & 0 \\ 4.5 & 3 & -6 \end{bmatrix}$

Exercise 14.5

In questions 1 through 5, use matrix methods to find the coordinates of the image points.

Pentagon ABCDE has vertices A(2, 4), B(4, 6), C(8, 5), D(8, 3), and E(4, 1). Use this pentagon for questions 1 through 5.

1. Find the vertices of the image of ABCDE under a reflection over the line $y = -x$.
2. ABCDE is transformed by a reflection over the x-axis followed by a reflection over the line $y = x$. Find the vertices of the final image.
3. Find the image of ABCDE under a rotation of 90° clockwise about the origin.
4. Find the image of ABCDE under a dilation centered at the origin with a scale factor of 1.5.
5. Pentagon ABCDE is rotated 180° about the origin. The result is reflected over the line $y = -x$, and that result is subjected to a dilation centered at the origin with a scale factor of 0.2. Find the vertices of the final image.

Use the algebra of matrices and your experience with transformations to solve questions 6 through 10.

6. A segment $\overline{XY}$ with endpoints X(2, 4) and Y(−1, −1) is reflected across the line $x = 5$ to produce an image segment $\overline{X'Y'}$ with endpoints X′(8, 4) and Y′(11, −1). A matrix $\begin{bmatrix} a & b \\ 0 & 1 \end{bmatrix}$ exists for which $\begin{bmatrix} a & b \\ 0 & 1 \end{bmatrix}\begin{bmatrix} 2 & -1 \\ 4 & -1 \end{bmatrix} = \begin{bmatrix} 8 & 11 \\ 4 & -1 \end{bmatrix}$. Find the values of a and b.
7. Does the matrix you found in the previous question correctly reflect $\overline{ZY}$ with Z(−2, 2) and Y(−1, −1) across the line $x = 5$?
8. If ΔRST is rotated 90°counterclockwise about the origin, the image ΔR′S′T′ has vertices R′(3, 5), S′(0, 2), and T′(2, −3). Find the vertices of ΔRST.
9. If you wished to rotate ΔRST 90°counterclockwise and then reflect the intermediate image across the y-axis, you might find a single matrix that accomplishes both transformations by multiplying the associated matrices. Find the product of those two square matrices and verify that it has the desired effect. What do you notice about the matrix you produced? How could you have accomplished the same transformation more quickly?
10. If you multiply the two associated matrices in the previous question in the opposite order, what single transformation is equivalent?

15

Rotating Conics

In the previous chapter, you applied transformation to polygons and graphs, including the graphs of some conics. You saw that, for example, rotating an ellipse which is centered at the origin 90° about the origin results in an ellipse but exchanges the lengths of the horizontal and vertical axes. Writing the equation of that new ellipse is a familiar process.

Other rotations may not be so friendly, however. What if that ellipse were rotated 30° or 45° or 72°? The results are more difficult to predict and the equation of the new ellipse is harder to write. It's not impossible, as you'll see, but it may not be intuitive.

Step 1. Create the Equation that Rotates a Conic

The matrix we used to create a rotation of 90° counterclockwise, $\begin{bmatrix} 0 & -1 \\ 1 & 0 \end{bmatrix}$, was fairly easy to create by just observing some rotations, but it is not a collection of random numbers. If you track the points (1, 0) and (0, 1) through a rotation counterclockwise about the origin, you can deduce the general form.

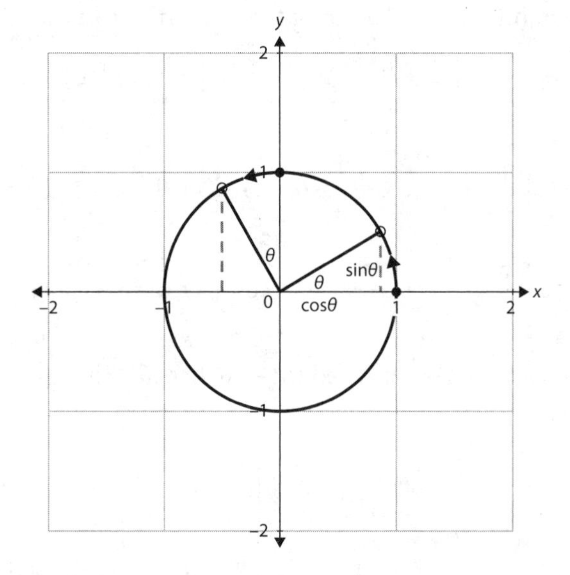

We can measure the amount of rotation in degrees or in radians, but let's agree to denote the amount of rotation as θ, and follow the points (1, 0) and (0, 1) around the unit circle through that much of a rotation. If we look first at the image of (1, 0), we can see that it has moved to a point whose coordinates are $(\cos\theta, \sin\theta)$. This tells us that the rotation matrix must include $\cos\theta$ and $\sin\theta$, placed to produce this change. That means $\begin{bmatrix} \cos\theta & ? \\ \sin\theta & ? \end{bmatrix}\begin{bmatrix} 1 \\ 0 \end{bmatrix} = \begin{bmatrix} \cos\theta \\ \sin\theta \end{bmatrix}$ but the zero in $\begin{bmatrix} 1 \\ 0 \end{bmatrix}$ obscures what those two other elements might be. So, look at what happens to (0,1).

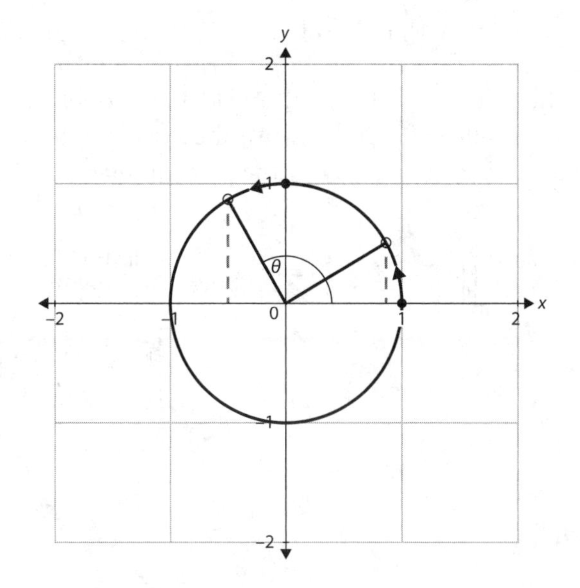

The point (0, 1) slides along the unit circle to a point whose coordinates are $\left(\cos\left(\frac{\pi}{2}+\theta\right), \sin\left(\frac{\pi}{2}+\theta\right)\right)$. Applying a pair of identities will simplify that.

$$\cos\left(\frac{\pi}{2}+\theta\right) = \cos\frac{\pi}{2}\cos\theta - \sin\frac{\pi}{2}\sin\theta = 0\cdot\cos\theta - 1\cdot\sin\theta = -\sin\theta$$

$$\sin\left(\frac{\pi}{2}+\theta\right) = \sin\frac{\pi}{2}\cos\theta + \cos\frac{\pi}{2}\sin\theta = 1\cdot\cos\theta + 0\cdot\sin\theta = \cos\theta$$

The point (0, 1) is mapped to $(-\sin\theta, \cos\theta)$. That means our matrix equation becomes

$$\begin{bmatrix} \cos\theta & -\sin\theta \\ \sin\theta & \cos\theta \end{bmatrix}\begin{bmatrix} 1 & 0 \\ 0 & 1 \end{bmatrix} = \begin{bmatrix} \cos\theta & -\sin\theta \\ \sin\theta & \cos\theta \end{bmatrix}.$$

To rotate a point or points counterclockwise about the origin by θ, multiply the matrix of coefficients by $\begin{bmatrix} \cos\theta & -\sin\theta \\ \sin\theta & \cos\theta \end{bmatrix}$.

Moving Between Worlds

It's probably a good idea to have a way to distinguish between coordinates of the original point before rotation and those of the image point after rotation. We'll use the traditional (x, y) for the pre-image, and (X, Y) for the image after rotation:

Each point (x, y) rotates to $(X, Y) = (x\cos\theta - y\sin\theta,\ x\sin\theta + y\cos\theta)$

Use the points (–2, 0), (0, –4), and (2, 0) to explore a rotation of the parabola $y = x^2 - 4$ by 60° counterclockwise about the origin. Create a 2×3 matrix of the coefficients and multiply by the rotation matrix.

$$\begin{bmatrix} \cos\theta & -\sin\theta \\ \sin\theta & \cos\theta \end{bmatrix}\begin{bmatrix} -2 & 0 & 2 \\ 0 & -4 & 0 \end{bmatrix} = \begin{bmatrix} \cos 60^\circ & -\sin 60^\circ \\ \sin 60^\circ & \cos 60^\circ \end{bmatrix}\begin{bmatrix} -2 & 0 & 2 \\ 0 & -4 & 0 \end{bmatrix}$$

$$= \begin{bmatrix} \frac{1}{2} & -\frac{\sqrt{3}}{2} \\ \frac{\sqrt{3}}{2} & \frac{1}{2} \end{bmatrix}\begin{bmatrix} -2 & 0 & 2 \\ 0 & -4 & 0 \end{bmatrix} = \begin{bmatrix} -1 & 2\sqrt{3} & 1 \\ -\sqrt{3} & -2 & \sqrt{3} \end{bmatrix}$$

The vertex of (0, −4) maps to $\left(2\sqrt{3},-2\right)$ and the x-intercepts are sent to $\left(-1,-\sqrt{3}\right)$ and $\left(1,\sqrt{3}\right)$. Once you see the image of the vertex and the images of the intercepts, you can imagine the rotated parabola.

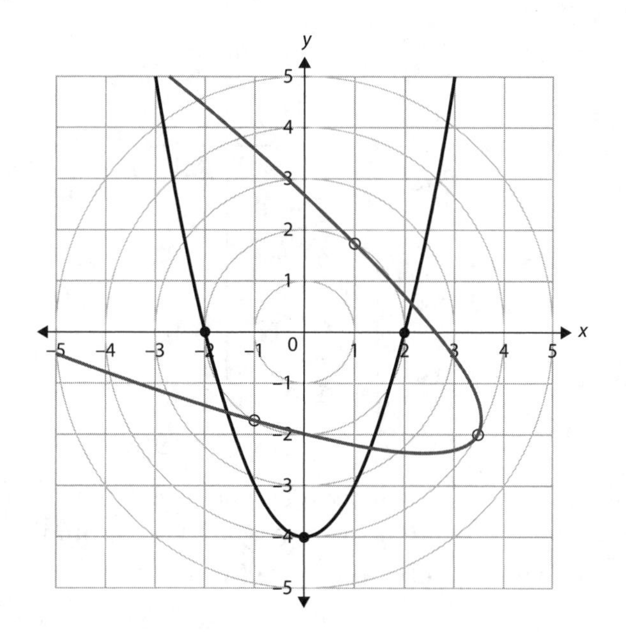

Rotated Axes

If we strip away the original parabola and add a couple of lines you can see that the rotated parabola is symmetric about a line through the image of the vertex, and the images of the x-intercepts sit on another line, perpendicular to the first. If you know how to find these rotated versions of the x-axis and y-axis, and focus on them rather than the original axes, you could sketch the graph of this rotated parabola just as you've sketched other parabolas, except perhaps with a crick in your neck from tilting your head to one side.

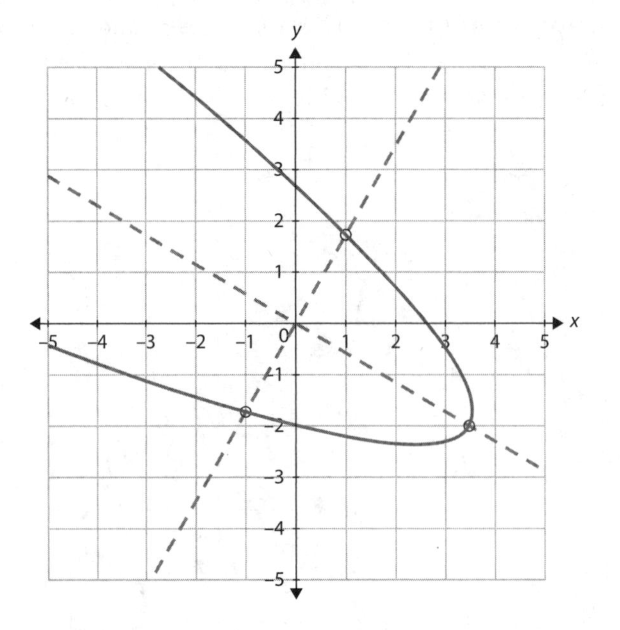

Identifying those rotated axes is not too difficult. If the amount of rotation is θ, the line $y = x\tan\theta$ is the X-axis and $y = -x\cot\theta$ is the Y-axis. For example, under a rotation of $\frac{\pi}{3}$, the rotated axes will be $Y = X\tan\frac{\pi}{3} = \sqrt{3}X \approx 1.732X$ and $Y = -X\cot\frac{\pi}{3} = -\frac{\sqrt{3}}{3}X \approx -0.577X$. Neither of those is tough to calculate, but they could be difficult to draw, so a sketch is probably the best you can hope for.

Finding the Equation

Eventually, you'll want to have an equation for the rotated form of the parabola. If $y = x^2 - 4$ denotes the parabola that opens up, how do we indicate a parabola that looks like that but rotated 60° counterclockwise? To map points from the original coordinate system into that rotated system that we denote as the (X, Y) system, we multiplied by $\begin{bmatrix} \cos\theta & -\sin\theta \\ \sin\theta & \cos\theta \end{bmatrix}$. To go from the (X, Y) system back to the (x, y) system, we would multiply the matrix of (X, Y) coordinates by the inverse matrix, $\begin{bmatrix} \cos\theta & \sin\theta \\ -\sin\theta & \cos\theta \end{bmatrix}$, and we would find that

$$x = X\cos\theta + Y\sin\theta$$

$$y = Y\cos\theta - X\sin\theta$$

We have a graph of a rotated parabola. In (X, Y) world, it looks like $Y = X^2 - 4$. To find an equation in (x, y) world, we're going to replace X with $x\cos\theta + y\sin\theta$ and replace Y with $y\cos\theta - x\sin\theta$. When that substitution is made, there's going to be a lot of simplifying to do. Patience and precision will be important.

$$Y = X^2 - 4$$
$$y\cos\theta - x\sin\theta = (x\cos\theta + y\sin\theta)^2 - 4$$

The parabola was rotated 60° or $\frac{\pi}{3}$ radians, so we can replace θ with the appropriate value.

$$Y = X^2 - 4$$
$$y\cos 60° - x\sin 60° = \left(x\cos 60° + y\sin 60°\right)^2 - 4$$
$$\tfrac{1}{2}y - \tfrac{\sqrt{3}}{2}x = \left(\tfrac{1}{2}x + \tfrac{\sqrt{3}}{2}y\right)^2 - 4$$
$$\tfrac{1}{2}y - \tfrac{\sqrt{3}}{2}x = \tfrac{1}{4}x^2 + 2\cdot\tfrac{1}{2}\cdot\tfrac{\sqrt{3}}{2}xy + \tfrac{3}{4}y^2 - 4$$
$$\tfrac{1}{2}y - \tfrac{\sqrt{3}}{2}x = \tfrac{1}{4}x^2 + \tfrac{\sqrt{3}}{2}xy + \tfrac{3}{4}y^2 - 4$$

Multiplying through by 4 will at least clear the fractions.

$$4\left(\tfrac{1}{2}y - \tfrac{\sqrt{3}}{2}x\right) = 4\left(\tfrac{1}{4}x^2 + \tfrac{\sqrt{3}}{2}xy + \tfrac{3}{4}y^2 - 4\right)$$
$$2y - 2\sqrt{3}x = x^2 + 2\sqrt{3}xy + 3y^2 - 4$$

Let's bring everything to one side and order the terms.

$$2y - 2\sqrt{3}x = x^2 + 2\sqrt{3}xy + 3y^2 - 4$$
$$x^2 + 2\sqrt{3}xy + 3y^2 + 2\sqrt{3}x - 2y - 4 = 0$$

Recognizing the Conic from the Equation

That's not an equation whose graph you could sketch with just a quick look, as you could with $y = x^2 - 4$, but it is the equation that relates the x- and y-coordinates of every point on that rotated parabola. It also gives us a chance to talk about the general form, not just for a particular type of conic, but for all conics.

$$Ax^2 + Bxy + Cy^2 + Dx + Ey + F = 0$$

In the conics you investigated earlier, B was equal to zero; there was no xy term. If either A or C was equal to zero, you had only one square term, and the graph was a parabola. If A and C were both non-zero, and one of them was negative, the graph was a hyperbola. If both were positive, it was an ellipse. A circle is a special case of an ellipse. The rotation of the conics brings in the xy-term. In a later section, we'll talk about identifying the rotated conics.

Exercise 15.1

In questions 1 through 3, you are given a point in the (x, y) coordinate system, and an angle θ. Find the image of the point under a rotation of θ.

1. $(2, -4);\ \theta = 45°$
2. $(-1, 6);\ \theta = 30°$
3. $(-4, -4);\ \theta = 60°$

In questions 4 through 6, you are given an angle θ, and a point (X, Y) that is the image of a point (x, y) under a rotation of θ. Find the pre-image (x, y).

4. $\theta = 30°;\ \left(4+4\sqrt{3}, 4-4\sqrt{3}\right)$
5. $\theta = 60°;\ \left(-5-3\sqrt{3}, 3-5\sqrt{3}\right)$
6. $\theta = 45°;\ \left(4\sqrt{2}, 0\right)$

In questions 7 through 10, you are given an angle θ and the equation of a conic. Find the equation that describes the graph of that conic under a rotation of θ counterclockwise about the origin.

7. $\theta = 60°;\ y = 9 - x^2$
8. $\theta = 45°;\ 4x^2 + y^2 = 4$
9. $\theta = 90°;\ 2x^2 - 4y^2 = 4$
10. $\theta = 30°;\ x = y^2 - 1$

Step 2. Recognize Rotation in Equations

The flag that signals that the equation you're looking at is a rotated conic is the presence of the xy term. The conversion of coordinates that occurs during a rotation means that both the x-coordinate and the y-coordinate are mapped onto values that depend on both x and y. $X = x\cos\theta - y\sin\theta$ and $Y = x\sin\theta + y\cos\theta$. Squaring either of those introduces the xy term.

$$X^2 = (x\cos\theta - y\sin\theta)^2 = x^2\cos^2\theta - 2\underline{\underline{xy}}\sin\theta\cos\theta + y^2\sin^2\theta$$

$$Y^2 = (x\sin\theta + y\cos\theta)^2 = x^2\sin^2\theta + 2\underline{\underline{xy}}\sin\theta\cos\theta + y^2\cos^2\theta$$

Magnitude of Rotation

When you look at the entire general form, $Ax^2 + Bxy + Cy^2 + Dx + Ey + F = 0$, you can identity how much of a rotation is going on by using the following rule:

$$\cot(2\theta) = \frac{A-C}{B} \text{ or } \tan(2\theta) = \frac{B}{A-C}$$

The cotangent version is guaranteed to have a non-zero denominator, because we know B ≠ 0. The tangent version gives you a value you're more likely to recognize.

The equation $5x^2 + 3xy + y^2 = 5$ shows A = 5, B = 3, and C = 1. The two square terms indicate this is a quadratic relation and the presence of the xy term tells us that the conic is rotated. We can find the angle of rotation using $\tan(2\theta) = \frac{B}{A-C}$. $\tan(2\theta) = \frac{3}{5-1} = \frac{3}{4}$. Calculators are helpful in exercises like this, allowing you to skip a lot of work with half-angle identities. The calculator will tell you that $2\theta \approx 37°$ and therefore $\theta \approx 18.5°$ but often the size of the angle is less important than the value of various trig ratios and those can often be deduced without calculator help.

Earlier we saw that under a rotation of θ, the x-axis rotates to the line $y = x\tan\theta$ and the y-axis rotates to the line $y = -x\cot\theta$. Once the value of θ has been identified from A, B, and C, these lines can be identified. If $\tan(2\theta) = \frac{3}{4}$, $\theta = \frac{\tan^{-1}(3/4)}{2}$ so $\tan\theta = \tan\left(\frac{\tan^{-1}(3/4)}{2}\right) = \frac{1}{3}$. In this case, the lines that form the rotated axes are approximately $y = \frac{1}{3}x$ and $y = -3x$.

Exercise 15.2

In questions 1 through 5, identify the amount of rotation, θ, to the nearest tenth of a degree for each equation.

1. $3x^2 + 2xy - y^2 + x - 1 = 0$
2. $5x^2 + 4xy + 2y^2 - 3x + 4y - 7 = 0$
3. $x^2 + 6xy + y^2 - 8 = 0$
4. $2x^2 + 2xy + 7 = 0$
5. $2\sqrt{3}x^2 + 3xy + \sqrt{3}y^2 = 10\sqrt{3}$

In questions 6 through 10, give the equations of the rotated axes. Round to the nearest hundredth if necessary.

6. $2x^2 + 6xy + y^2 - 8 = 0$
7. $xy = 4$
8. $x^2 - xy + 2y^2 = 5$
9. $7x^2 + 6xy + 3y^2 = 9$
10. $\sqrt{3}xy - y^2 = 1$

Step 3. Graph a Rotated Conic

To graph a rotated conic, you need to start with an equation containing an xy-term, which is an equation from (x, y) world, and change it to an equation in (X, Y) world with no xy-term, which can be graphed easily. Then you can focus on the X-axis and Y-axis and sketch the conic.

- Make sure the equation is in general form.
- Use $\cot(2\theta) = \dfrac{A-C}{B}$ or $\tan(2\theta) = \dfrac{B}{A-C}$ to find the angle of rotation, θ.
- Identify and sketch the rotated axes $y = x\tan\theta$ and $y = -x\cot\theta$.
- Determine the values of $\sin\theta$ and $\cos\theta$ in simplest form.
- Substitute for $\sin\theta$ and $\cos\theta$ in $X = x\cos\theta - y\sin\theta$ and $Y = x\sin\theta + y\cos\theta$.
- Substitute in the equation and simplify.
- Sketch the rotated conic, using the converted equation.

> The frequency of errors, especially sign errors, when simplifying is frustrating. Work in a neat, well-organized manner. This is not something to cram into a margin. Take your time, double check every few steps, and keep the goal in mind.

The equation $3x^2 + 4\sqrt{3}xy - y^2 = 7$ is in general form with A = 3, B = $4\sqrt{3}$, C = −1, and D = E = F = 0. $\tan(2\theta) = \dfrac{B}{A-C} = \dfrac{4\sqrt{3}}{3-(-1)} = \sqrt{3}$ so $2\theta = 60°$ and θ = 30°. We know $\sin(30°) = \frac{1}{2}$, $\cos(30°) = \frac{\sqrt{3}}{2}$, and $\tan(30°) = \frac{\sqrt{3}}{3}$. The rotated axes are $y = \frac{\sqrt{3}}{3}x$ and $y = -\sqrt{3}x$. $X = x\cos\theta - y\sin\theta = \frac{\sqrt{3}}{2}x - \frac{1}{2}y$ and $Y = y\cos\theta + x\sin\theta = \frac{\sqrt{3}}{2}y + \frac{1}{2}x$. Take a deep breath and a sharp pencil and make the conversion.

$$3x^2 + 4\sqrt{3}xy - y^2 = 7$$

$$3\left(\tfrac{\sqrt{3}}{2}x - \tfrac{1}{2}y\right)^2 + 4\sqrt{3}\left(\tfrac{\sqrt{3}}{2}x - \tfrac{1}{2}y\right)\left(\tfrac{\sqrt{3}}{2}y + \tfrac{1}{2}x\right) - \left(\tfrac{\sqrt{3}}{2}y + \tfrac{1}{2}x\right)^2 = 7$$

$$3\left(\tfrac{3}{4}x^2 - \tfrac{\sqrt{3}}{2}xy + \tfrac{1}{4}y^2\right) + 4\sqrt{3}\left(\tfrac{3}{4}xy + \tfrac{\sqrt{3}}{4}x^2 - \tfrac{\sqrt{3}}{4}y^2 - \tfrac{1}{4}xy\right) - \left(\tfrac{3}{4}y^2 + \tfrac{\sqrt{3}}{2}xy + \tfrac{1}{4}x^2\right) = 7$$

$$\left(\tfrac{9}{4}x^2 - \tfrac{3\sqrt{3}}{2}xy + \tfrac{3}{4}y^2\right) + 4\sqrt{3}\left(\tfrac{\sqrt{3}}{4}x^2 + \tfrac{1}{2}xy - \tfrac{\sqrt{3}}{4}y^2\right) - \left(\tfrac{3}{4}y^2 + \tfrac{\sqrt{3}}{2}xy + \tfrac{1}{4}x^2\right) = 7$$

$$\tfrac{9}{4}x^2 - \tfrac{3\sqrt{3}}{2}xy + \tfrac{3}{4}y^2 + \left(3x^2 + 2\sqrt{3}xy - 3y^2\right) - \tfrac{3}{4}y^2 - \tfrac{\sqrt{3}}{2}xy - \tfrac{1}{4}x^2 = 7$$

$$\left(3x^2 + \tfrac{9}{4}x^2 - \tfrac{1}{4}x^2\right) - \tfrac{3\sqrt{3}}{2}xy + 2\sqrt{3}xy - \tfrac{\sqrt{3}}{2}xy + \left(\tfrac{3}{4}y^2 - 3y^2 - \tfrac{3}{4}y^2\right) = 7$$

$$5x^2 - 3y^2 = 7$$

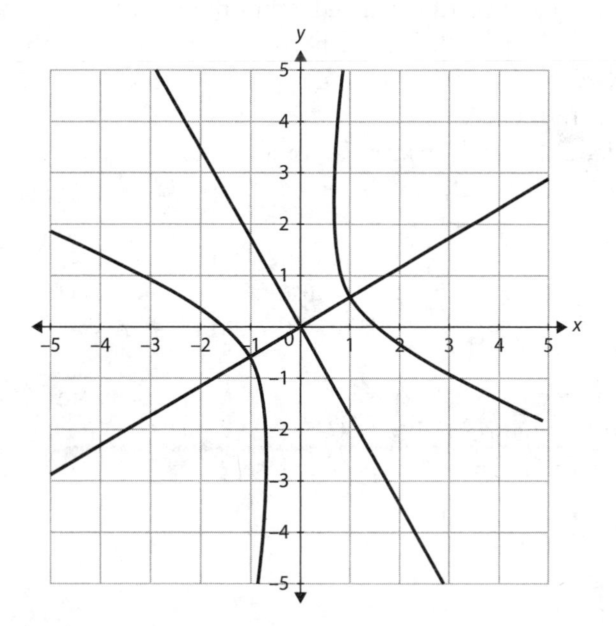

Exercise 15.3

In each question, you are given the equation of a rotated conic. Transform each equation by eliminating the xy-term, draw the rotated axes, and sketch the rotated conic.

1. $x^2 + 4xy + y^2 = 3$
2. $x^2 - xy + y^2 = 2$
3. $5x^2 + 26xy + 5y^2 + 72 = 0$
4. $17x^2 - 30\sqrt{3}xy + 51y^2 = 64$
5. $x^2 - 3xy + y^2 = 5$
6. $3x^2 + 2xy + 3y^2 = 19$
7. $x^2 + 2xy + y^2 - 3\sqrt{2} - 5\sqrt{2} + 8 = 0$
8. $7x^2 - 8xy + y^2 = 9$
9. $2x^2 + \sqrt{3}xy + y^2 = 5$
10. $x^2 - 2xy + y^2 - \sqrt{2}x - \sqrt{2}y = 0$

Step 4: Classify Rotated Conics without Graphing

All of that converting can be frustrating if all you want to know is what type of conic the equation describes. The discriminant, the formula that helped us determine the number and type of solutions for a quadratic function, will help with this task.

If the general form of a conic is $Ax^2 + Bxy + Cy^2 + Dx + Ey + F = 0$, the graph of the equation will be:

$$\begin{aligned} &\text{an ellipse if } B^2 - 4AC < 0 \\ &\text{a parabola if } B^2 - 4AC = 0 \\ &\text{a hyperbola if } B^2 - 4AC > 0 \end{aligned}$$

The equation $x^2 - xy + y^2 = 2$ has A = 1, B = −1, and C = 1. The discriminant for this equation is $B^2 - 4AC = (-1)^2 - 4(1)(1) = 1 - 4 = -3$. The negative discriminant indicates this is an ellipse.

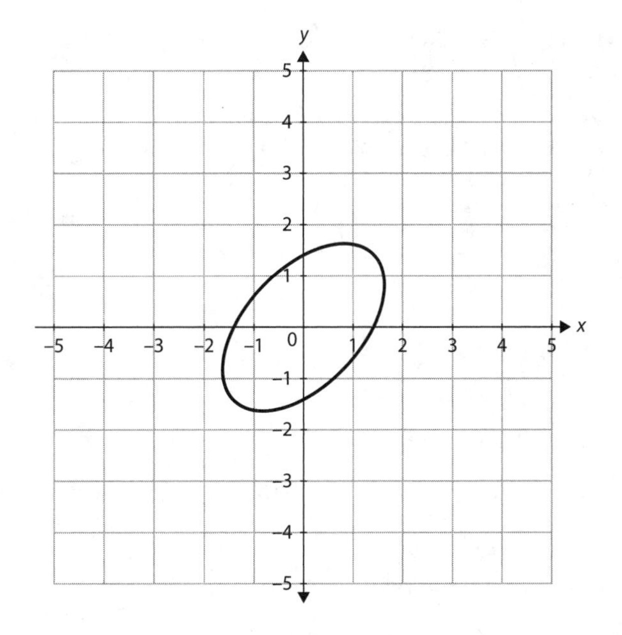

The equation $x^2 + 4xy + y^2 = 3$ has a discriminant of $B^2 - 4AC = (4)^2 - 4(1)(1) = 16 - 4 = 12$. The positive discriminant tells us this is a hyperbola.

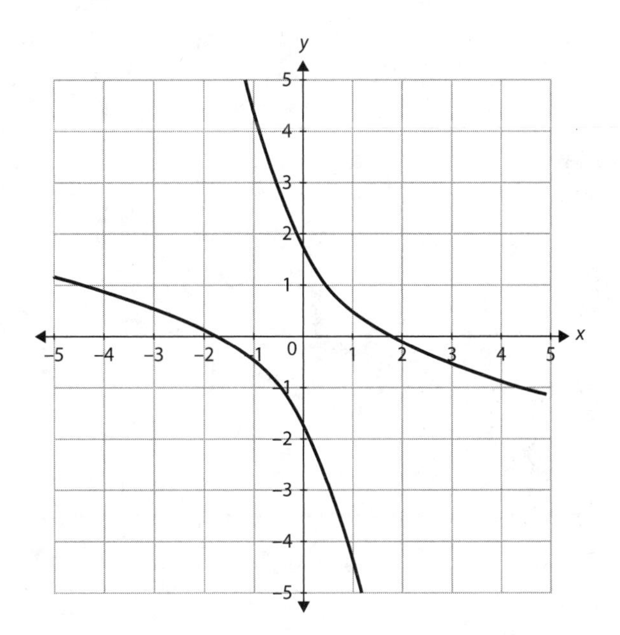

Exercise 15.4

Use the discriminant to classify each equation as an ellipse, a parabola, or a hyperbola without converting the equation.

1. $x^2 + xy + y^2 = 1$
2. $x^2 - 2xy + y^2 - x - y = 0$
3. $x^2 + \sqrt{2}xy = 1$
4. $x^2 - xy + y^2 = 1$
5. $3x^2 - 12\sqrt{5}xy + 6y^2 + 9 = 0$
6. $\sqrt{3}xy + y^2 = 1$
7. $xy = 4$
8. $5x^2 - 3xy + 2y^2 + 3x + 4y + 2 = 0$
9. $3x^2 + 12xy + 2y^2 - 3x - 2y + 5 = 0$
10. $x^2 + 12xy + 36y^2 - 4x + 3y - 10 = 0$

16

Complex Numbers

When you solved quadratic equations, you had a brief introduction to complex numbers. The idea that a whole group of equations would have to be abandoned as having no solution is not something mathematicians take lightly. Yes, there are some equations that cannot be true, no matter what you substitute for the variable, but the definition of $i = \sqrt{-1}$ and the number system that can be built from that allow the solution of many equations you might once have declared unsolvable.

In order to solve quadratic equations, you only need to know a little bit about complex numbers. You need to know how to simplify the quadratic formula down to the simplest radical form, and if the radicand is negative, that leads to a complex number in $a + bi$ form. But there are other forms in which complex numbers can be written, and there is a whole system of arithmetic for working with complex numbers. This chapter will lead you through that.

Step 1. Change Between Forms

There are three forms in which complex numbers can be expressed, and which form you choose depends on the work you want to do. The $a + bi$ or rectangular form is probably the most common. It is formed from a real part, a, and an imaginary part, bi. If $a = 0$ the number is purely imaginary. If $b = 0$, the number is a real number. The Real Numbers and the Imaginary Numbers are subsets of the Complex Numbers.

Rectangular Form

Complex numbers can be represented by points in the plane, by placing the reals on the horizontal axis and the imaginaries on the vertical axis. The complex number $a + bi$ is represented by the point (a, b).

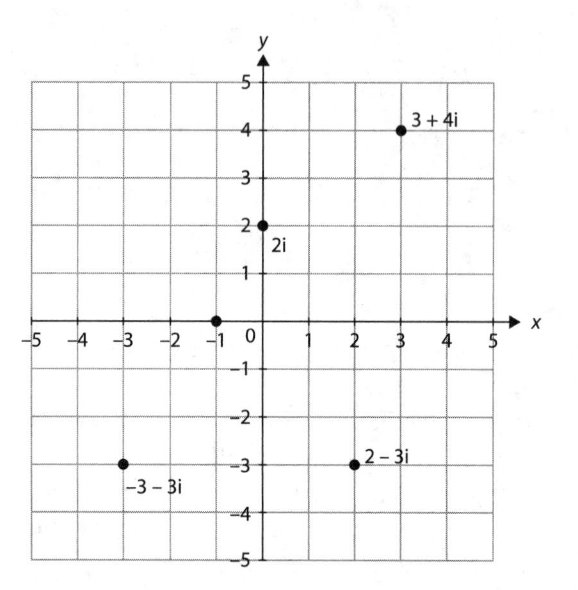

Polar Form

The second form is polar, or trigonometric form $r(\cos\theta + i\sin\theta)$. Trigonometric form is longer to write, which is why it's sometimes abbreviated as $r(cis\theta)$, but does make some calculations much easier. For this form, imagine the point on the polar plane that corresponds to the complex number and a line segment connecting that point to the origin. The number r is the modulus of the complex number, its distance from the origin, and θ is the angle between the horizontal axis and the line segment connecting the point to the origin.

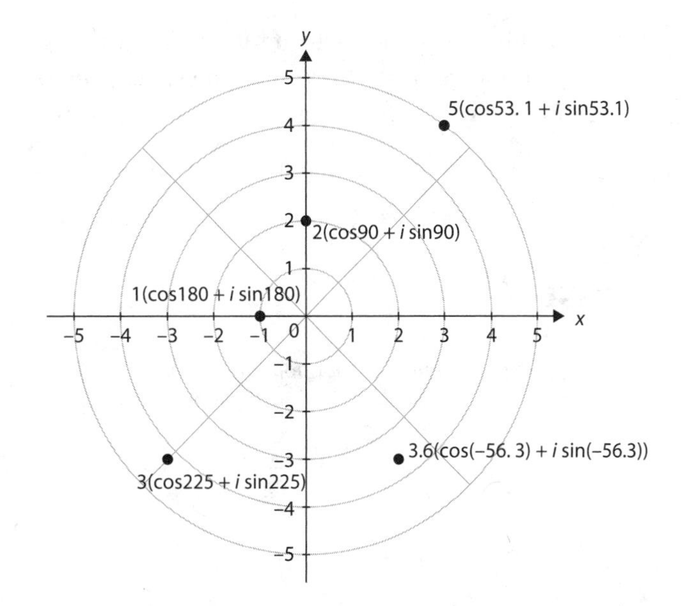

To convert a number in $a+bi$ form to trigonometric form, calculate $r=\sqrt{a^2+b^2}$ and $\theta=\tan^{-1}\left(\dfrac{b}{a}\right)$.

To convert $3-4i$ to trig form, first consider that the point falls in quadrant IV. This will help with the determination of the angle. Calculate $r=\sqrt{(-3)^2+4^2}=\sqrt{9+16}=\sqrt{25}=5$ and $\theta=\tan^{-1}\left(\dfrac{4}{-3}\right)\approx -53.1^\circ$ or approximately -0.93 radians. In trig form, the number is $5\,(\cos(-53.1^\circ)+i\sin(-53.1^\circ))$.

To convert a number in polar form to $a+bi$ form, remember that $r(\cos\theta+i\sin\theta)=r\cos\theta+i\cdot r\sin\theta$ and substitute $x=r\cos\theta$ and $y=r\sin\theta$.

The complex number $6\left(\cos\dfrac{5\pi}{4}+i\sin\dfrac{5\pi}{4}\right)$ has $r=6$ and $\theta=\dfrac{5\pi}{4}$ so $x=6\cos\left(\frac{5\pi}{4}\right)=6\left(-\frac{\sqrt{2}}{2}\right)=-3\sqrt{2}$ and $y=6\sin\left(\frac{5\pi}{4}\right)=6\left(-\frac{\sqrt{2}}{2}\right)=-3\sqrt{2}$. The rectangular form of the number is $-3\sqrt{2}-3i\sqrt{2}$.

Exponential Form

The third form, known as exponential form, expresses complex numbers as a power of the natural base e, which you first met in your study of logarithms. The form is built from an identity known as Euler's Formula: $e^{\theta i}=\cos\theta+i\sin\theta$. There's a bit of calculus involved in deriving that, so you'll have to take it on faith for now, but if $e^{\theta i}=\cos\theta+i\sin\theta$, then $r(\cos\theta+i\sin\theta)=re^{\theta i}$. Expressing complex numbers in that form means you get all the benefits of the laws of exponents: adding exponents to multiply, subtracting exponents to divide, and multiplying exponents to raise a power to a power. We'll focus on rectangular and polar form for the most part, with just an occasional mention of exponential form. From the example above, $6\left(\cos\frac{5\pi}{4}+i\sin\frac{5\pi}{4}\right)=6e^{\frac{5\pi}{4}i}$.

Exercise 16.1

Convert each complex number to trigonometric form.

1. $4-4i$
2. $1+\sqrt{3}i$
3. $-5i$
4. $-5\sqrt{3}+5i$
5. $-3+4i$

Convert each number to a + bi form.

6. $6\left(\cos\frac{\pi}{3}+i\sin\frac{\pi}{3}\right)$
7. $8\left(\cos\frac{\pi}{4}+i\sin\frac{\pi}{4}\right)$
8. $5\left(\cos\frac{\pi}{6}+i\sin\frac{\pi}{6}\right)$
9. $2\left(\cos\frac{\pi}{2}+i\sin\frac{\pi}{2}\right)$
10. $12(\cos 210°+i\sin 210°)$
11. $\sqrt{3}(\cos 60°+i\sin 60°)$
12. $\sqrt{3}(\cos 420°+i\sin 420°)$
13. $\sqrt{3}(\cos 780°+i\sin 780°)$
14. $6\left(\cos\frac{\pi}{4}+i\sin\frac{\pi}{4}\right)$
15. $6\left(\cos\frac{9\pi}{4}+i\sin\frac{9\pi}{4}\right)$

Any complex number has infinite polar representations.

Step 2. Add and Subtract Complex Numbers

To add complex numbers in $a+bi$ form, add the real parts and add the imaginary parts.

> To avoid combining unlike terms and prevent sign errors, consider showing the rearranging of terms.

$$(a+bi)+(c+di)=(a+c)+(b+d)i$$

$$(7-3i)+(-2+5i)=(7-2)+(-3+5)i=5+2i$$

To subtract complex numbers in $a+bi$ form, remember that subtracting is adding the opposite. Change the signs of both the real and the imaginary term of the second number and then add.

> Subtraction is especially vulnerable to sign errors so take the time to assure you have changed signs correctly.

$$(a+bi)-(c+di)=a+bi-c-di=(a-c)+(b-d)i$$

$$(11+4i)-(6-8i)=11+4i-6+8i=(11-6)+(4+8)i=5+12i$$

Trying to add complex numbers in trigonometric form will require the law of cosines, the law of sines, and a lot of rounding, so it is not recommended. Convert to rectangular form, add or subtract, and if necessary convert back to polar form.

Exercise 16.2

Perform each calculation. Give your answers in rectangular form.

1. $(5-7i)+(2+9i)$
2. $(-6+4i)+(2-4i)$
3. $(12+11i)-(7+19i)$
4. $(2-9i)-(5+3i)$
5. $(6+5i)+(8-3i)-(11-4i)$
6. $(6+5i)-(-7+2i)+(3-4i)$
7. $(-1+3i)-(5+4i)+(12-8i)$
8. $(4+3i)+(-9+7i)-(-5-2i)$
9. $(13+8i)-(9+6i)-(7+2i)$
10. $(3+2i)+(5-8i)+(-6+i)$

Step 3. Multiply and Divide Complex Numbers

Multiplication and division of complex numbers can be done with numbers in rectangular or polar form. The methods for rectangular form can become cumbersome, especially for multistep problems, but are applications of familiar algebraic techniques. The methods for polar form are based on trigonometric identities, but it's not necessary to work through the whole rationale each time and you can jump right to the simple arithmetic.

To multiply a complex number in $a+bi$ by a real number or by a pure imaginary, apply the distributive property. To multiply two complex numbers in $a+bi$ form, apply the FOIL rule and simplify using the fact that $i^2=-1$.

> Because $i^2=-1$ you should always be able to combine two real terms and two imaginary terms.

$$\begin{aligned}(a+bi)(c+di) &= ac+adi+bci+bdi^2\\ &= ac+(adi+bci)+bd(-1)\\ &= (ac-bd)+(ad+bc)i\end{aligned}$$

$$\begin{aligned}(3+5i)(-2-3i) &= -6-9i-10i-15i^2\\ &= -6-19i-15(-1)\\ &= -6+15-19i\\ &= 9-19i\end{aligned}$$

To divide complex numbers in rectangular form, use the techniques for rationalizing denominators.

> In general form, it looks more complicated than it actually is. Multiply numerator and denominator by the conjugate of the denominator, using the FOIL rule, and then simplify.

$$\frac{a+bi}{ci}=\frac{a+bi}{ci}\cdot\frac{i}{i}=\frac{ai+bi^2}{ci^2}=\frac{ai-b}{-c}$$

$$\frac{a+bi}{c+di}=\frac{a+bi}{c+di}\cdot\frac{c-di}{c-di}=\frac{ac-adi+bci-bdi^2}{c^2-d^2i^2}=\frac{(ac+bd)+(bc-ad)i}{c^2+d^2}$$

$$\frac{8+4i}{1+2i}\cdot\frac{1-2i}{1-2i}=\frac{8-16i+4i-8i^2}{1-2i+2i-4i^2}=\frac{8+8-12i}{1+4}=\frac{16-12i}{5}=3.2-2.4i$$

To multiply complex numbers in trigonometric form, multiply the moduli, r_1 and r_2 and add the angle measures $\theta_1 + \theta_2$.

You can easily multiply the *r*'s, the modulus of the number. The work is in finding the angle, but luckily the identities for sin and cos of a sum do the job.

$$r_1(\cos\theta_1 + i\sin\theta_1)\cdot r_2(\cos\theta_2 + i\sin\theta_2) = r_1\cdot r_2(\cos(\theta_1+\theta_2) + i\sin(\theta_1+\theta_2))$$

$$3\left(\cos\frac{\pi}{3} + i\sin\frac{\pi}{3}\right)\cdot 5\left(\cos\frac{\pi}{4} + i\sin\frac{\pi}{4}\right) = 15\left(\cos\left(\frac{\pi}{3}+\frac{\pi}{4}\right) + i\sin\left(\frac{\pi}{3}+\frac{\pi}{4}\right)\right)$$

$$= 15\left(\cos\left(\frac{7\pi}{12}\right) + i\sin\left(\frac{7\pi}{12}\right)\right)$$

To divide complex numbers in polar form, divide the moduli and subtract the angle measurements.

As with multiplication, the *r*'s are simple. The angles take work and the application of the identities for the sin of a difference and cos of a difference of two angles.

$$\frac{r_1(\cos\theta_1 + i\sin\theta_1)}{r_2(\cos\theta_2 + i\sin\theta_2)}\cdot = \frac{r_1}{r_2}\cdot(\cos(\theta_1-\theta_2) + i\sin(\theta_1-\theta_2))$$

$$\frac{15\left(\cos\frac{\pi}{2} + i\sin\frac{\pi}{2}\right)}{3\left(\cos\frac{\pi}{6} + i\sin\frac{\pi}{6}\right)}\cdot = 5\left(\cos\left(\frac{\pi}{2}-\frac{\pi}{6}\right) + i\sin\left(\frac{\pi}{2}-\frac{\pi}{6}\right)\right)$$

$$= 5\left(\cos\left(\frac{\pi}{3}\right) + i\sin\left(\frac{\pi}{3}\right)\right)$$

Exponential form also makes multiplication and division easier. To multiply two complex numbers in exponential form, multiply the moduli $r_1 \cdot r_2$ and add the exponents.

Being able to add the exponents makes the work easy. The *i* just goes along, and you add the θ's.

$$r_1e^{\theta_1 i}\cdot r_2e^{\theta_2 i} = r_1\cdot r_2\cdot e^{(\theta_1+\theta_2)i}$$

$$4e^{\frac{\pi}{3}i}\cdot 5e^{\frac{\pi}{6}i} = 20e^{\frac{\pi}{2}i}$$

To divide two complex numbers in exponential form, divide the moduli $r_1 \div r_2$ and subtract the exponents.

As with multiplication, let the i tag along as you subtract the θ's.

$$\frac{r_1 e^{\theta_1 i}}{r_2 e^{\theta_2 i}} = \frac{r_1}{r_2} e^{(\theta_1 - \theta_2)i}$$

$$\frac{24e^{\frac{\pi}{3}i}}{6e^{\frac{\pi}{6}i}} = 4e^{\frac{\pi}{6}i}$$

Exercise 16.3

Perform each calculation. Simplify your answers as much as possible, but it is not necessary to change the form (rectangular, polar, or exponential).

1. $(2+5i)(3-4i)$
2. $4\left(\cos\frac{\pi}{3}+i\sin\frac{\pi}{3}\right)\cdot 2\left(\cos\frac{\pi}{6}+i\sin\frac{\pi}{6}\right)$
3. $(-3i)(5-9i)$
4. $12(\cos 40° + i\sin 40°)\cdot 5(\cos 70° + i\sin 70°)$
5. $(-2-3i)(-2+3i)$
6. $\frac{24(\cos 100° + i\sin 100°)}{8(\cos 55° + i\sin 55°)}$
7. $\frac{2+3i}{1+2i}$
8. $3e^{\frac{5\pi}{4}i}\cdot 7e^{\frac{3\pi}{2}i}$
9. $\frac{18(\cos 225° + i\sin 225°)}{3(\cos 135° + i\sin 135°)}$
10. $\frac{16e^{\frac{3\pi}{4}i}}{4e^{\frac{2\pi}{3}i}}$
11. $\frac{8+4i}{2i}$
12. $7\left(\cos\frac{\pi}{6}+i\sin\frac{\pi}{6}\right)\cdot 9\left(\cos\frac{\pi}{4}+i\sin\frac{\pi}{4}\right)$

13. $\dfrac{18(\cos 120° + i\sin 120°)}{6(\cos 30° + i\sin 30°)}$

14. $\dfrac{2+5i}{3-3i}$

15. $4(\cos 70° + i\sin 70°)\cdot 7(\cos 80° + i\sin 80°)$

Polar (trig) form looks like it would be more cumbersome, and if you actually had to raise $\cos\theta + i\sin\theta$ to the nth power by FOIL, it would be. The shortcut is a tremendous time saver.

Step 4. Raise a Complex Number to a Power

Raising a complex number in rectangular form to a power can become tedious if the exponent is greater than 2 or 3, because it calls for repeated application of the FOIL rule. Polar form or exponential form will make the task quicker and easier.

To raise a complex number in trigonometric form to the nth power, raise r to the nth power and multiply θ by n.

$$[r(\cos\theta + i\sin\theta)]^n = r^n(\cos(n\theta) + i\sin(n\theta))$$

$$\begin{aligned}[2(\cos 40° + i\sin 40°)]^5 &= 2^5(\cos(5\cdot 40°) + i\sin(5\cdot 40°)) \\ &= 32(\cos(200°) + i\sin(200°))\end{aligned}$$

To raise a complex number in exponential form to the nth power, raise r to the nth power and multiply θ by n.

Raising a power to a power is a simple matter so raising a complex number in exponential form to a power can be accomplished easily.

$$\left(re^{\theta i}\right)^n = r^n e^{n\theta i}$$

$$\left(3e^{\frac{\pi}{4}i}\right)^3 = 3^3 e^{3\cdot\frac{\pi}{4}i} = 27e^{\frac{3\pi}{4}i}$$

Exercise 16.4

Raise each complex number to the power shown. Work within the form given. Simplify answers as completely as possible.

1. $(2-3i)^3$
2. $(2i)^8$
3. $[5(\cos 20° + i\sin 20°)]^4$

4. $\left[2(\cos 60^\circ + i\sin 60^\circ)\right]^5$

5. $\left[10\left(\cos\frac{\pi}{4} + i\sin\frac{\pi}{4}\right)\right]^6$

6. $\left[4\left(\cos\frac{\pi}{6} + i\sin\frac{\pi}{6}\right)\right]^5$

7. $\left[2\left(\cos\frac{\pi}{3} + i\sin\frac{\pi}{3}\right)\right]^9$

8. $\left[\frac{1}{2}\left(\cos\frac{3\pi}{2} + i\sin\frac{3\pi}{2}\right)\right]^6$

9. $\left(3e^{\frac{\pi}{3}i}\right)^4$

10. $\left(2e^{\frac{3\pi}{2}i}\right)^{12}$

Step 5. Find the Roots of a Complex Number

Soon after you learned about square roots, you learned that every non-zero real number has both a positive and a negative square root, but you probably didn't investigate whether every non-zero real number had three cube roots or five fifth roots. That's because two of the three cube roots are complex and aren't easily found until you know more about the arithmetic of complex numbers. When you can express complex numbers in polar or exponential form, that task is easier.

To find the nth root of a complex number in polar form:

- Write n versions of the radicand.

  As you saw in earlier exercises, there are infinitely many ways to represent a complex number in polar form. If θ is measured in degrees, adding 360° brings you around the circle back to the same point. If θ is measured in radians, adding 2π will have the same effect. Whatever the value of n, it is possible to find n representations of the number whose roots are sought.

$$r(\cos\theta + i\sin\theta)$$
$$r(\cos(\theta + 2\pi) + i\sin(\theta + 2\pi))$$
$$r(\cos(\theta + 4\pi) + i\sin(\theta + 4\pi))$$
$$r(\cos(\theta + 6\pi) + i\sin(\theta + 6\pi))$$

This pattern can be continued until the necessary number of representations have been found.

- For each version, take the nth root of r and divide θ by n.

  To find the three cube roots of −8, first express −8 + 0i in polar form: $8(\cos 180° + i\sin 180°)$. Then find two more representations to give a total of three, for the third root.

$$8(\cos 180° + i\sin 180°)$$

$$8(\cos(180° + 360°) + i\sin(180° + 360°)) = 8(\cos(540°) + i\sin(540°))$$

$$8(\cos(180° + 720°) + i\sin(180° + 720°)) = 8(\cos(900°) + i\sin(900°))$$

- Finally, take the third root of each representation.

$$\sqrt[3]{8}\left(\cos\frac{180°}{3} + i\sin\frac{180°}{3}\right) = 2(\cos 60° + i\sin 60°)$$

$$\sqrt[3]{8}\left(\cos\left(\frac{540°}{3}\right) + i\sin\left(\frac{540°}{3}\right)\right) = 2(\cos 180° + i\sin 180°)$$

$$\sqrt[3]{8}\left(\cos\left(\frac{900°}{3}\right) + i\sin\left(\frac{900°}{3}\right)\right) = 2(\cos 300° + i\sin 300°)$$

If plotted in the complex plane, the cube roots form the vertices of an equilateral triangle, and, in general, the nth roots of a complex number are the vertices of a regular n-gon.

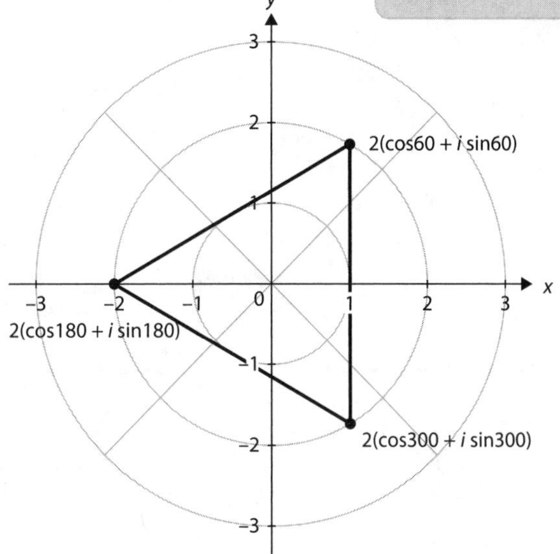

Exercise 16.5

Find the indicated roots of each of the numbers given. Begin by writing the number in polar (trig) form. It may help to locate the point in the complex plane that represents the number. Then apply the procedure described in this section. Plot the roots in the complex plane to check your results.

1. Find the 3 cube roots of 1.
2. Find the 5 fifth roots of $-32\boldsymbol{i}$.
3. Find the 2 square roots of $16\boldsymbol{i}$.
4. Find the 4 fourth roots of -81.
5. Find the 6 sixth roots of 1,000,000.
6. Find the 2 square roots of $49\boldsymbol{i}$.
7. Find the 3 cube roots of -64.
8. Find the 5 fifth roots of 243.
9. Find the 3 cube roots of $1{,}000\boldsymbol{i}$.
10. Find the 4 fourth roots of $625\boldsymbol{i}$.

Limits

If you set out from home to walk to school to meet a friend, you have a clear destination. You know whether or not you have reached your goal. Even if you arrive and don't find your friend, you've reached the designated spot. If you set out from home to take a walk, just for the exercise or the enjoyment of the out-of-doors, you have no such goal (except probably to get back home eventually).

There's a rigorous and formal definition of what mathematicians mean by a limit, but for working purposes, you can think about limits as where the function is headed on its travels. When you see $\lim_{x \to 4}(x^2 - 1)$, which you'd read aloud as "the limit of $x^2 - 1$ as x approaches 4," you can follow the graph of $f(x) = x^2 - 1$ toward the point on the graph where $x = 4$. What y-value will you reach? Does the answer change if you approach from above 4 instead of below? Or is it possible that you can't get there at all? These are the kinds of questions limits ask.

> Traditionally called the epsilon-delta definition of a limit, the formal definition uses two small numbers, designated as epsilon and delta, and presents the finding of a limit almost as a challenge. If you propose that $\lim_{x \to a} f(x) = L$, I pick a small number (ε) and insist that all the y-values of the function be within the number of units to either side of L. If you can find a small number (δ) so that when we consider values of x between $a - \delta$ and $a + \delta$, all the y-values are in my required strip, then L is the limit.

To be a little more formal, if a function $f(x)$ gets extremely close to, or even equal to, a number L when x gets close to some value a, we say that $\lim_{x \to a} f(x) = L$. So $\lim_{x \to 4}(x^2 - 1) = 15$, $\lim_{x \to -3} \frac{1}{x} = -\frac{1}{3}$, and $\lim_{x \to 2} \frac{x^2 - 4}{x - 2} = 4$. Even though the expression $\frac{x^2 - 4}{x - 2}$ is undefined for $x = 2$, when you get close to $x = 2$, you're heading for $y = 4$. But $\lim_{x \to 0} \frac{1}{x}$ doesn't exist. You can't get there, because when you approach from one side, you go up to ∞, and when you approach from the other side you go down to $-\infty$.

Step 1. Evaluate Limits

How do you find a limit? If a function is well behaved—smooth and continuous—calculating a limit is just a matter of evaluating the function. So $\lim_{x \to 4}(x^2 - 1) = 15$ because you can evaluate $x^2 - 1$ when $x = 4$ and it equals 15. Even if the function has discontinuities, as $f(x) = \frac{1}{x}$ does, you may still be able to plug in if you're looking for the limit as x approaches a value where the function is continuous. That's why you can say that $\lim_{x \to -3} \frac{1}{x} = -\frac{1}{3}$, but you can't evaluate $\lim_{x \to 0} \frac{1}{x}$. The value $x = 0$ is not in the domain of the function.

When you're looking for a limit of a function that's not behaving well for you, there are strategies you can use to find the limit.

Investigate Numerically

You're looking for a number that the function gets extremely close to, so take advantage of your calculator to find the function values close to the x-value you're approaching. If you're trying to find $\lim_{x \to 3} \frac{x-3}{x^2 - x - 6}$, you won't be able to just plug in 3 for x, because you'll get a 0 denominator, but you can look at the values of $f(x) = \frac{x-3}{x^2 - x - 6}$ for values of x close to 3 and see if there's a pattern to where they're going. Use values both above and below 3, and organize them in order so you can see a trend.

2.5	2.9	2.99	2.999	3	3.001	3.01	3.1	3.5
0.22222	0.20408	0.2004	0.20004	?	0.19996	0.1996	0.19608	0.18182

$\rightarrow\rightarrow\rightarrow\rightarrow\rightarrow\rightarrow\rightarrow\rightarrow\rightarrow\rightarrow\rightarrow\rightarrow\rightarrow\rightarrow\rightarrow\rightarrow\rightarrow\rightarrow \qquad \leftarrow\leftarrow\leftarrow\leftarrow\leftarrow\leftarrow\leftarrow\leftarrow\leftarrow\leftarrow\leftarrow\leftarrow\leftarrow\leftarrow$

The values seem to be heading toward 0.2, so you can say $\lim_{x\to 3}\frac{x-3}{x^2-x-6}=0.2$.

Explore Graphically

The same kind of investigation of the values of the function near $x = a$ that you can do by computing values can be done by looking at the graph of a function.

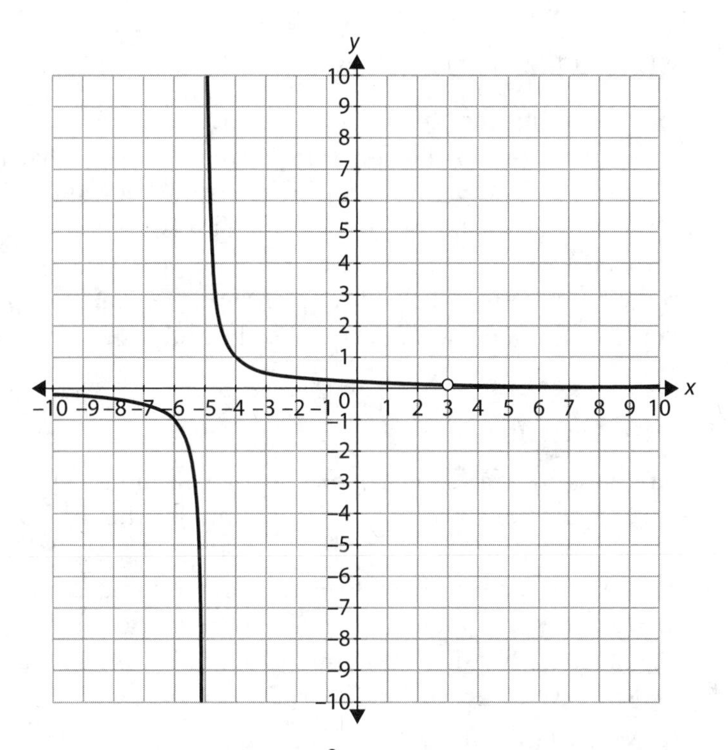

The graph of $f(x)=\frac{x-3}{x^2+2x-15}$ has an essential discontinuity at $x=-5$ and a hole at $x=3$

When you look at the graph of the function $f(x)=\frac{x-3}{x^2+2x-15}$ in the previous figure, you can see that for values of x close to -4, the value of the function is close to 1 (and in fact, $f(-4) = 1$). You can say that $\lim_{x\to -4} f(x)=1$. If you look closely at the neighborhood around $x = 3$, you'll see that the function is not defined for $x = 3$, but for all the values of x close to 3, the values of the function are close to $\frac{1}{8}$. You can conclude that $\lim_{x\to3} f(x) = \frac{1}{8}$. But if you look at values of x near -5, not only is the function not defined there, but you can't say that the function values are heading toward a particular value as x gets close to -5. In fact, as x approaches -5 from below, the function is heading toward $-\infty$, but as x approaches -5 from above, the function is approaching ∞. It's not possible to say what the limit is.

Manipulate Algebraically

If you're looking for the limit of a function as x approaches a certain value—call it a—and the function is well behaved, that is, smooth and continuous, around a, you can just substitute the value of a and evaluate the function. The value of the function at a is the limit of the function as x approaches a.

If the function isn't defined at a, however, you can't just plug in, but don't give up. A little bit of algebra may solve the problem.

Factor and Cancel

If you're looking at a rational expression, an algebraic fraction, and you can't plug in, it's probably because you're looking for the limit as x approaches a value for which the function is not defined, a value at which the denominator would be 0. Check to see if the numerator and denominator can be factored. You may find that the numerator and denominator have a factor in common. Canceling that common factor leaves you with a function that is equivalent to yours at every point except $x = a$.

If you need to find $\lim_{x\to1} \frac{x-1}{x^2-1}$, you won't be able to plug in $x = 1$, because, if you try, you'll have $\frac{0}{0}$, an indeterminate form. Notice, however, that the denominator is factorable. Because $\frac{x-1}{x^2-1} = \frac{\overset{1}{\cancel{x-1}}}{(x+1)\cancel{(x-1)}}$, you know that $\frac{x-1}{x^2-1}$ matches $\frac{1}{x+1}$ everywhere except at $x = 1$. When $x = 1$, $\frac{1}{x+1}$ is defined but $\frac{x-1}{x^2-1}$ is not. Because they agree everywhere else, both of them will have the same limit as you get close to $x = 1$: $\lim_{x\to1} \frac{x-1}{x^2-1} = \lim_{x\to1} \frac{1}{x+1} = \frac{1}{2}$.

Rationalize

If your problem finding a limit is caused by a radical, you may be able to find a function that matches yours at all but one point by rationalizing either the denominator or the numerator.

If you're asked for $\lim\limits_{x\to 2}\frac{x-2}{\sqrt{x^2-4}}$, you'll find that $\frac{x-2}{\sqrt{x^2-4}}$ becomes the indeterminant $\frac{0}{0}$ if you try to plug in. Instead, rationalize the denominator.

$$\lim_{x\to 2}\frac{x-2}{\sqrt{x^2-4}}=\lim_{x\to 2}\frac{(x-2)\sqrt{x^2-4}}{\left(\sqrt{x^2-4}\right)}$$

$$=\lim_{x\to 2}\frac{(x-2)\sqrt{x^2-4}}{x^2-4}$$

$$=\lim_{x\to 2}\frac{\cancel{(x-2)}\sqrt{x^2-4}}{(x+2)\cancel{(x-2)}}$$

$$=\lim_{x\to 2}\frac{\sqrt{x^2-4}}{x+2}=0$$

While you're accustomed to rationalizing denominators, it's also possible to rationalize a numerator, and often that will help you to find limits as well. To find $\lim\limits_{x\to 3}\frac{\sqrt{x+1}-2}{x-3}$, first rationalize the numerator by multiplying the numerator and denominator by the conjugate of the numerator, $\sqrt{x+1}-2$.

$$\lim_{x\to 3}\frac{\sqrt{x+1}-2}{x-3}=\lim_{x\to 3}\frac{\left(\sqrt{x+1}-2\right)\left(\sqrt{x+1}+2\right)}{(x-3)\left(\sqrt{x+1}+2\right)}$$

$$=\lim_{x\to 3}\frac{x+1-4}{(x-3)\left(\sqrt{x+1}+2\right)}$$

$$=\lim_{x\to 3}\frac{\cancel{x-3}}{\cancel{(x-3)}\left(\sqrt{x+1}+2\right)}$$

$$=\lim_{x\to 3}\frac{1}{\sqrt{x+1}+2}=\frac{1}{4}$$

Exercise 17.1

Find each limit. Use the following figure for questions 1 through 4.

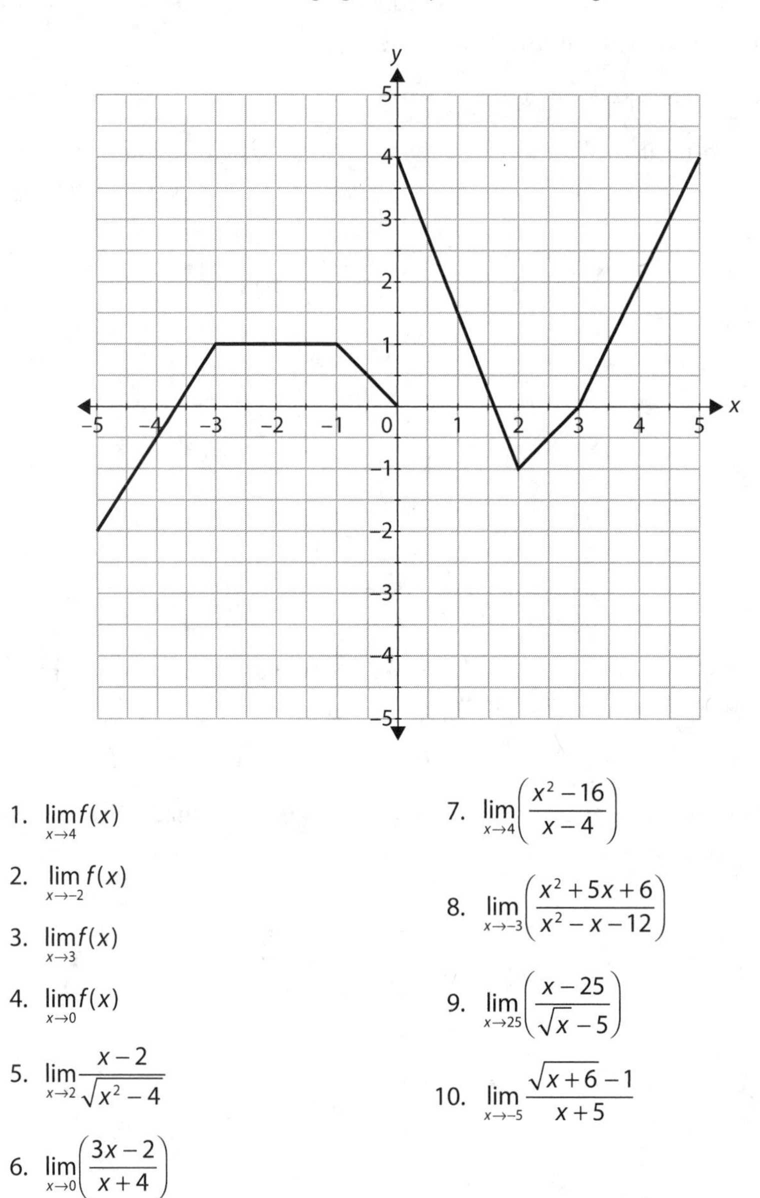

1. $\lim_{x\to 4} f(x)$
2. $\lim_{x\to -2} f(x)$
3. $\lim_{x\to 3} f(x)$
4. $\lim_{x\to 0} f(x)$
5. $\lim_{x\to 2} \frac{x-2}{\sqrt{x^2-4}}$
6. $\lim_{x\to 0} \left(\frac{3x-2}{x+4}\right)$
7. $\lim_{x\to 4} \left(\frac{x^2-16}{x-4}\right)$
8. $\lim_{x\to -3} \left(\frac{x^2+5x+6}{x^2-x-12}\right)$
9. $\lim_{x\to 25} \left(\frac{x-25}{\sqrt{x}-5}\right)$
10. $\lim_{x\to -5} \frac{\sqrt{x+6}-1}{x+5}$

Step 2. Deal with Problems

Once you understand what finding a limit means and you've had some experience determining limits or deciding that there is no limit, the next thing to consider is under what conditions you have to say the limit does not exist. Before trying to address that, you need to introduce the notion of a one-sided limit.

One-Sided Limits

When you looked at the limit of a well-behaved function as x approached some value a, you noticed that it didn't matter whether x was approaching a by coming up to a from smaller values of x or coming down to a from larger values. The limit was the same whether x approached a from the left or the right. If you want to talk about the behavior of the function as x approaches a from just one side, you can use the notation $\lim_{x \to a^+} f(x)$ to denote the limit of $f(x)$ as x approaches a from the right, or from higher values of x. Think of the + as saying that you're coming from the positive side of the number line. The expression $\lim_{x \to a^-} f(x)$ denotes the limit of $f(x)$ as x approaches a from the left, or below, or from the negative end of the number line.

Suppose f is a piecewise function defined by

$$f(x) = \begin{cases} x^2 - 4 & x < 2 \\ 3x - 1 & x \geq 2 \end{cases}$$

If you look for the limit of f as x approaches 2 from the right, you're looking at values of $3x - 1$ when x takes values close to but greater than 2: $\lim_{x \to 2^+} f(x) = 5$. When you approach 2 from the left, however, you're looking at the values $x^2 - 4$ takes for values of x just below 2: $\lim_{x \to 2^-} f(x) = 0$.

When Limits Do Not Exist

There are three situations in which you will find that the limit of a function as x approaches a value a does not exist.

Different from the Left and the Right. The first situation in which the limit does not exist is when the limit from the left and the limit from the right do not agree. Earlier, you saw that for the function

$$f(x)=\begin{cases} x^2-4 & x<2 \\ 3x-1 & x\geq 2 \end{cases}$$

$\lim\limits_{x\to 2^+} f(x)=5$ but $\lim\limits_{x\to 2^-} f(x)=0$. Because the limit from the left and the limit from the right are different, it's impossible to say what $\lim\limits_{x\to 2} f(x)$ is. For this function, $\lim\limits_{x\to 2} f(x)$ does not exist.

Unbounded Behavior. The limit of a function as x approaches a does not exist if, as x gets close to a either from the right or from the left, the function increases without bound or decreases without bound. Generally, this is written using the infinity symbol, as in $\lim\limits_{x\to a^-} f(x)=\infty$, but the definition of *limit* tells you that the limit is a number and ∞ is not a number. If the function is increasing without bound or decreasing without bound, there is no number that can be called the limit. The function $f(x)=\dfrac{1}{x-3}$ has a discontinuity, a break in the graph, at $x=3$, and as x gets close to 3, the values of the function become increasingly large if you approach from above or increasing small if you approach from below: $\lim\limits_{x\to 3^+}\dfrac{1}{x-3}=\infty$ and $\lim\limits_{x\to 3^-}\dfrac{1}{x-3}=-\infty$. Because the function is unbounded near $x=3$, $\lim\limits_{x\to 3}\dfrac{1}{x-3}$ does not exist.

Oscillation. When you look for a limit, you're looking for a trend or pattern to the behavior of the function in a particular little neighborhood. If the function is oscillating, or bouncing up and down rapidly, there is no such trend. The graph in the figure on page 273 shows a function f on the interval from -1 to 1.

If you were asked for $\lim\limits_{x\to \frac{1}{2}} f(x)$, you could use the graph to estimate the limit, but if you were asked to find $\lim\limits_{x\to 0} f(x)$, it would be impossible to say what that limit is, because the graph fluctuates so rapidly near 0; $\lim\limits_{x\to 0} f(x)$ does not exist.

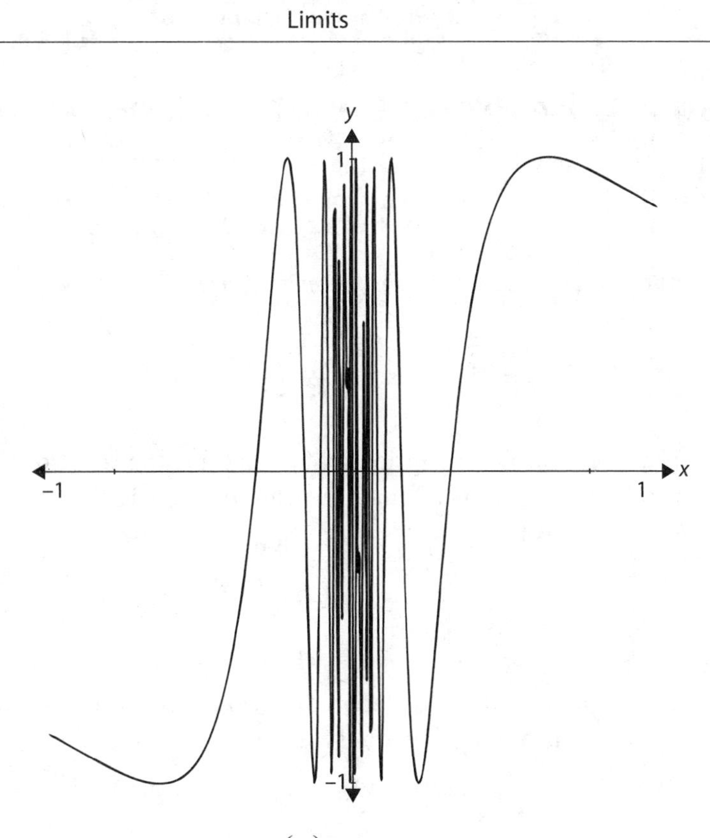

The graph of $f(x) = \sin\left(\frac{1}{x}\right)$ oscillates near $x = 0$

Exercise 17.2

Find each limit or indicate why it does not exist.

1. $\lim\limits_{x\to 3^+} \frac{1}{x-3}$
2. $\lim\limits_{x\to 2^+} \frac{x^2-4}{x+2}$
3. $\lim\limits_{x\to 2^+} \frac{x}{x-2}$
4. $\lim\limits_{x\to 2^-} \frac{x-2}{x}$
5. $\lim\limits_{x\to 3^+} \sqrt{x-3}$
6. $\lim\limits_{x\to 0}\left[\sin\left(\frac{1}{x}\right)\right]$
7. $\lim\limits_{x\to -4} \frac{|x+4|}{x+4}$
8. $\lim\limits_{x\to 0} \frac{x}{4-x}$
9. $\lim\limits_{x\to 4} \frac{4}{4-x}$
10. $\lim\limits_{x\to 0}(\sin|x|)$

Step 3. Use Properties of Limits

Once you know how to find the limit of a simple function, the properties of limits let you break down a more complicated function into manageable pieces.

- The limit of a constant is the constant itself.

$$\lim_{x \to a} c = c$$

$$\lim_{x \to 4} 5 = 5$$

- If a function is multiplied by a constant, the limit of the result is the product of the constant and the limit of the original function.

$$\lim_{x \to a} cf(x) = c \lim_{x \to a} f(x)$$

$$\begin{aligned} \lim_{x \to 3} 5(x-2) &= 5 \lim_{x \to 3}(x-2) \\ &= 5(1) = 5 \end{aligned}$$

- The limit of a sum or difference of two functions is the sum or difference of the limits of the individual functions.

$$\lim_{x \to a} f(x) + g(x) = \lim_{x \to a} f(x) + \lim_{x \to a} g(x)$$

$$\begin{aligned} \lim_{x \to 0} e^x + x^2 &= \lim_{x \to 0} e^x + \lim_{x \to 0} x^2 \\ &= 1 + 0 = 1 \end{aligned}$$

- The limit of a product of two functions is the product of the limits of the individual functions.

$$\lim_{x \to a} f(x) \cdot g(x) = \lim_{x \to a} f(x) \cdot \lim_{x \to a} g(x)$$

$$\begin{aligned} \lim_{x \to 3} x^2 \cdot 2^x &= \lim_{x \to 3} x^2 \cdot \lim_{x \to 3} 2^x \\ &= 9 \cdot 8 = 72 \end{aligned}$$

- The limit of a quotient of two functions is the quotient of the limits of two functions, provided that the limit of the denominator is not 0.

$$\lim_{x \to a} \frac{f(x)}{g(x)} = \frac{\lim_{x \to a} f(x)}{\lim_{x \to a} g(x)}$$

$$\begin{aligned} \lim_{x \to 3} \frac{x^2 + 2}{x^2 - 4} &= \frac{\lim_{x \to 3} x^2 + 2}{\lim_{x \to 3} x^2 - 4} \\ &= \frac{11}{5} \end{aligned}$$

- The limit of a power of a function is the limit of the function raised to that power.

$$\lim_{x\to a}[f(x)]^n = \left[\lim_{x\to a} f(x)\right]^n$$

$$\lim_{x\to 1}[2x-1]^3 = \left[\lim_{x\to 1} 2x-1\right]^3$$

$$= [1]^3 = 1$$

- The limit of the nth root of a function is the nth root of the limit of the function.

$$\lim_{x\to a}\sqrt[n]{f(x)} = \sqrt[n]{\lim_{x\to a} f(x)}$$

$$\lim_{x\to 3}\sqrt[3]{x^2-1} = \sqrt[3]{\lim_{x\to 3} x^2-1}$$

$$= \sqrt[3]{8} = 2$$

Exercise 17.3

Use properties of limits to help you find each limit.

1. $\lim_{x\to 2}(x^2-2x)$
2. $\lim_{x\to 1}(4x^2-3x+1)$
3. $\lim_{x\to 0}[(x-3)(2x+5)]$
4. $\lim_{x\to \frac{\pi}{2}}\left(\frac{\sin x}{x}\right)$
5. $\lim_{x\to 2}\left(\frac{x^2-5}{x+3}\right)$
6. $\lim_{x\to 4}\left(\frac{x-3}{5-x}\right)^{10}$
7. $\lim_{x\to 3}\sqrt{x^2-4x+7}$
8. $\lim_{x\to 1}\frac{x^2-1}{x}$
9. $\lim_{x\to 2}(6x^2-18)$
10. $\lim_{x\to 6}\sqrt[5]{x^2-4}$

Step 4. Evaluate Infinite Limits and Limits at Infinity

The definition of a limit talks about the function values approaching a number L, called the limit, when the values of x approach a number a. When limit notation is used to talk about unbounded behavior, as in $\lim_{x \to 0^+} \frac{1}{x} = \infty$, it takes liberties with the definition, but it's a convenient and compact way to describe unbounded increase or decrease.

Often, the use of limit notation to describe unbounded behavior involves one-sided limits, because the function is increasing without bound on one side of a vertical asymptote and decreasing without bound on the other. If you have a function that increases without bound on both sides, or decreases without bound on both sides of a vertical asymptote, such as $f(x) = \frac{1}{x^2}$, then you can say $\lim_{x \to 0^+} \frac{1}{x^2} = \infty$, $\lim_{x \to 0^-} \frac{1}{x^2} = \infty$, and $\lim_{x \to 0} \frac{1}{x} = \infty$. If the function turns in different directions on either side of a vertical asymptote, such as $\lim_{x \to 0^+} \frac{1}{x} = \infty$ and $\lim_{x \to 0^-} \frac{1}{x} = -\infty$, then $\lim_{x \to 0} \frac{1}{x}$ does not exist.

Using limit notation to describe end behavior of a function also diverges from the actual definition, but as with the infinite limit notation, it is useful. When you write $x \to \infty$ or $x \to -\infty$, you know that ∞ and $-\infty$ are not numbers but symbols indicating that x is increasing to large positive values or decreasing to extremely negative values. You're looking at the end behavior of the function.

If the function values level out approaching a constant as $x \to \infty$ or $x \to -\infty$, that is, if there is a horizontal asymptote, you can say that $\lim_{x \to \infty} f(x) = c$ or $\lim_{x \to -\infty} f(x) = c$. The function $f(x) = \frac{3x^2 - 7}{5 - 2x^2}$ has a horizontal asymptote of $y = -\frac{3}{2}$, and you can say that $\lim_{x \to \infty} \frac{3x^2 - 7}{5 - 2x^2} = -\frac{3}{2}$ and $\lim_{x \to -\infty} \frac{3x^2 - 7}{5 - 2x^2} = -\frac{3}{2}$.

If the function increases without bound or decreases without bound as x goes to infinity or negative infinity, you can cheat the limit definition in both ways to say that concisely: $\lim_{x \to \infty} x^3 - 3x^2 + 2x - 1 = \infty$ and $\lim_{x \to -\infty} x^3 - 3x^2 + 2x - 1 = -\infty$.

Exercise 17.4

Evaluate each limit, if possible.

1. $\lim_{x \to \infty} \frac{5x - 2}{x + 3}$

2. $\lim_{x \to \infty} \frac{x^4 - 1}{2x^3 + 1}$

3. $\lim\limits_{x\to\infty} \dfrac{4-3x^2}{2x+4x^2}$

4. $\lim\limits_{x\to\infty} \dfrac{5x^2-3x+1}{x^2+x+3}$

5. $\lim\limits_{x\to\infty} 2^{-x}$

6. $\lim\limits_{x\to-\infty} (2+e^x)$

7. $\lim\limits_{x\to 3} \dfrac{1}{(x-3)^2}$

8. $\lim\limits_{x\to 0} (\ln x)$

9. $\lim\limits_{x\to 0} \left(\dfrac{8}{x^2}\right)$

10. $\lim\limits_{x\to \frac{\pi}{2}} (\tan x)$

18

Sequences and Series

Most people think of a sequence as simply a list of numbers that have a pattern, but from a mathematical point of view, a sequence is a function with a domain of nonnegative integers. The sequence $\left\{1, \frac{1}{2}, \frac{1}{4}, \frac{1}{8}, \ldots\right\}$, for example, can be seen as the function $f(n) = \frac{1}{2^n}$, defined for integers greater than or equal to 0. Because the domain of positive integers is an infinite set, the sequence is an infinite sequence.

A series is a summation of the terms of a sequence. The series $1 + \frac{1}{2} + \frac{1}{4} + \frac{1}{8} + \cdots + \frac{1}{2^n}$ can be written using an uppercase Greek letter sigma (Σ) to mean "the summation of." The expression $\sum_{k=0}^{n} \frac{1}{2^k}$ indicates the sum of terms of the form $\frac{1}{2^k}$ as k goes from 0 to some value n. An infinite series is an expression of the form $a_1 + a_2 + a_3 + \cdots + a_n + \cdots = \sum_{k=1}^{\infty} a_k$, which adds the terms of an infinite sequence.

At the bottom of the sigma, you'll see the index variable and its starting value. The general form of the terms will be given in terms of the index variable. In this example, the index variable is *k*, and it starts at 0. Commonly, the index will start from 0 or 1, but it can start from any value. Above the sigma, you'll see the final value of the index, if the series is finite, or a ∞, indicating that it's an infinite series.

Step 1. Find Terms of Sequences

As with any other function, a sequence may be specified by an equation or rule. If a sequence is defined as $f(n) = \frac{2^n}{n}$, you can evaluate the 5th term of the sequence by substituting 5 for n: $f(5)=\frac{2^5}{5}=\frac{32}{5}$. The sequence defined by $a(n)=\frac{n^2}{n!}$ has a 6th term equal to $a(6)=\frac{6^2}{6!}=\frac{6\cdot 6}{6\cdot 5\cdot 4\cdot 3\cdot 2\cdot 1}=\frac{1}{20}$.

The symbol $n!$ is read "n factorial" and denotes the product of the integers from n down to 1. The symbol 6! means 6·5·4·3·2·1. It's not uncommon to see factorials in the definitions of sequences.

Some sequences alternate positive and negative terms and so are referred to as alternating sequences. The common shorthand for these changing signs is $(-1)^n$ or $(-1)^{n+1}$, depending on whether the odd or the even terms are negative.

A sequence is called arithmetic if consecutive terms have a common difference. If you begin with 7 and form successive terms by adding 3, to get the sequence {7, 10, 13, 16, 19, … }, you have an arithmetic sequence with a common difference of 3. The nth term of an arithmetic sequence with a common difference d and a first term a_1 is $a_n = a_1 + (n - 1)d$. The 20th term of the sequence {7, 10, 13, 16, 19, … } is $7 + (20 - 1) \cdot 3 = 7 + 19 \cdot 3 = 7 + 57 = 64$.

A sequence in which consecutive terms have a common ratio is called a geometric sequence. In the sequence {5, 10, 20, 40, … } the common ratio is 2. The nth term of a geometric sequence with a first term of a_1 and a common ratio of r is $a_n = a_1 r^{n-1}$. The 10th term of the sequence {5, 10, 20, 40, …} is $5 \cdot 2^{10-1} = 5 \cdot 2^9 = 5 \cdot 512 = 2560$.

Exercise 18.1

Determine whether the sequence is arithmetic, geometric, or neither.

1. {4, −20, 10, −50, 25, −125, …}
2. {4, 7, 10, 13, 16, …}
3. {3, 12, 48, 192, …}

Find the nth *term of the sequence.*

4. $f(n)=\frac{n+1}{n^2}$, $n=10$
5. $f(n)=\frac{1}{\sqrt{n-1}}$, $n=50$
6. $f(n)=\frac{n!}{(n-2)!}$, $n=20$
7. $f(n)=\frac{n(n-1)}{2^n}$, $n=10$

List the first five terms of the sequence.

8. $f(n) = 4 - \frac{1}{n}$

9. $f(n) = \left(\frac{3}{2}\right)^{n-1}$

10. $f(n) = (-1)^n \frac{2^n}{n!}$

Step 2. Find Limits of Sequences

If the terms of a sequence become arbitrarily close to a number, L, as the number of terms becomes large, then L is the limit of the sequence. If a sequence has a limit, if the terms of the sequence approach a number L, then the sequence converges to L. The sequence $\left\{1, \frac{1}{2}, \frac{1}{4}, \frac{1}{8}, \ldots, \frac{1}{2^n}, \ldots\right\}$ converges to 0, because as n becomes large, the fractions $\frac{1}{2^n}$ get close to 0. The sequence $\{1, 2, 4, 8, \ldots, 2^n, \ldots\}$, on the other hand, does not converge.

Because a sequence is actually a function defined on the domain of positive or nonnegative integers, talking about the limit of a sequence is looking at the end behavior of the function. The sequence $\left\{1, \frac{1}{2}, \frac{1}{4}, \frac{1}{8}, \ldots, \frac{1}{2^n}, \ldots\right\}$ can be viewed as a function $f(n) = \frac{1}{2^n}$ for $n \in \{0,1,2,3,\ldots\}$, so $\lim\limits_{n\to\infty} f(n) = \lim\limits_{n\to\infty} \frac{1}{2^n} = \lim\limits_{n\to\infty} 2^{-n}$ describes the end behavior of the function. In this case, $\lim\limits_{n\to\infty} f(n) = \lim\limits_{n\to\infty} \frac{1}{2^n} = \lim\limits_{n\to\infty} 2^{-n} = 0$. The techniques you used to describe the end behavior of functions can be used to find the limit of the sequence.

Exercise 18.2

Find the limit of each sequence, if a limit exists.

1. $\left\{1, -\frac{1}{2}, \frac{1}{4}, -\frac{1}{8}, \ldots\right\}$

2. $\{6, 18, 30, 42, \ldots\}$

3. $\{10, 2, 0.4, 0.08, 0.016, 0.0032, \ldots\}$

4. $\{1.8, 1.4, 1.0, 0.6, \ldots\}$

5. $f(n) = \frac{n^2}{n!}$

6. $f(n) = \frac{n}{2n-1}$

7. $f(n) = \frac{n-3}{(n-1)(n-2)}$

8. $f(n) = 100 - 2(n-1)$

9. $f(n) = \frac{(-1)^{n-1}}{3^n}$

10. $f(n) = \frac{1}{n}$

Step 3. Find Sums of Series

Remember that a series is the sum of the terms of a sequence. If you add a finite number of terms, it's a finite series. If you let it go on forever, it's an infinite series.

For a finite series, the sum of the series is just what it sounds like: the result of adding the terms of the series. $\sum_{k=1}^{5} \frac{1}{2^k} = \frac{1}{2} + \frac{1}{4} + \frac{1}{8} + \frac{1}{16} + \frac{1}{32} = \frac{31}{32}$. For a small number of terms, you can simply make the list and add, but for larger series, you'll need more efficient methods. For an infinite series, the nth partial sum is the result of adding the first n terms.

When you look for the sum of a series, there are several properties of summations that may be helpful.

- The summation of a sum is the sum of the summations.

$$\sum (a_n + b_n) = \sum a_n + \sum b_n$$

- The summation of a difference is the difference of the summations.

$$\sum (a_n - b_n) = \sum a_n - \sum b_n$$

- If each term has a common constant multiple, it can be factored out.

$$\sum ca_n = c\sum a_n$$

The sum of the series

$$\sum_{n=1}^{\infty} \left[5\left(\frac{1}{2}\right)^n - 3\left(\frac{1}{n!}\right) \right] = \sum_{n=1}^{\infty} 5\left(\frac{1}{2}\right)^n - \sum_{n=1}^{\infty} 3\left(\frac{1}{n!}\right)$$

$$= 5\sum_{n=1}^{\infty} \left(\frac{1}{2}\right)^n - 3\sum_{n=1}^{\infty} \left(\frac{1}{n!}\right)$$

If you sum the terms of an arithmetic sequence, you get an arithmetic series. If you add the terms of a geometric sequence, the result is a geometric series. Beyond simple vocabulary, this is important because for arithmetic and geometric series, there are methods of finding summations that are more efficient than just listing terms and adding.

The nth partial sum of an arithmetic sequence with a first term of a_1 and nth term a_n is $S_n = \frac{n}{2}(a_1 + a_n)$. If the common difference of the terms is d, you know that $a_n = a_1 + (n - 1)d$, so the nth partial sum can be expressed $S_n = \frac{n}{2}(a_1 + a_1 + (n-1)d)$ or $S_n = \frac{n}{2}(2a_1 + (n-1)d)$.

The series 7 + 10 + 13 + 16 + 19 + ⋯ or $\sum_{n=1}^{\infty}(7+3(n-1))$ has a 50th term equal to $7 + 3 \cdot 49 = 154$. The 50th partial sum can be written as $\sum_{n=1}^{50}(7+3(n-1))$ and calculated using the formula $S_{50} = \frac{50}{2}(a_1 + a_{50}) = \frac{50}{2}(7+154) = 25(161) = 4025$ or by $S_{50} = \frac{50}{2}(2 \cdot 7 + 49 \cdot 3) = 25(14+147) = 25(161) = 4025$.

The nth partial sum of a geometric series with a first term of a_1 and a common ratio of r is $S_n = \frac{a_1(1-r^n)}{1-r}$. The series 5 + 10 + 20 + 40 + ⋯ can be written as $\sum_{n=1}^{\infty} 5 \cdot 2^{n-1}$. The 12th partial sum of this series can be denoted by $\sum_{n=1}^{12} 5 \cdot 2^{n-1}$ and calculated by the formula

$$S_{12} = \frac{5(1-2^{12})}{1-2} = \frac{5(1-4096)}{-1} = \frac{5(-4095)}{-1} = 20{,}475.$$

Exercise 18.3

Find the nth partial sum of the series.

1. $\sum_{i=1}^{10}[8+12(n-1)]$
2. $\sum_{i=1}^{10}\frac{1}{2^n}$
3. $\sum_{i=1}^{20}3\left(\frac{1}{2}\right)^{n-1}$
4. $\sum_{i=1}^{10}(-2)^n$
5. $\sum_{i=1}^{20}(1.55-.05n)$
6. $\sum_{i-1}^{10}1024\left(\frac{1}{4}\right)^{n-1}$

7. $\sum_{i=1}^{12}\left(\frac{2}{3}\right)^n$

8. $\sum_{i=1}^{100} n$

9. $\sum_{i=1}^{15}(100-2n)$

10. $\sum_{i=1}^{20}\left(10-\frac{2}{n}\right)$

Step 4. Find Sums of Infinite Series

How can you find the sum of an infinite series if it goes on forever? It would seem that it just keeps getting bigger and bigger. For some infinite series, that is true. For others, however, it is possible to get a sum of the infinite series, and the distinction has to do with limits.

To determine if you can find the sum of an infinite series, you need to look at the partial sums. Build a sequence of the partial sums of the series. If a limit, S, of the sequence of partial sums $\{S_1, S_2, S_3, \ldots, S_n, \ldots\}$ exists, the series converges to S. If the sum of an infinite series exists, the series converges to that sum.

The series $1+\frac{1}{2}+\frac{1}{4}+\frac{1}{8}+\cdots+\frac{1}{2^{n-1}}+\cdots=\sum_{k-0}^{n}\frac{1}{2^{k-1}}$ is a convergent series. The partial sums form the sequence $\left\{1,\frac{3}{2},\frac{7}{4},\frac{15}{8},\frac{31}{16},\ldots,\frac{2^n-1}{2^{n-1}},\ldots\right\}$ and $\lim_{n\to\infty}\frac{2^n-1}{2^{n-1}}=2$. The sum of the infinite series $\sum_{k=0}^{n}\frac{1}{2^{k-1}}$ is 2. The series $\sum_{k=0}^{n}\frac{1}{2^{k-1}}$ converges to 2.

If a series converges, the limit of the sequence of partial sums is the sum of the series, but the limit of the terms of the series is $\lim_{k\to\infty} a_k = 0.$ The terms of the series get smaller as k increases. An infinite geometric series will converge when the constant ratio r has an absolute value less than 1, that is, if $|r| < 1$. If the infinite geometric series converges, the sum is $S=\frac{a}{1-r}$.

The infinite geometric series

$$2+\frac{2}{3}+\frac{2}{9}+\frac{2}{27}+\cdots+\frac{2}{3^k}+\cdots=\sum_{n=1}^{\infty}2\left(\frac{1}{3}\right)^{n-1}$$

converges because $r=\frac{1}{3}<1$, and it converges to the sum $\frac{2}{1-\frac{1}{3}}=\frac{2}{\frac{2}{3}}=3.$

Exercise 18.4

Determine whether the series converges, and if it does, find the sum of the series.

1. $\sum_{n=1}^{\infty} \frac{1}{2^{n-1}}$

2. $\sum_{n=1}^{\infty} \frac{1}{n}$

3. $\sum_{n=1}^{\infty} 6\left(\frac{1}{2}\right)^n$

4. $\sum_{n=1}^{\infty} \frac{3^{n-1}}{243}$

5. $\sum_{n=1}^{\infty} 8\left(\frac{3}{4}\right)^{n-1}$

6. $\sum_{n=1}^{\infty} n^2$

7. $\sum_{n=1}^{\infty} 3^n$

8. $\sum_{n=1}^{\infty} \frac{2}{4n^2 - 1}$

9. $\sum_{n=1}^{\infty} \left(\frac{1}{4}\right)^n$

10. $\sum_{n=1}^{\infty} (10 - n)$

Answer Key

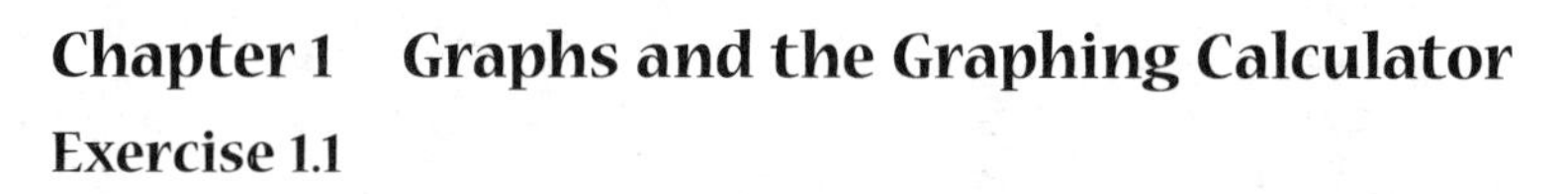

Chapter 1 Graphs and the Graphing Calculator

Exercise 1.1

1.

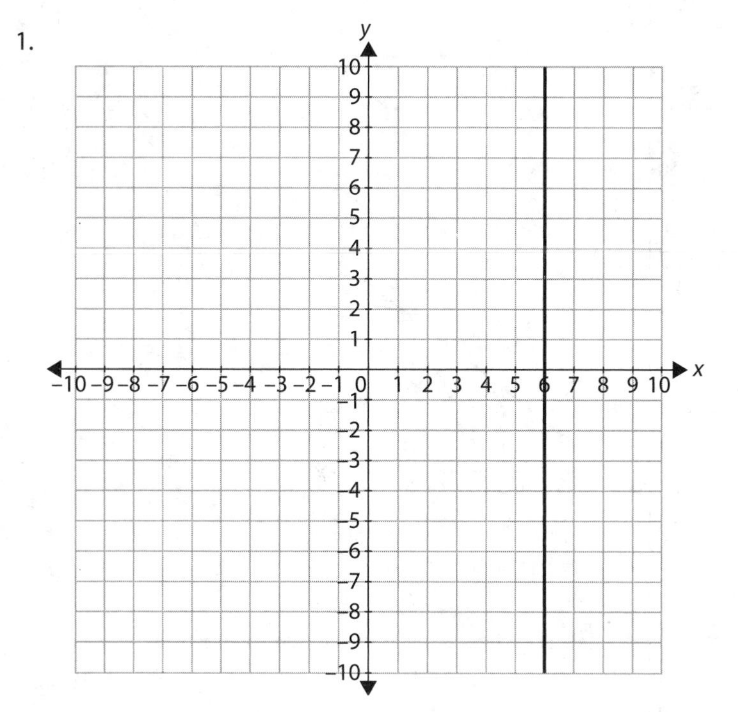

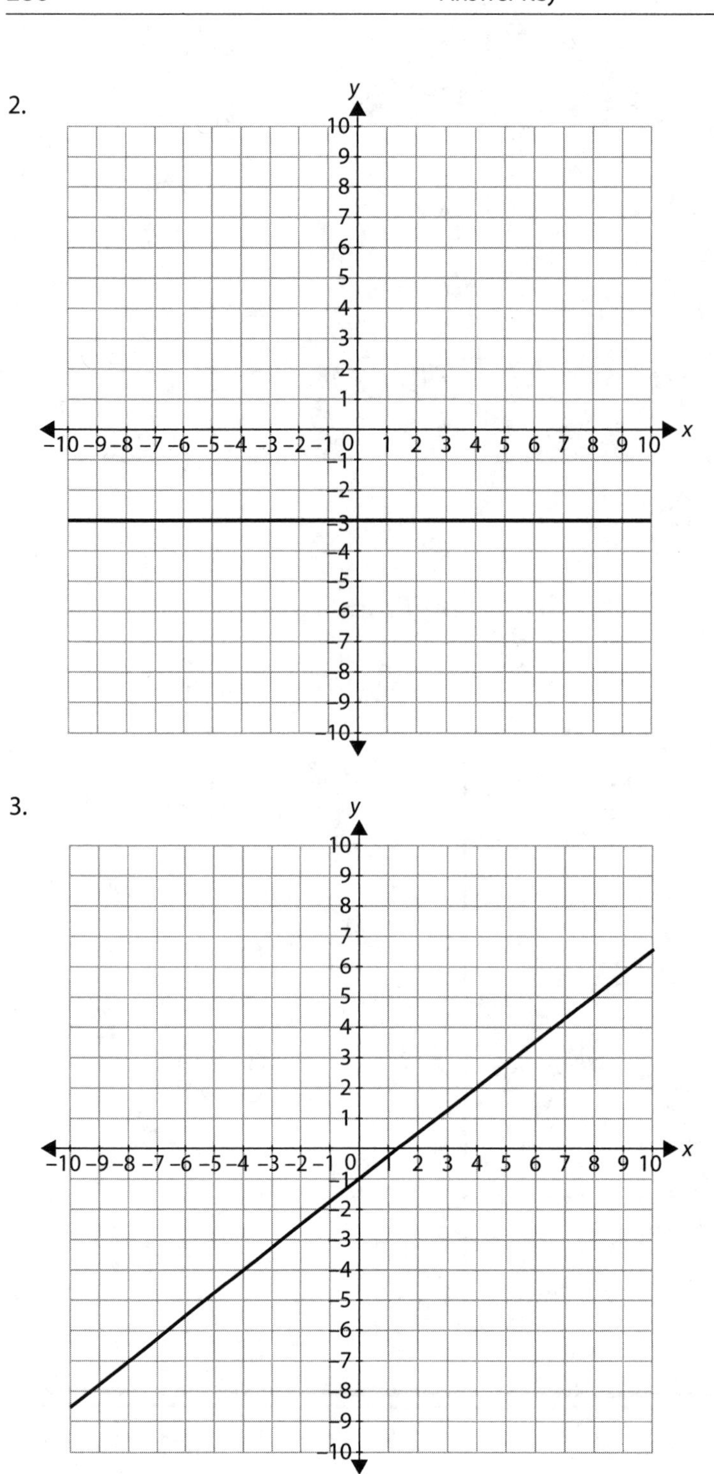
2.
y
x
3.
y
x

4.

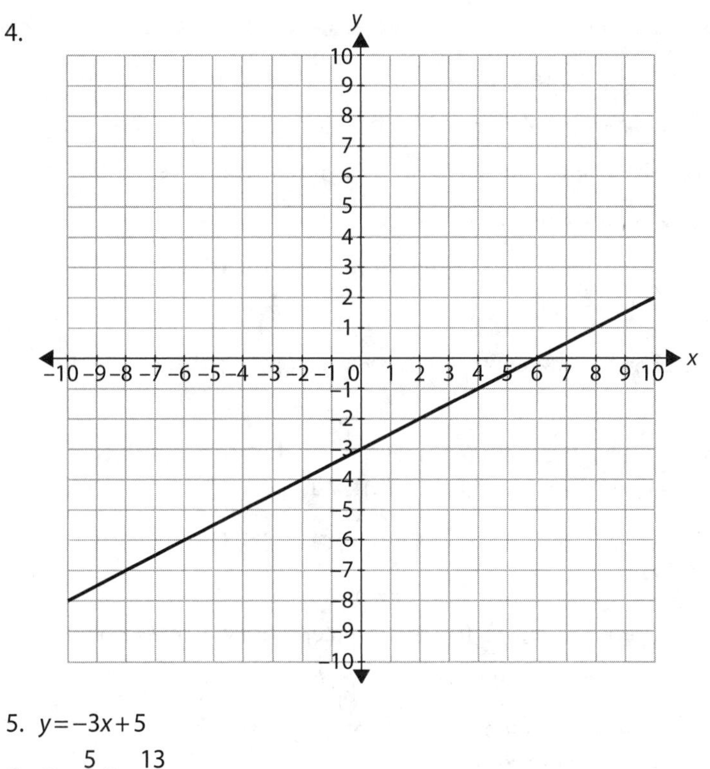

5. $y=-3x+5$
6. $y=\frac{5}{3}x-\frac{13}{3}$
7. $y=x+6$
8. $y=\frac{5}{7}x+9$
9. $y=-\frac{1}{2}x+4$
10. The midpoint of the segment is $\left(\frac{3-5}{2},\frac{-2+6}{2}\right)=(-1,2)$. The slope of the segment is $m=\frac{-2-6}{3+5}=-1$. The equation of the perpendicular bisector of the segment is $y=x+3$.

Exercise 1.2

1. Quadratic
2. Cubic
3. Logarithmic
4. Square root
5. Exponential
6. Rational
7. Cube root
8. Linear
9. Constant
10. Quadratic

Exercise 1.3

1. Domain: $(-\infty,\infty)$; range: $(-\infty,\infty)$; x-intercept: $\left(\frac{9}{2},0\right)$; y-intercept: $(0,9)$; end behavior: $x \to \infty$, $f(x) \to -\infty$; $x \to -\infty$, $f(x) \to \infty$
2. Domain: $(-\infty,\infty)$; range: $[-4,\infty)$; x-intercept: $\left(\pm\frac{2\sqrt{3}}{3},0\right)$; y-intercept: $(0,-4)$; end behavior: $x \to \infty$, $f(x) \to \infty$; $x \to -\infty$, $f(x) \to \infty$
3. Domain: $(-\infty,\infty)$; range: $(-\infty,\infty)$; x-intercept: $(9,0)$; y-intercept: $(0,-36)$; end behavior: $x \to \infty$, $f(x) \to \infty$; $x \to -\infty$, $f(x) \to -\infty$
4. Domain: $\left[-\frac{5}{2},\infty\right)$; range: $[0,\infty)$; x-intercept: $\left(-\frac{5}{2},0\right)$; y-intercept: $(0,\sqrt{5})$; end behavior: $x \to \infty$, $f(x) \to \infty$
5. Domain: $(-\infty,\infty)$; range: $(-\infty,9)$; x-intercept (approximate): $(-1.83,0)$; y-intercept: $(0,-23)$; end behavior: $x \to \infty$, $f(x) \to -\infty$; $x \to -\infty$, $f(x) \to 9$
6. Domain: $(3,\infty)$; range: $(-\infty,\infty)$; x-intercept: $(4,0)$; y-intercept: none; end behavior: $x \to \infty$, $f(x) \to \infty$; $x \to 3$, $f(x) \to -\infty$
7. Domain: $\left(-\infty,\frac{1}{3}\right) \cup \left(\frac{1}{3},\infty\right)$; range: $(-\infty,0) \cup (0,\infty)$; x-intercept: none; y-intercept: $(0,-2)$; end behavior: $x \to \infty$, $f(x) \to 0$; $x \to -\infty$, $f(x) \to 0$
8. Domain: $(-\infty,\infty)$; range: $(-\infty,\infty)$; x-intercept: $(72,0)$; y-intercept: $(0,6)$; end behavior: $x \to \infty$, $f(x) \to -\infty$; $x \to -\infty$, $f(x) \to \infty$
9. Domain: $(-\infty,-4) \cup (-4,1) \cup (1,\infty)$; range $(-\infty,0) \cup (0,\infty)$: $(-\infty,0.21467) \cup (0.74533,\infty)$; x-intercept: $\left(\frac{3}{2},0\right)$; y-intercept: $\left(0,\frac{3}{4}\right)$; end behavior: $x \to \infty$, $f(x) \to 0$; $x \to -\infty$, $f(x) \to 0$
10. Domain: $(-\infty,\infty)$; range: $(-\infty,-6.25]$; x-intercepts: $(2, 0)$ and $(7, 0)$; y-intercept: $(0,-14)$; end behavior: $x \to \infty$, $f(x) \to -\infty$; $x \to -\infty$, $f(x) \to -\infty$

Exercise 1.4

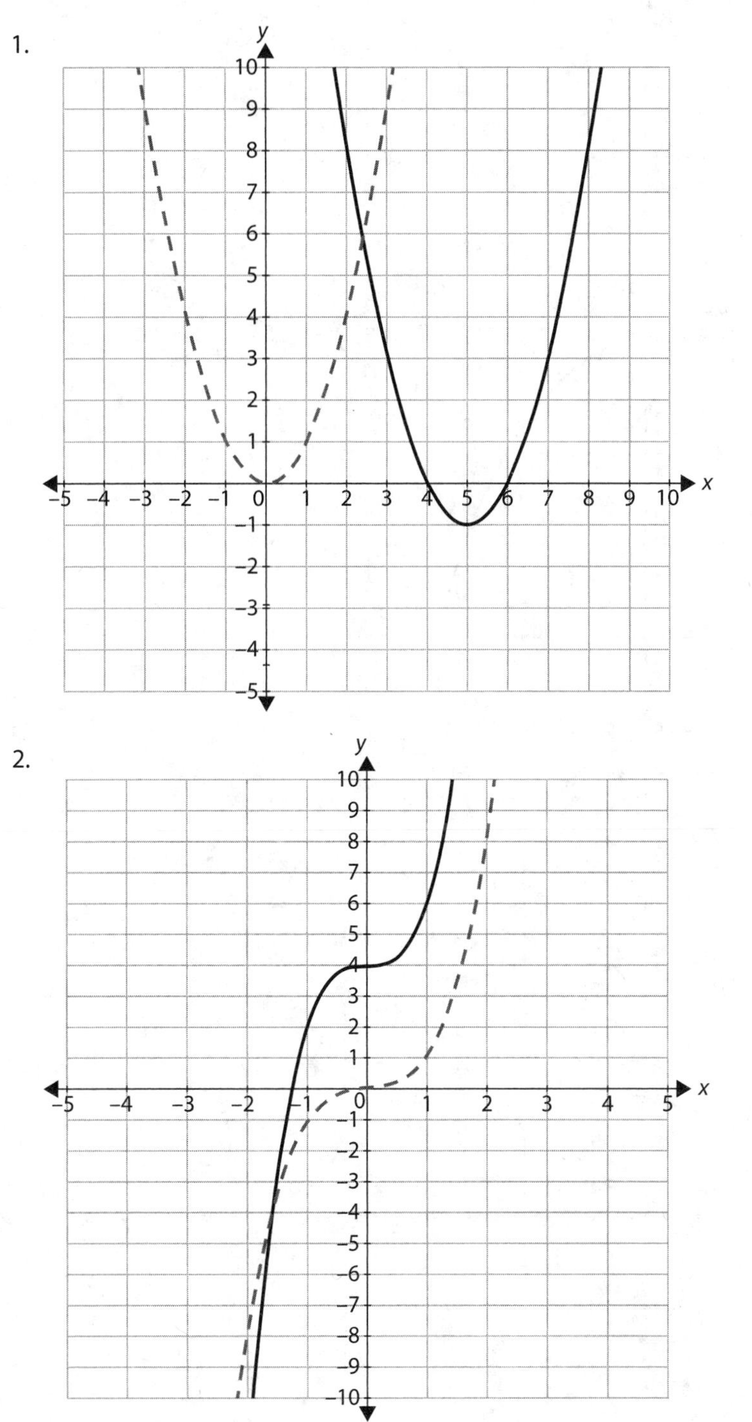

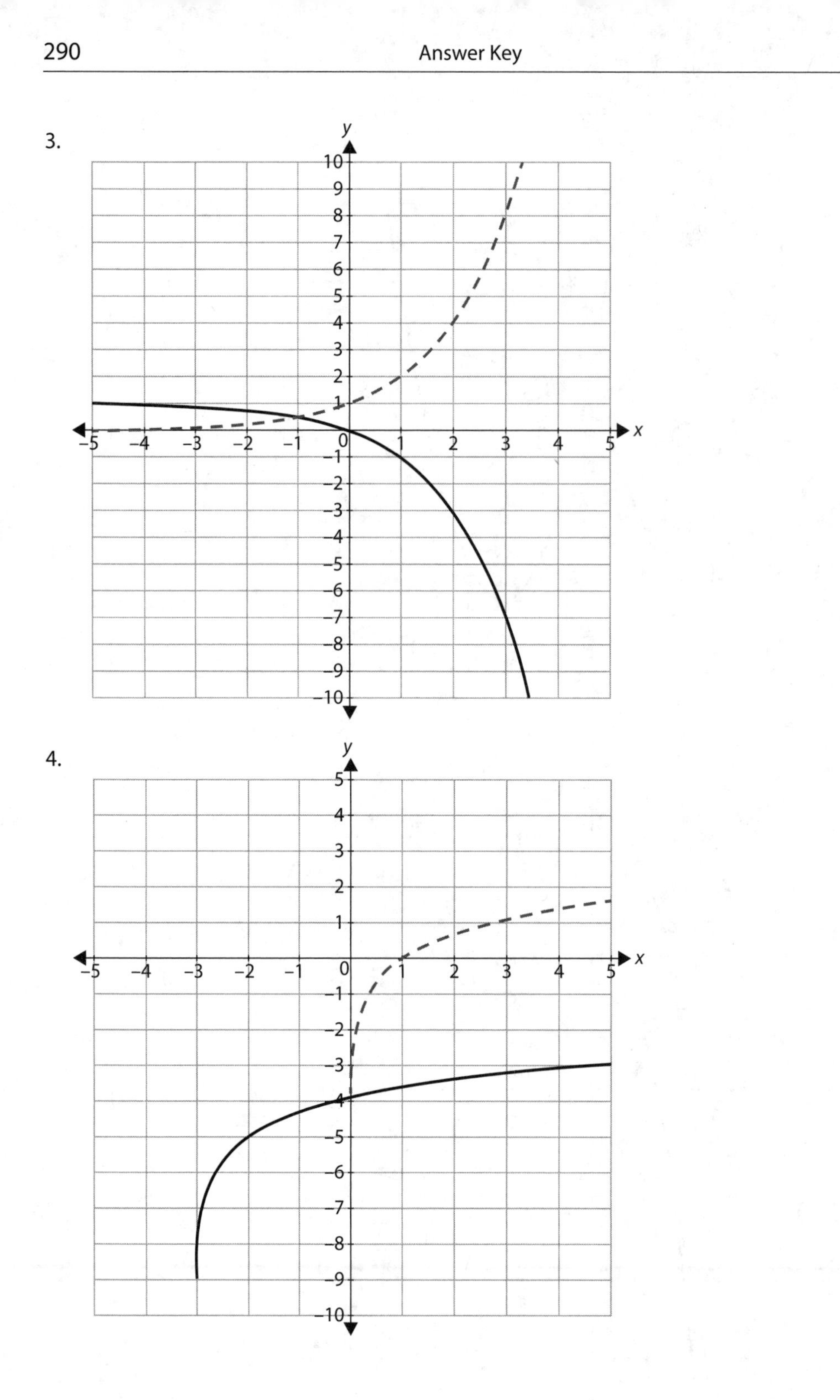
3.
y
x
10
9
8
7
6
5
4
3
2
1
0
−1
−2
−3
−4
−5
−6
−7
−8
−9
−10
−5
−4
−3
−2
−1
1
2
3
4
5
4.
y
x
5
4
3
2
1
0
−1
−2
−3
−4
−5
−6
−7
−8
−9
−10
−5
−4
−3
−2
−1
1
2
3
4
5

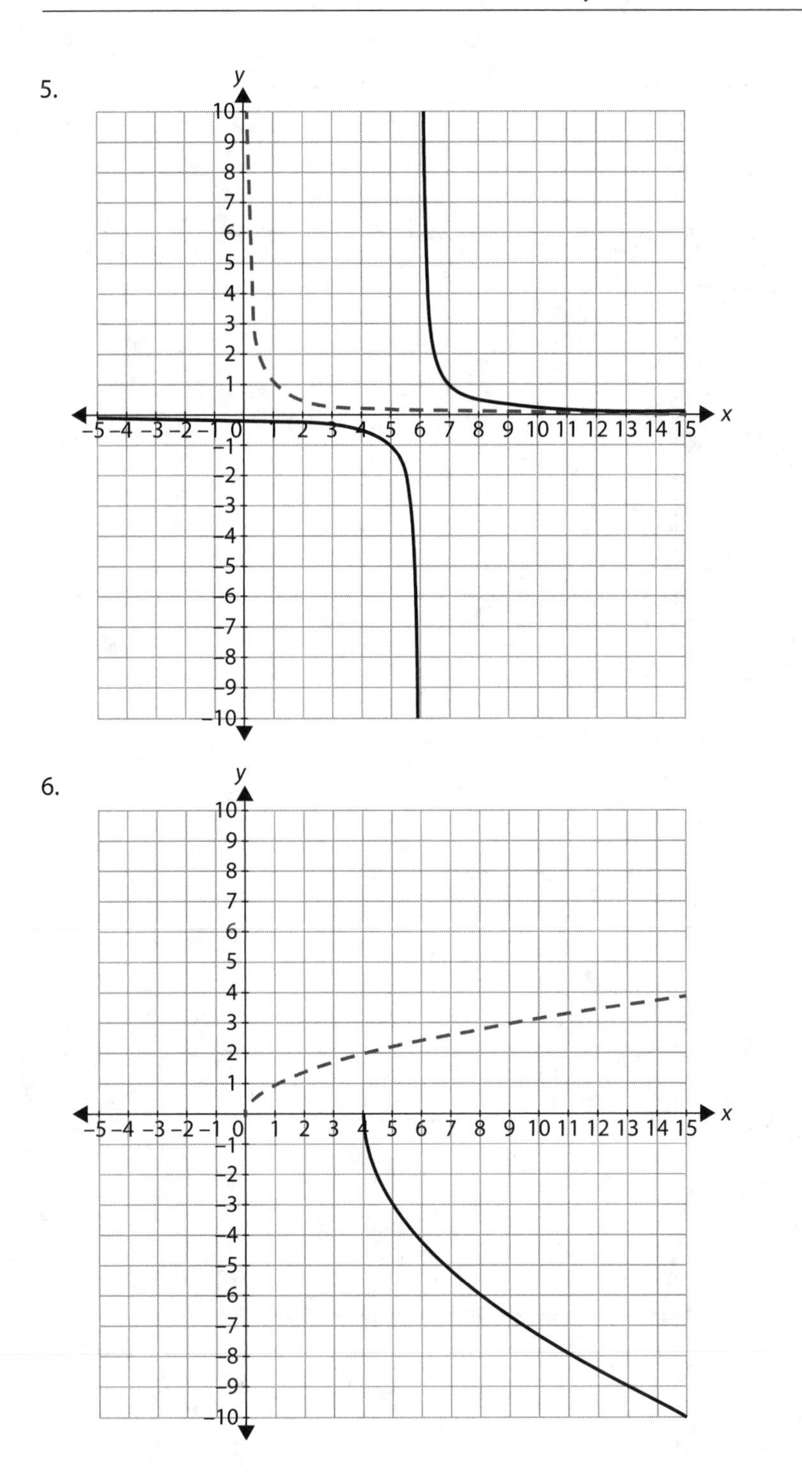
5.
y
x
6.
y
x

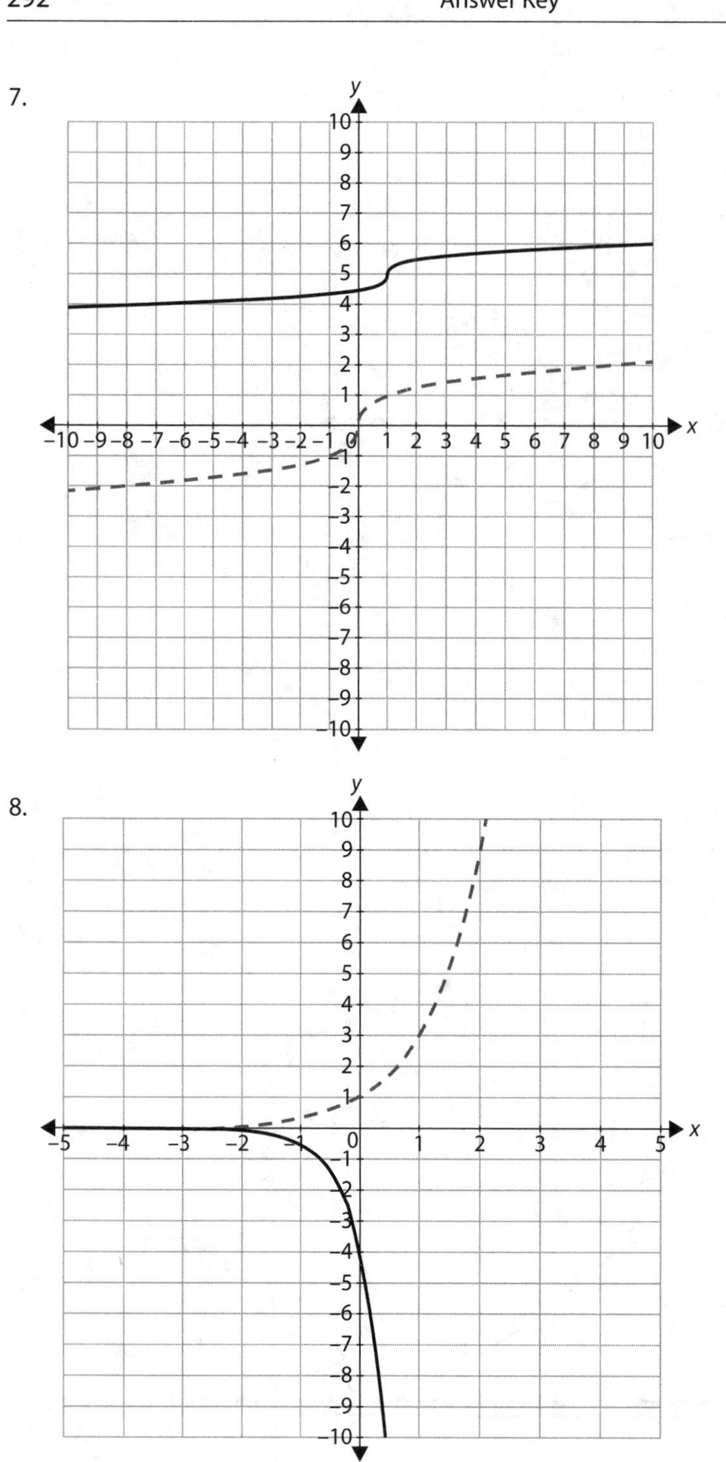
7.
y
x
−10 −9 −8 −7 −6 −5 −4 −3 −2 −1 0 1 2 3 4 5 6 7 8 9 10
10 9 8 7 6 5 4 3 2 1 −1 −2 −3 −4 −5 −6 −7 −8 −9 −10
8.
y
x
−5 −4 −3 −2 −1 0 1 2 3 4 5
10 9 8 7 6 5 4 3 2 1 −1 −2 −3 −4 −5 −6 −7 −8 −9 −10

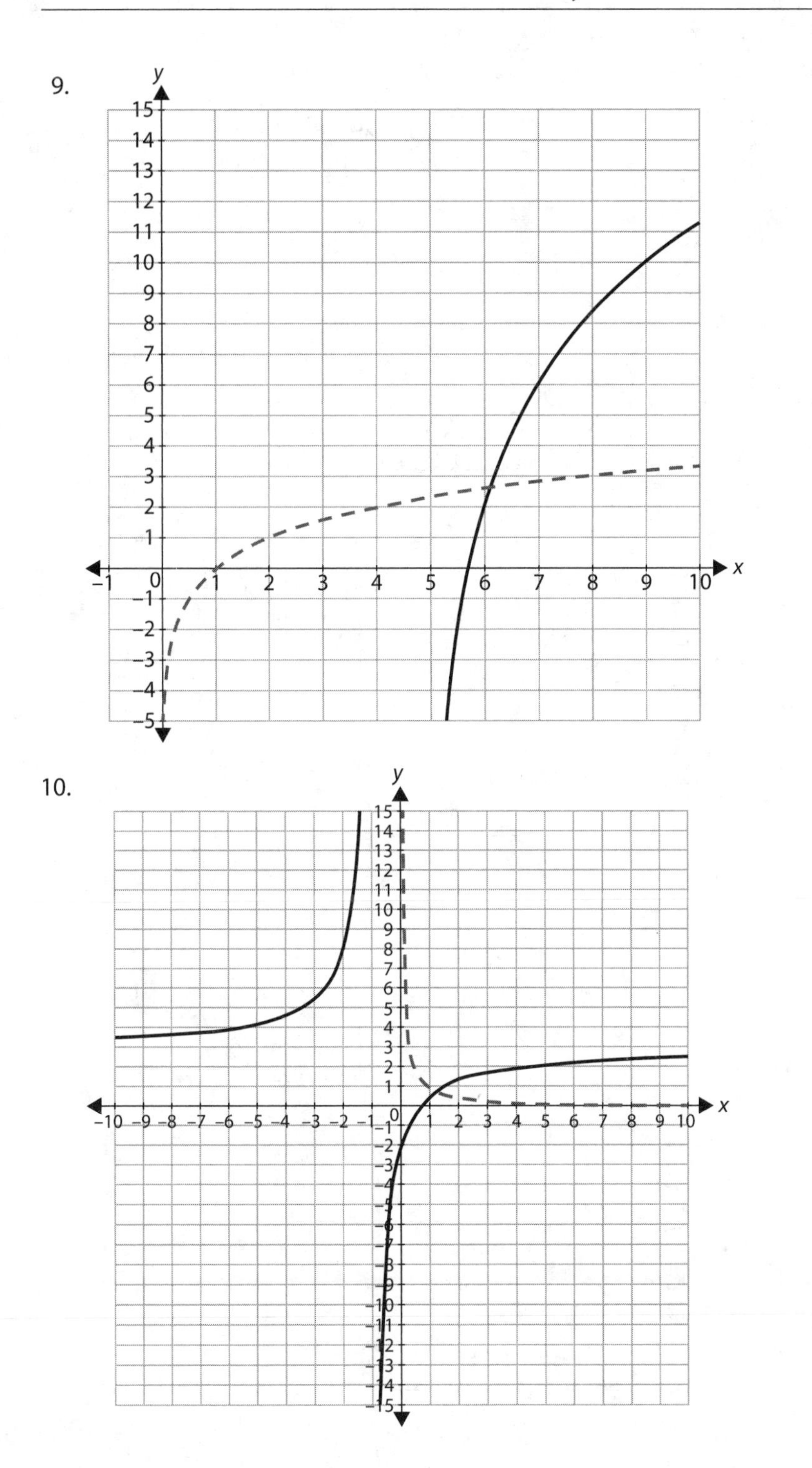
9.
y
x
10.
y
x

Exercise 1.5

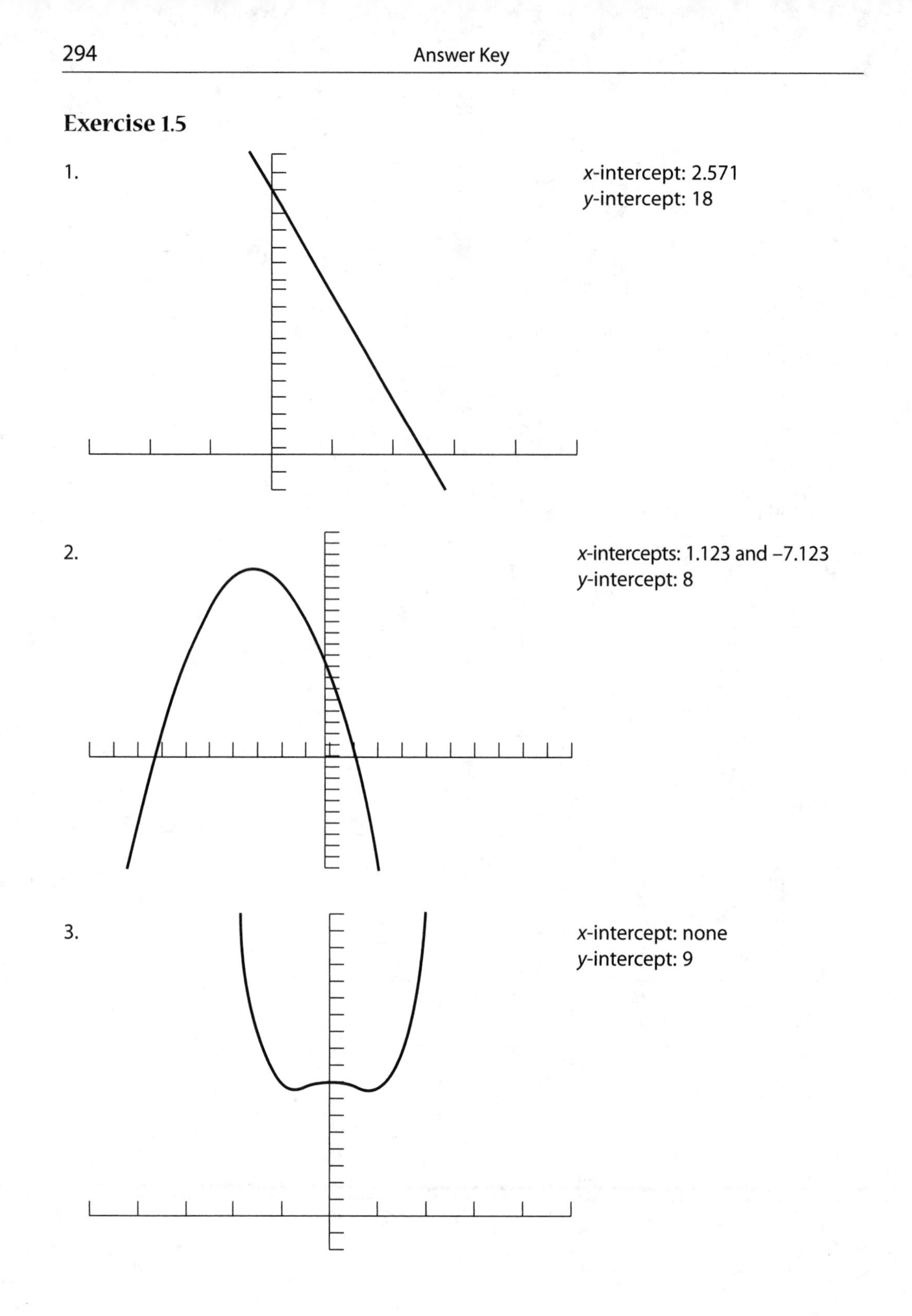

1. x-intercept: 2.571
   y-intercept: 18

2. x-intercepts: 1.123 and −7.123
   y-intercept: 8

3. x-intercept: none
   y-intercept: 9

4.

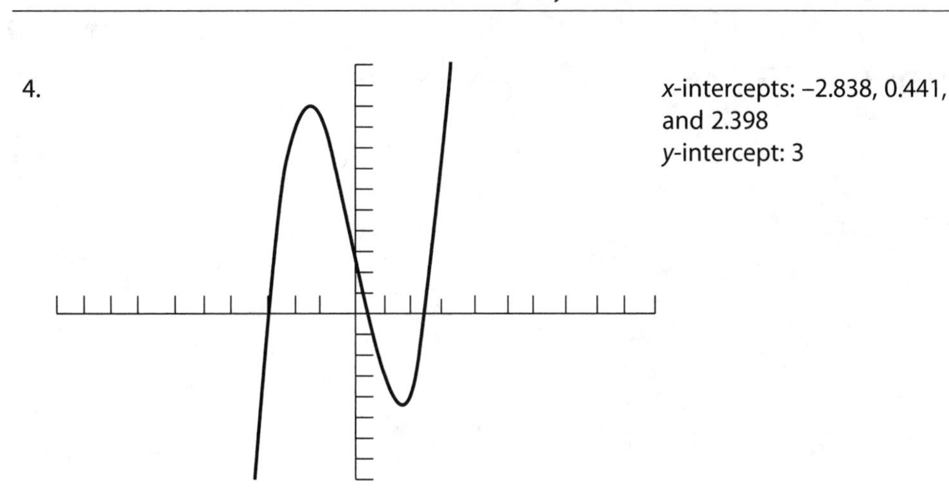

x-intercepts: –2.838, 0.441, and 2.398
y-intercept: 3

5. (14.32,7.15)
6. (2.85,–2.66)
7. (4.125,1.8125)
8. Minimum: $(1.1\overline{6}, 0.91\overline{6})$, no maximum
9. Maximum: (0.451,–1.369), minimum: (2.215,–4.113)
10. Maximum: (0,19), no minimum

Chapter 2 Functions

Exercise 2.1

1. Domain: $(-\infty,\infty)$; range: $(-\infty,\infty)$
2. Domain: $\left[\frac{9}{4},\infty\right]$; range: $[0,\infty)$
3. Domain: $(-\infty,-5)\cup(-5,1)\cup(1,\infty)$; range: $(-\infty,0)\cup(0,\infty)$
4. Minimum: $\left(2,-\frac{1}{3}\right)$; maximum: $\left(-2,\frac{31}{3}\right)$
5. Minima: (2,–25), (–2,–25); maximum: (0,–9)
6. Maximum: (0,9), no minimum
7. Increasing: $(-\infty,0)$; decreasing: $(0,\infty)$
8. Increasing: $(-\infty,-2)$, $(2, \infty)$; decreasing: $(-2,2)$
9. Decreasing: $(-\infty,\infty)$
10. Increasing: $(-2,0)$, $(2,\infty)$; decreasing: $(-\infty,-2)$, $(0,2)$

Exercise 2.2

1. $(f+g)(-2)=1$
2. $(f-g)(1)=-5$
3. $\dfrac{f}{g}(2)$ is undefined.
4. $(f\cdot g)(0)=2$
5.

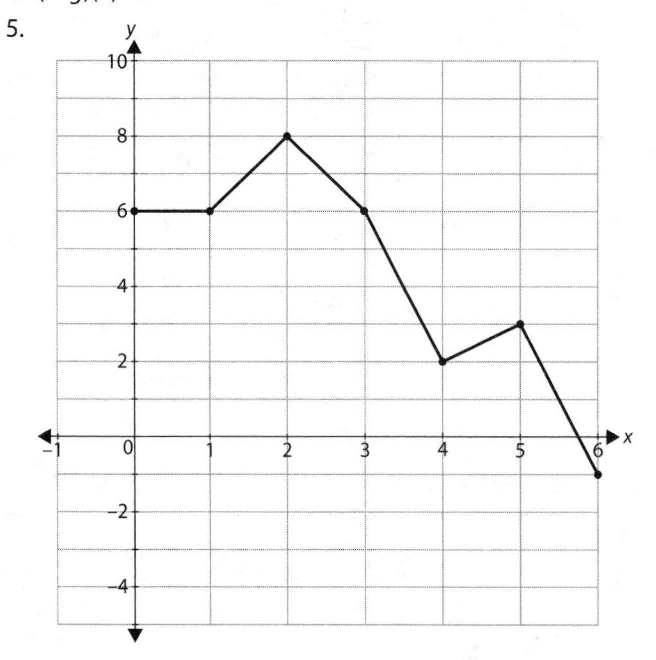

6.

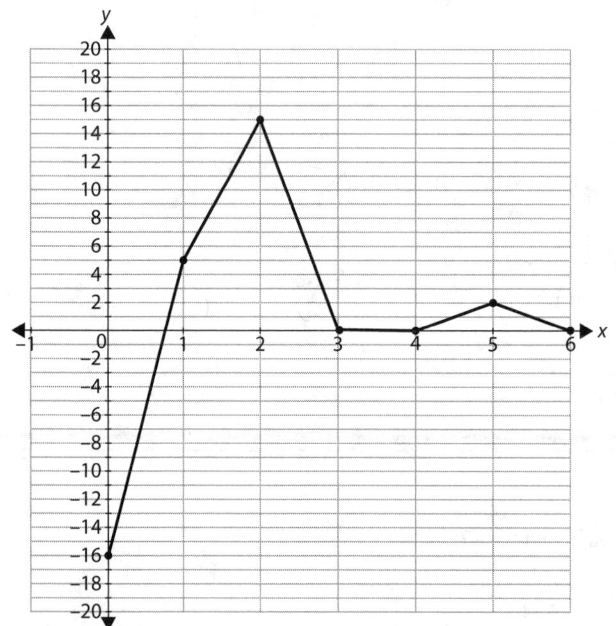

7. $\frac{f}{g}(x) = x - 2$, domain: $(-\infty,-2) \cup (-2,\infty)$

8. $(f-g)(x) = x^2 - x + 8$, domain: $(-\infty,\infty)$

9. $(f+g)(x) = \frac{2x}{x^2-1}$, domain: $(-\infty,-1) \cup (-1,1) \cup (1,\infty)$

10. $(f \cdot g)(x) = x\sqrt{x^2-9}$, domain: $(-\infty,-3) \cup (3,\infty)$

Exercise 2.3

1. $f(g(2)) = f(-1) = 1$
2. $f(g(-2)) = f(-2) = 3$
3. $g(f(1)) = g(-2) = -2$
4. $g(f(-1)) = g(1) = 1$
5. $g(g(0)) = g(2) = -1$
6. $f(g(x)) = x - 6$, domain: $[4,\infty)$
7. $g(f(x)) = \frac{x-1}{3}$, domain: $(-\infty,1) \cup (1,\infty)$
8. $f(g(x)) = x$, domain: $[0, \infty)$
9. $g(f(x)) = \left(x + \frac{1}{x}\right) - \frac{2}{x + \frac{1}{x}} = \frac{(x^2+1)^3 - 2x^3}{x(x^2+1)}$, domain: $(-\infty,0] \cup [0,\infty)$
10. $f(g(f(x))) = 1 - 2x + x^2$, domain: $(-\infty,\infty)$

Exercise 2.4

1. Not inverses
2. Inverses
3. Not inverses
4. Not 1–1
5. 1–1
6. $f^{-1}(x) = \frac{x+1}{3}$, domain: (∞,∞)
7. $f^{-1}(x) = \sqrt{x+4}$, domain: $[-4,\infty)$
8. $f^{-1}(x) = \frac{-1}{3x-2}$, domain: $\left(-\infty, \frac{2}{3}\right) \cup \left(\frac{2}{3}, \infty\right)$
9. $f^{-1}(x) = \frac{x^2+9}{5}$, domain: $[0,\infty)$

10.

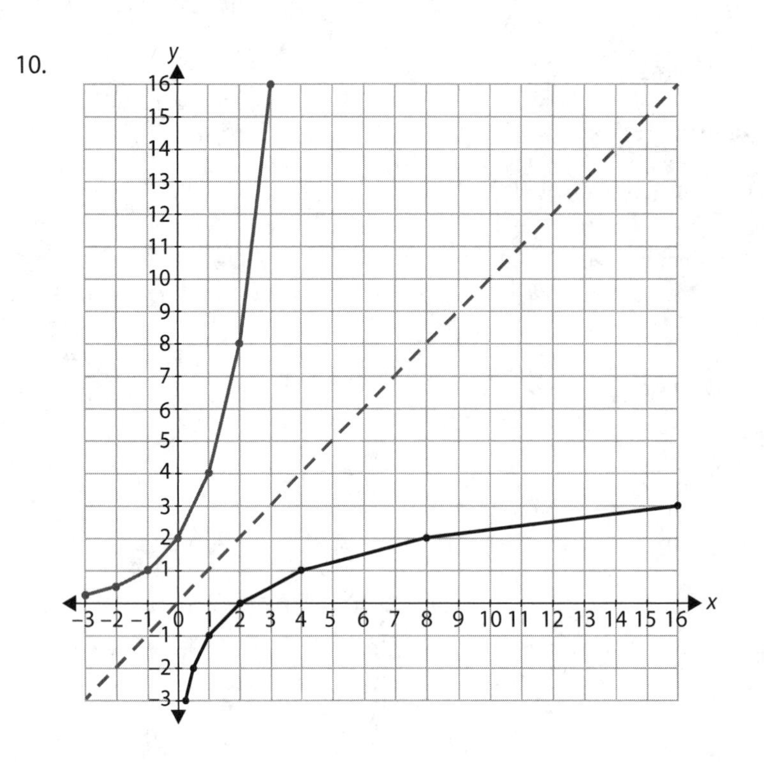

Chapter 3 Quadratic Functions

Exercise 3.1

1. $(2x-5)^2 = 16$
   $2x-5 = \pm 4$
   $2x = 9 \quad 2x = 1$
   $x = \frac{9}{2} \quad x = \frac{1}{2}$

2. $2(5x-7)^2 - 1 = 15$
   $(5x-7)^2 = 8$
   $5x-7 = \pm 2\sqrt{2}$
   $x = \frac{7 \pm 2\sqrt{2}}{5}$

3. $x^2 - 12x - 9 = 0$
   $x^2 - 12x + 36 = 9 + 36$
   $(x-6)^2 = 45$
   $x - 6 = \pm 3\sqrt{5}$
   $x = 6 \pm 3\sqrt{5}$

4. $2x^2 + 8x - 3 = 0$
   $x^2 + 4x = \frac{3}{2}$
   $x^2 + 4x + 4 = \frac{3}{2} + 4$
   $(x+2)^2 = \frac{11}{2}$
   $x + 2 = \pm\sqrt{\frac{11}{2}} = \pm\frac{\sqrt{22}}{2}$
   $x = \frac{-4 \pm \sqrt{22}}{2}$

5. $5x^2+15x+11=0$

$$x^2+3x=-\frac{11}{5}$$

$$x^2+3x+\frac{9}{4}=-\frac{11}{5}+\frac{9}{4}$$

$$\left(x+\frac{3}{2}\right)^2=\frac{1}{20}$$

$$x+\frac{3}{2}=\pm\frac{2\sqrt{5}}{20}=\pm\frac{\sqrt{5}}{10}$$

$$x=\frac{-15\pm\sqrt{5}}{10}$$

6. $3x^2-7x+2=0$

$$x=\frac{7\pm\sqrt{49-4(3)(2)}}{2(3)}=\frac{7\pm\sqrt{25}}{6}=\frac{7\pm5}{6}$$

$$x=\frac{7+5}{6}=2,\ x=\frac{7-5}{6}=\frac{1}{3}$$

7. $8x^2-9=4x$

$$8x^2-4x-9=0$$

$$x=\frac{4\pm\sqrt{16-4(8)(-9)}}{2(8)}=\frac{4\pm\sqrt{16+288}}{16}$$

$$x\approx1.340,\ x\approx-0.840$$

8. $3-7x^2=12x$

$$7x^2+12x-3=0$$

$$x=\frac{-12\pm\sqrt{144-4(7)(-3)}}{2(7)}$$

$$=\frac{-12\pm\sqrt{144+84}}{14}$$

$$x\approx0.221,\ x\approx-1.936$$

9. $2x^2-5x-18=0$

$$(2x-9)(x+2)=0$$

$$x=\frac{9}{2},\ x=-2$$

10. $15x^2-13x-20=0$

$$(5x+4)(3x-5)=0$$

$$x=-\frac{4}{5},\ x=\frac{5}{3}$$

Exercise 3.2

1. $(3-7i)+(2+5i)=5-2i$
2. $(11-2i)-(9+8i)=2-10i$
3. $(4-7i)\ (6+2i)=38-34i$
4. $\dfrac{4-9i}{2+3i}$

$$\frac{4-9i}{2+3i}\frac{(2-3i)}{(2-3i)}=\frac{-19-30i}{13}$$

5. $2i\left(\dfrac{5+3i}{2-i}\right)-(4+3i)\left(\dfrac{6-5i}{2+i}\right)$

$$\frac{10i-6}{2-i}-\frac{39-2i}{2+i}$$

$$\frac{-22+14i}{(2-i)(2+i)}-\frac{76-43i}{(2-i)(2+i)}$$

$$\frac{-98+57i}{5}$$

6. Two nonreal zeros
7. Two real, irrational zeros
8. One real zero with a multiplicity of 2
9. Two real, irrational zeros
10. Two real, rational zeros

Exercise 3.3

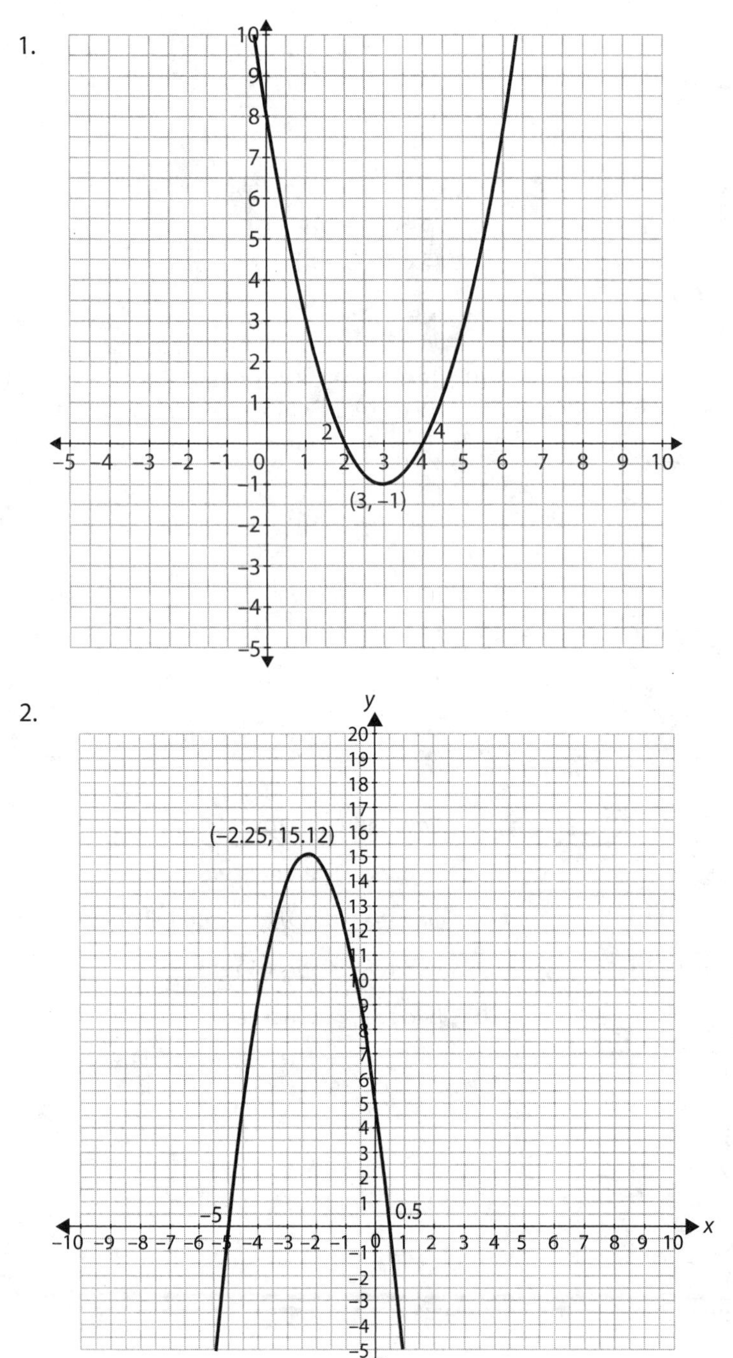

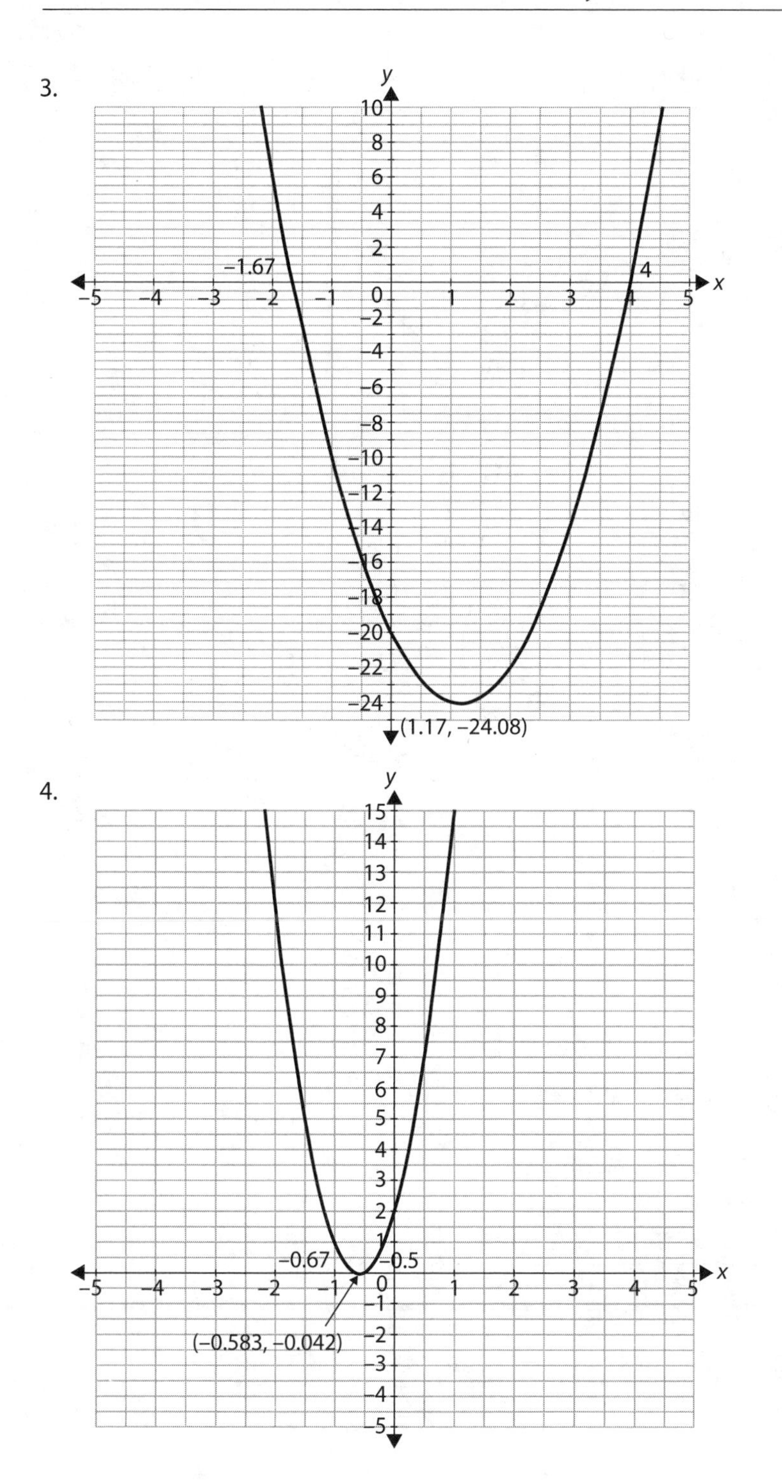
3.
y
x
−1.67
4
(1.17, −24.08)
4.
y
x
−0.67
−0.5
(−0.583, −0.042)

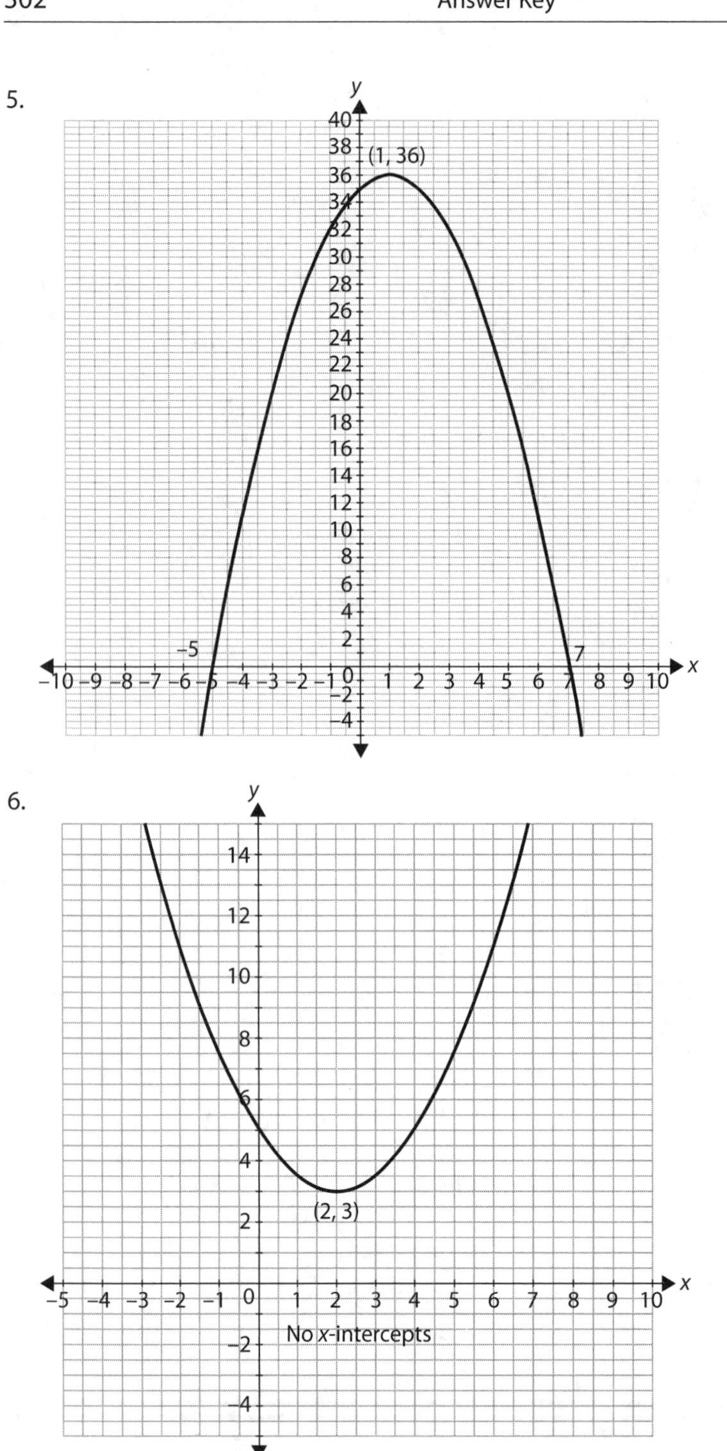
5.
(1, 36)
−5
7
6.
(2, 3)
No x-intercepts

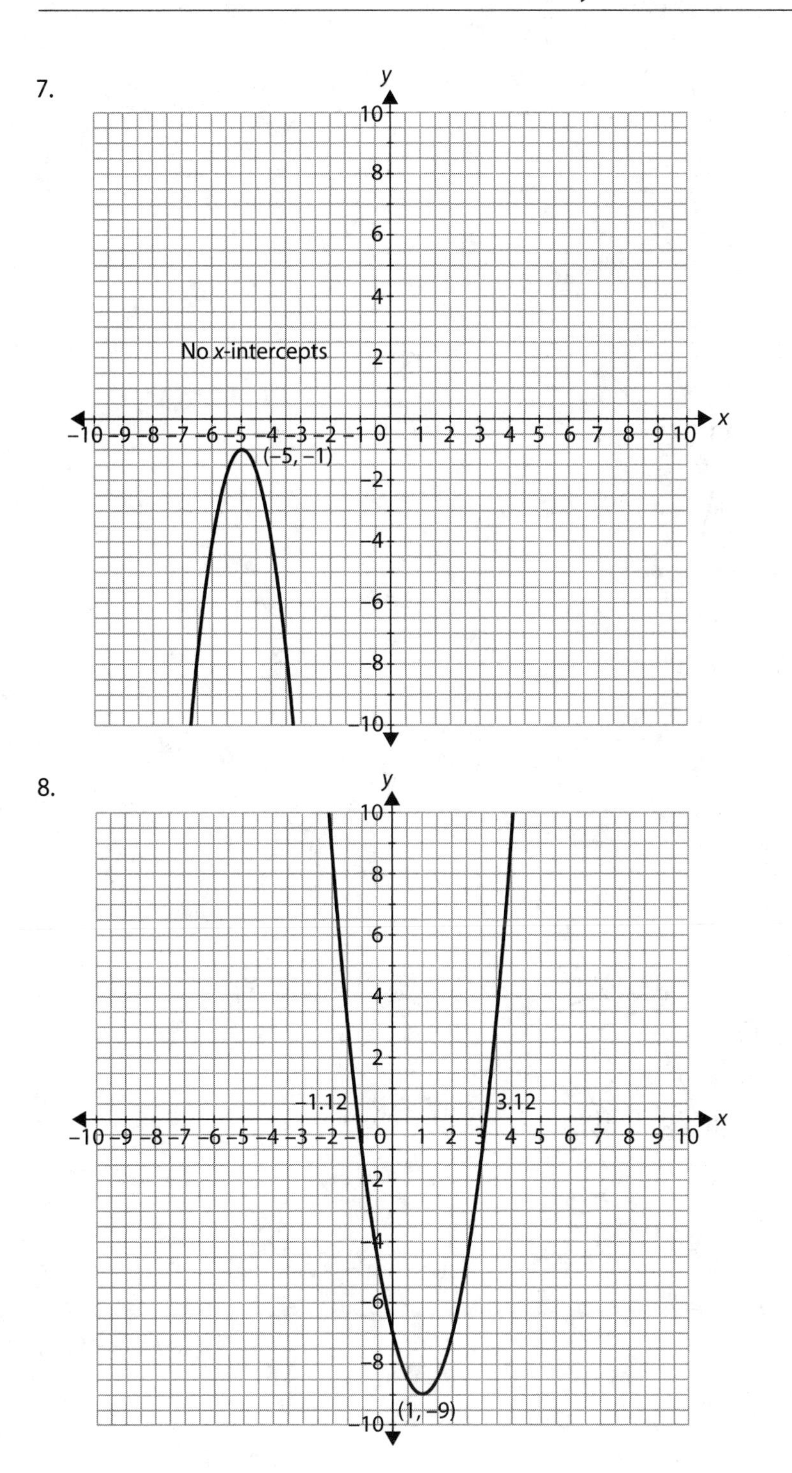
7.
No x-intercepts
(−5, −1)
8.
−1.12
3.12
(1, −9)

9.

10.

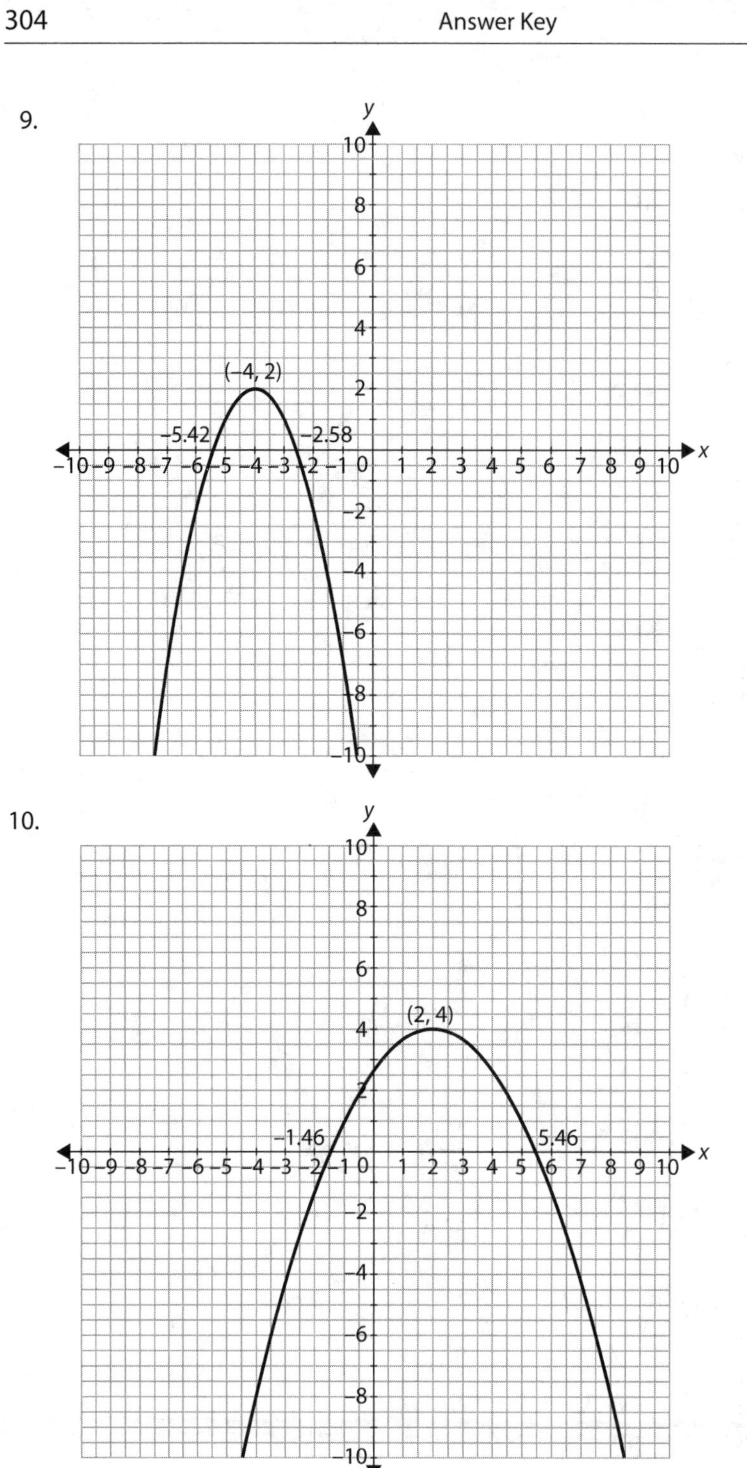

Exercise 3.4

1. $h(t) = -16t^2 + 120$

   $0 = -16t^2 + 120$

   $t^2 = \frac{120}{16} = \frac{15}{2}$

   $t \approx 2.739$ seconds

2. $h(t) = -16t^2 + 40t$; maximum height is 25 feet.
3. $h(t) = -16t^2 + 33t + 7$; ball hits the ground after approximately 2.26 seconds.
4. $A(x) = x(400 - 2x) = 400x - 2x^2$; maximum area is achieved when dimensions are 100 feet by 200 feet.
5. $V(x) = 6x(30 - x) = 180x - 6x^2$; maximum volume is achieved by a carton with a square base 15 inches on a side.
6. Area of the patio is $A(x) = (60 + 2x)(40 + 2x) - 2400$. Solve $(60 + 2x)(40 + 2x) - 2400 = 2400$ to find that the width of the patio is 10 feet.
7. $y = -\frac{1}{7}(x - 4)^2 + 5$
8. $y = \frac{1}{7}(x + 3)^2 - 2$
9. $y = 3x^2 - 5x + 2$
10. $y = -0.464x^2 + 7.857x - 2.821$

Chapter 4 Polynomial Functions

Exercise 4.1

1. $x^3 - 5x^2 + 9x - 45$

   $x^2(x - 5) + 9(x - 5)$

   $(x - 5)(x^2 + 9)$

2. $8x^3 - 27$

   $(2x - 3)(4x^2 + 6x + 9)$

3. $3x^3 + 21x^2 - 54x$

   $3x(x^2 + 7x - 18)$

   $3x(x + 9)(x - 2)$

4. $x^4 - 5x^2 - 14$

   $(x^2 - 7)(x^2 + 2)$

5. $10x^2 - 13x - 3$

   $(5x + 1)(2x - 3)$

6. $125 + 64x^3$

   $(5 + 4x)(25 - 20x + 16x^2)$

7. $x^3 + 5x^2 - 4x - 20$

   $x^2(x + 5) - 4(x + 5)$

   $(x + 5)(x^2 - 4)$

   $(x + 5)(x + 2)(x - 2)$

8. $3x^7 + 21x^4 - 24x$

   $3x(x^6 + 7x^3 - 8)$

   $3x(x^3 + 8)(x^3 - 1)$

   $3x(x + 2)(x^2 - 2x + 4)(x - 1)(x^2 + x + 1)$

9. $9x^2 - 25$

   $(3x + 5)(3x - 5)$

10. $16x^2 - 8x + 1$

    $(4x - 1)^2$

Exercise 4.2

1. Quotient: x^2-2x+1, remainder: 0
2. Quotient: $x^4-x^3+5x^2-8x+8$, remainder: -3
3. Quotient: $x^3+5x^2+25x+125$, remainder: 600
4. Not a factor, remainder $=15$
5. Factor
6. Factor
7. Not a factor, remainder $=129$
8. Not a factor, remainder $=-1056$
9. $p(-2)=-95$
10. $p(4)=139$

Exercise 4.3

1. $p(x)=2x^3-3x^2+4x-3=(2x-3)(x^2-5)+(14x+12)$
2. $p(x)=x^3-9x+1=x(x^2+2)+(-11x+1)$
3. $p(x)=4x^5-6x^4+2x^3-5x^2+2x-1$
   $=(2x^3-3x^2+2x-4)(2x^2-1)+(4x-5)$
4. $p(x)=9x^4+7x^2+8=\left(3x^2+\frac{1}{3}\right)(3x^2+2)+7\frac{1}{3}$
5. $p(x)=x^6-x^4+x^2+1=(x^3-x-1)(x^3+1)+(x^2+x+2)$
6. Factor
7. Not a factor
8. Not a factor
9. Factor
10. Factor

Exercise 4.4

1. Three zeros; possible rational zeros: $\pm1, \pm5, \pm25, \pm\frac{1}{3}, \pm\frac{5}{3}, \pm\frac{25}{3}$
2. Five zeros; possible rational zeros: ±1, ±2, ±4, ±5, ±8, ±10, ±16, ±20, ±40, ±80
3. Four zeros; possible rational zeros: $\pm1, \pm2, \pm\frac{1}{3}, \pm\frac{2}{3}$
4. At most three positive real zeros; no negative real zeros
5. At most two positive real zeros; at most two negative real zeros

6. At most three positive real zeros; at most one negative real zero

7. $f(x) = x^3 - 5x^2 - 15x + 27$

$$= (x+3)(x^2 - 8x + 9)$$
$$= (x+3)(x - 4 - \sqrt{7})(x - 4 + \sqrt{7})$$

8. $f(x) = 2x^3 - 5x^2 + 12x - 5$

$$= \left(x - \frac{1}{2}\right)(x - 1 - 2i)(x - 1 + 2i)$$

9. $f(x) = x^4 + 10x^2 + 9$

$$= (x^2 + 9)(x^2 + 1)$$
$$= (x + 3i)(x - 3i)(x + i)(x - i)$$

10. $f(x) = x^4 - 4x^3 - 2x^2 + 12x - 16$

$$= (x+2)(x-4)(x^2 - 2x + 2)$$
$$= (x+2)(x-4)(x - 1 - i)(x - 1 + i)$$

Exercise 4.5

1.

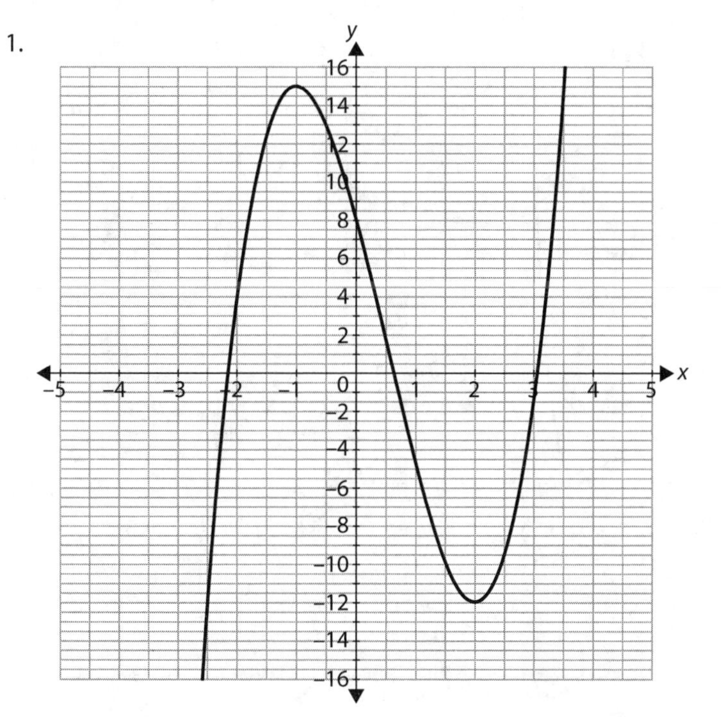

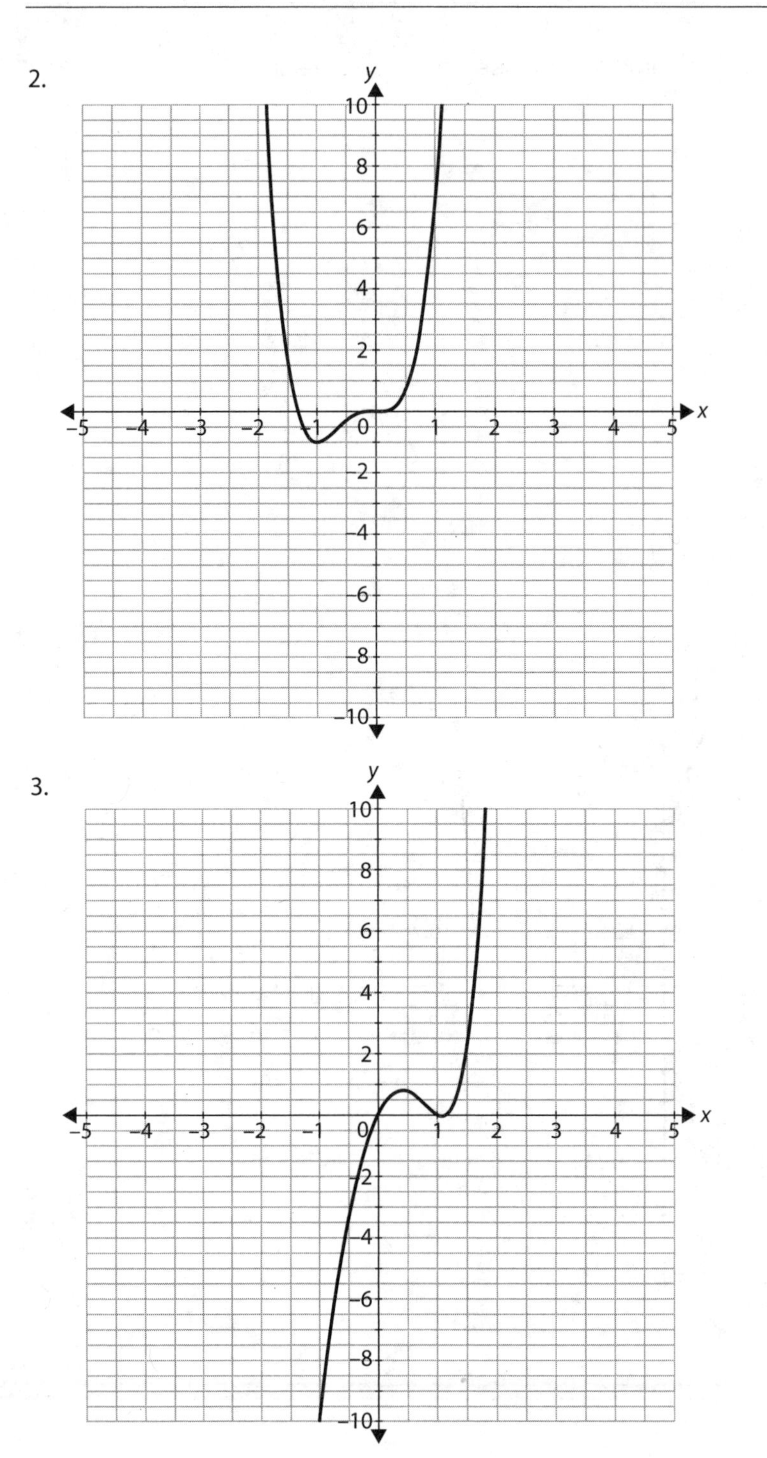
2.
y
x
3.
y
x

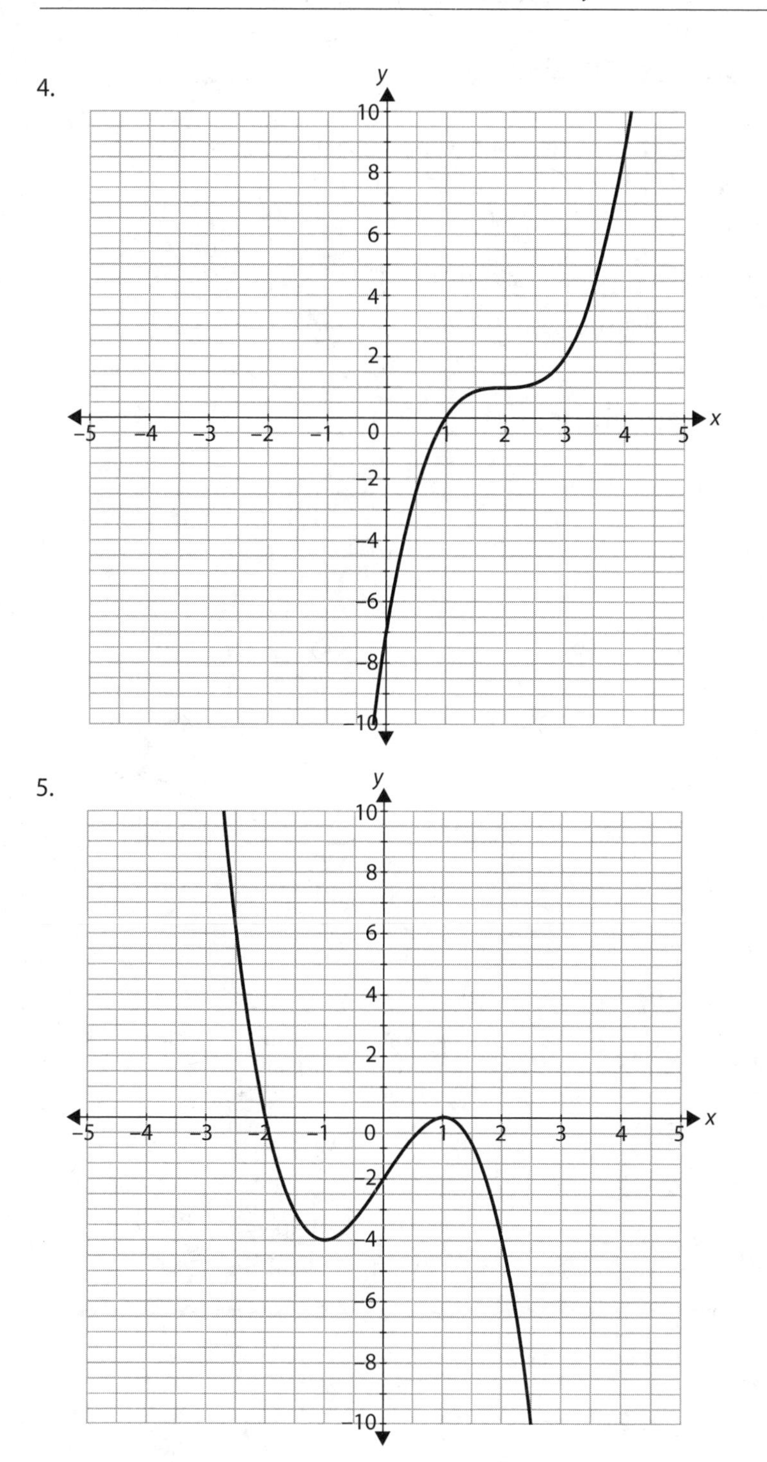
4.
y
10
8
6
4
2
0
−2
−4
−6
−8
−10
x
−5
−4
−3
−2
−1
1
2
3
4
5
5.
y
10
8
6
4
2
0
−2
−4
−6
−8
−10
x
−5
−4
−3
−2
−1
1
2
3
4
5

6. x-intercepts: (1,0), (−1,0); y-intercept: (0,1); end behavior: $x \to \infty$, $f(x) \to -\infty$; $x \to -\infty$, $f(x) \to -\infty$

7. x-intercepts: (0,0), $(\sqrt{2},0)$, $(-\sqrt{2},0)$; y-intercept: (0, 0); end behavior: $x \to \infty$, $f(x) \to \infty$; $x \to -\infty$, $f(x) \to -\infty$

8. Maximum: (0,0); minimum: (2,–4)

9. Maximum: (1.155,3.079); minimum: (−1.155,–3.079)

10. Maximum: (0.5,0.0625); minima: (0,0), (1,0)

Chapter 5 Rational Functions

Exercise 5.1

1. $\dfrac{x-3}{(x+1)(x+2)}$

2. $x+1$

3. $\dfrac{3x^2-5x+5}{(x+5)(x-2)}$

4. $\dfrac{x^2+4x+9}{x(x+3)}$

5. $\dfrac{x^2+3x-4}{3x^2+5x-12}$

6.
$$\frac{3x}{x-2}+\frac{5}{x+5}=\frac{7x}{x^2+3x-10}$$
$$3x(x+5)+5(x-2)=7x$$
$$3x^2+15x+5x-10=7x$$
$$3x^2+13x-10=0$$
$$(3x-2)(x+5)=0$$
$$3x-2=0 \qquad x+5=0$$
$$x=\frac{2}{3} \qquad \cancel{x=-5}$$

7.
$$\frac{3}{x+2}-\frac{39}{x^2+2x}=\frac{-2}{x}$$
$$3x-39=-2(x+2)$$
$$3x-39=-2x-4$$
$$5x=35$$
$$x=7$$

8.
$$\frac{4x}{x+4}+\frac{3}{x-1}=\frac{15}{x^2+3x-4}$$
$$4x(x-1)+3(x+4)=15$$
$$4x^2-4x+3x+12=15$$
$$4x^2-x-3=0$$
$$(4x+3)(x-1)=0$$
$$4x+3=0 \qquad x-1=0$$
$$x=-\frac{3}{4} \qquad \cancel{x=1}$$

9.
$$\frac{2x}{4-x}=\frac{x^2}{x-4}$$
$$\frac{-2x}{x-4}=\frac{x^2}{x-4}$$
$$-2x=x^2$$
$$x^2+2x=0$$
$$x(x+2)=0$$
$$x=0 \qquad x=-2$$

10.
$$\frac{2}{x-10}-\frac{3}{x-2}=\frac{6}{x^2-12x+20}$$
$$2(x-2)-3(x-10)=6$$
$$2x-4-3x+30=6$$
$$-x+26=6$$
$$-x=-20$$
$$x=20$$

Exercise 5.2

1. $(-\infty,3)\cup(3,\infty)$
2. $(-\infty,0)\cup(0,\infty)$
3. $(-\infty,\infty)$
4. $x=6, x=-5$
5. $x=\frac{9}{4}$
6. $x=1$: essential, $x=-1$: essential
7. $x=3$: essential, $x=-3$: essential
8. $x=3$: removable, $x=-4$: essential
9. $g(x)=\frac{2}{3x+2}$
10. $g(x)=\frac{x-4}{x+4}$

Exercise 5.3

1. $y=0$
2. $y=0$
3. $y=-2$
4. $y=x-6$
5. $y=x$
6. $y=2x-3$
7. As $x\to\infty$, $f(x)\to\infty$, and as $x\to-\infty$, $f(x)\to\infty$.
8. As $x\to\infty$, $f(x)\to-\infty$, and as $x\to-\infty$, $f(x)\to-\infty$.
9. As $x\to\infty$, $f(x)\to\infty$, and as $x\to-\infty$, $f(x)\to-\infty$.
10. As $x\to\infty$, $f(x)\to 0$, and as $x\to-\infty$, $f(x)\to 0$.

Exercise 5.4

1. x-intercepts: none; y-intercept: $\left(0,\frac{1}{2}\right)$
2. x-intercept: $\left(\frac{5}{3},0\right)$; y-intercept: $\left(0,-\frac{5}{2}\right)$
3. x-intercepts: $(-3,0)$, $(7,0)$; y-intercept: $(0,-3)$
4. x-intercept: $\left(\frac{1}{2},0\right)$; y-intercept: none
5. x-intercepts: $(0,0)$, $(1,0)$; y-intercept: $(0,0)$

Exercise 5.5

1. Vertical asymptote: $x=1$; horizontal asymptote: $y=0$; x-intercepts: none; y-intercept: -3

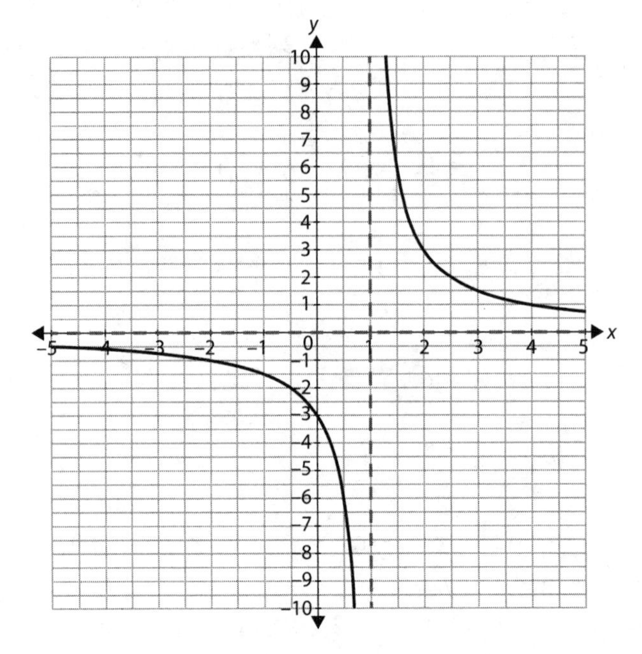

2. Vertical asymptote: $x=-5$; horizontal asymptote: $y=0$; x-intercepts: none; y-intercept: -0.2

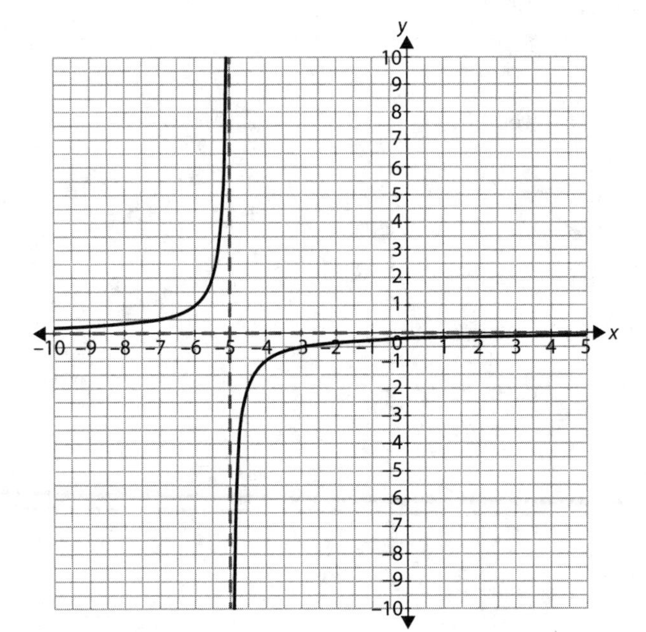

3. Vertical asymptote: $x = 2$; horizontal asymptote: $y = 1$; x-intercept: -3; y-intercept: -1.5

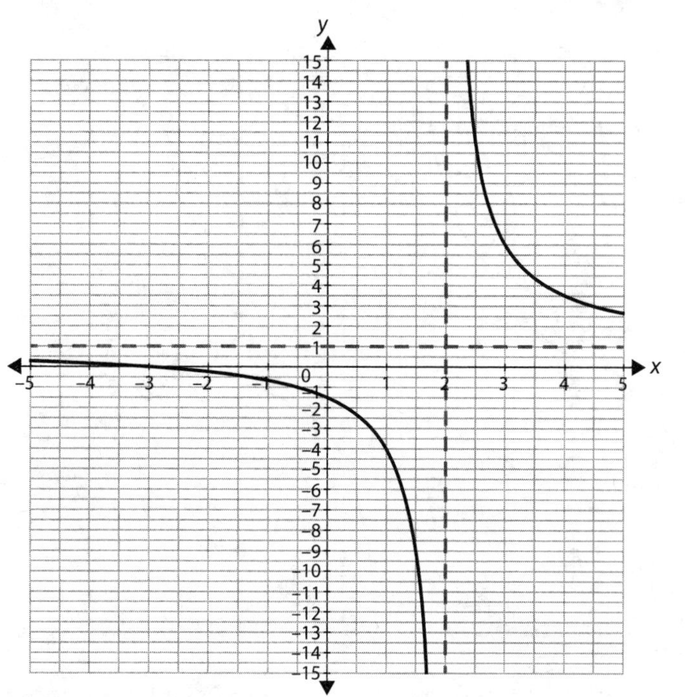

4. Vertical asymptote: $x = -3$; horizontal asymptote: $y = 1$; x-intercept: 0; y-intercept: 0

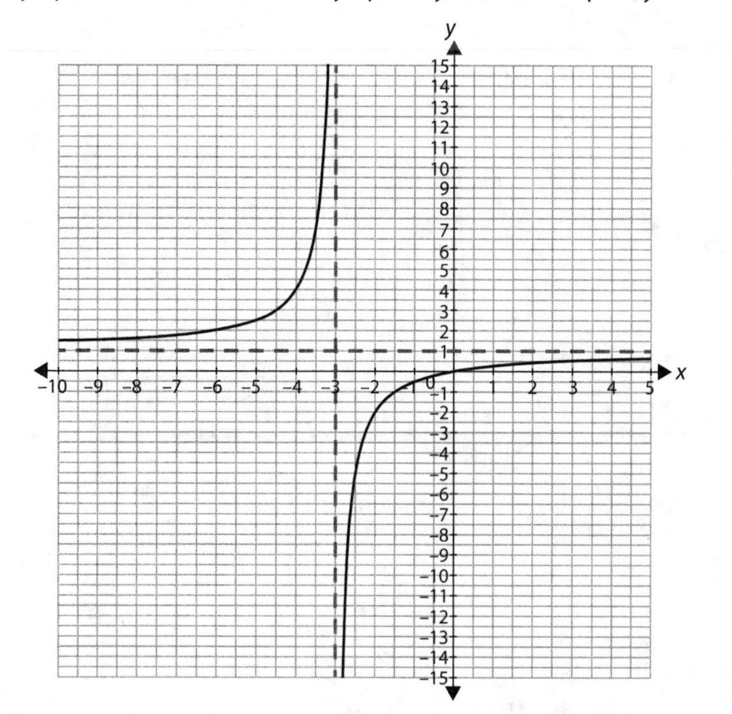

5. Vertical asymptote: $x = -1$; horizontal asymptote: $y = 5$; x-intercept: 0.2; y-intercept: -1

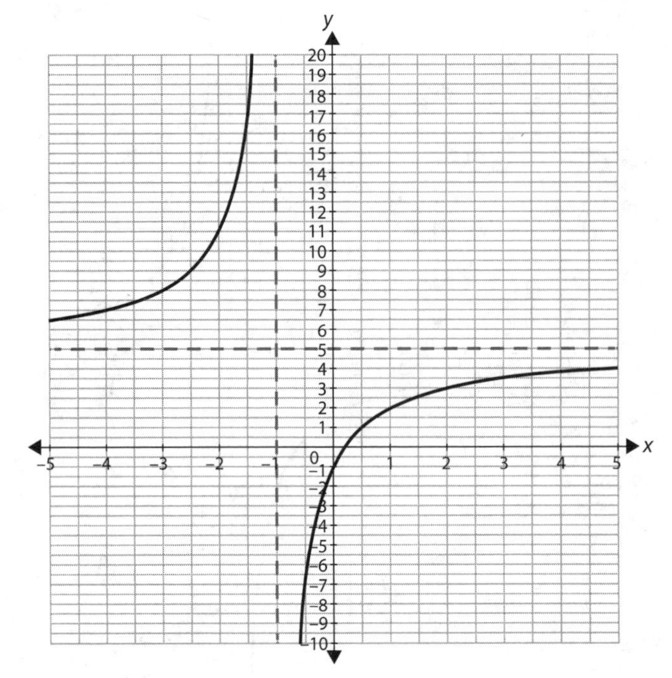

6. Vertical asymptote: $x = \pm 3$; horizontal asymptote: $y = 0$; x-intercepts: none; y-intercept: $-\frac{4}{9}$

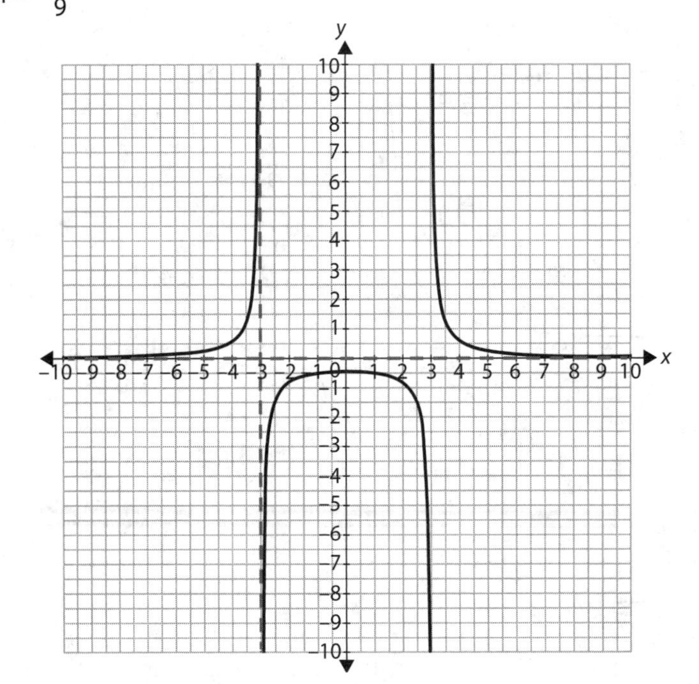

7. Vertical asymptotes: $x=-1$, $x=-2$; horizontal asymptote: $y=0$, x-intercepts: none; y-intercept: -1.5

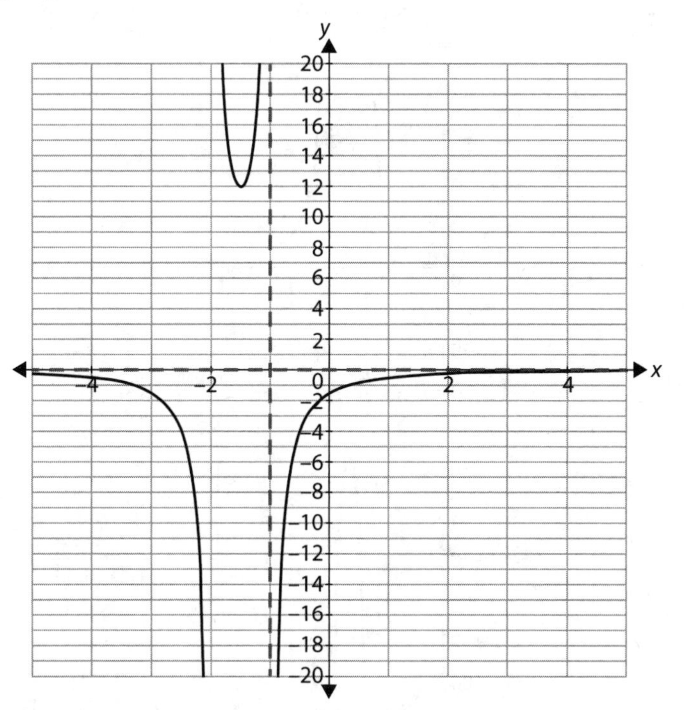

8. Vertical asymptote: $x=-2$; horizontal asymptote: $y=0$; x-intercept: 0; y-intercept: 0

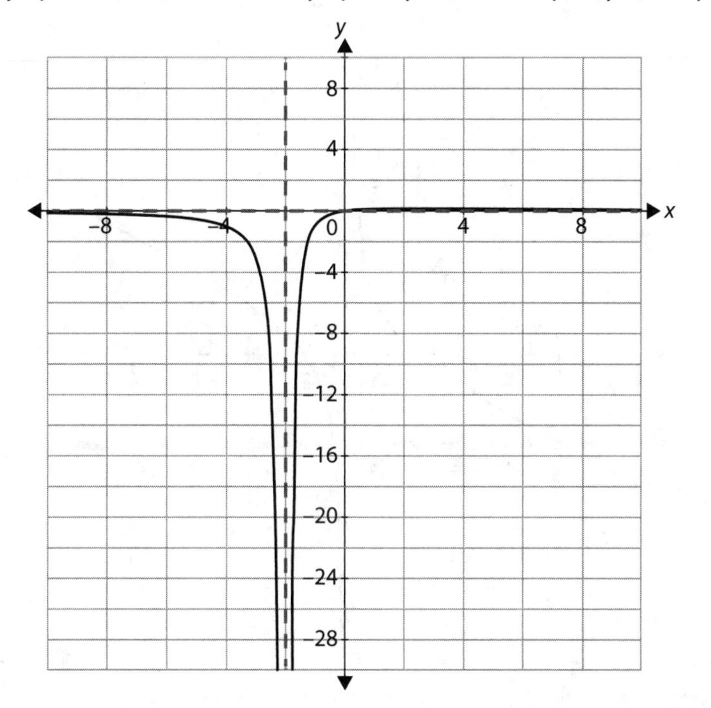

9. Vertical asymptotes: $x=2$, $x=3$; horizontal asymptote: $y=0$; x-intercept: $-\frac{4}{3}$; y-intercept: $\frac{2}{3}$

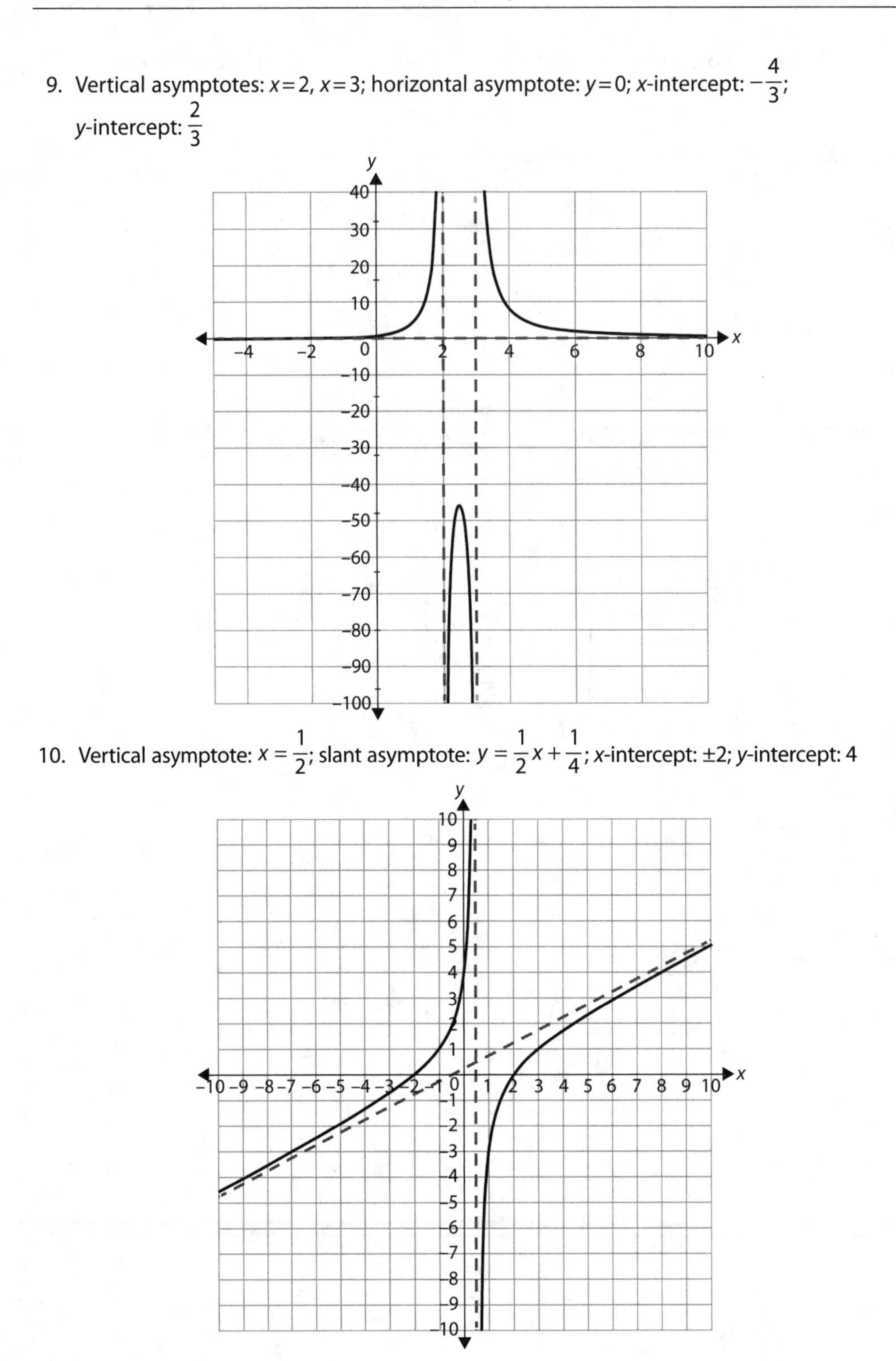

10. Vertical asymptote: $x = \frac{1}{2}$; slant asymptote: $y = \frac{1}{2}x + \frac{1}{4}$; x-intercept: ± 2; y-intercept: 4

Chapter 6 Conic Sections

Exercise 6.1

1. Vertex: (2,–4) focus: $\left(2,-3\frac{31}{32}\right)$, directrix: $y=-4\frac{1}{32}$

2. Vertex: (3,–2), focus: (3.5, –2), directrix: $x=2.5$

3. $y-\frac{13}{2}=-\frac{1}{2}(x+4)^2$

4. $x-34=-\ (y-6)^2$

5. 

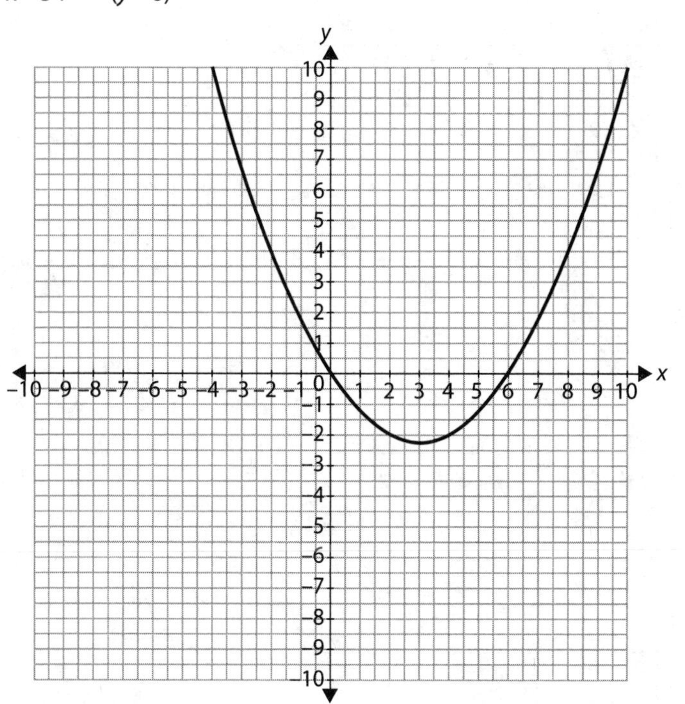

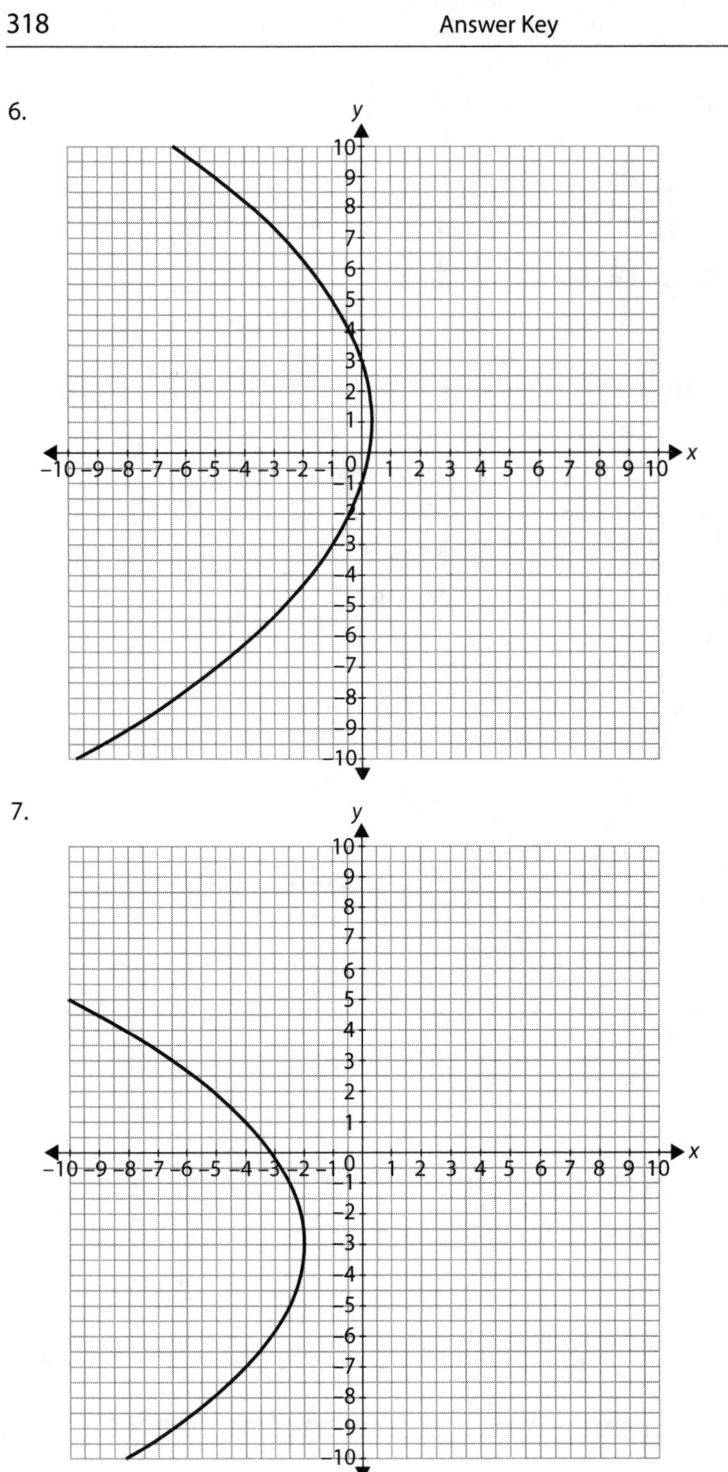
6.
y
x
10
9
8
7
6
5
4
3
2
1
0
−1
−2
−3
−4
−5
−6
−7
−8
−9
−10
−10 −9 −8 −7 −6 −5 −4 −3 −2 −1 1 2 3 4 5 6 7 8 9 10
7.
y
x
10
9
8
7
6
5
4
3
2
1
0
−1
−2
−3
−4
−5
−6
−7
−8
−9
−10
−10 −9 −8 −7 −6 −5 −4 −3 −2 −1 1 2 3 4 5 6 7 8 9 10

8.

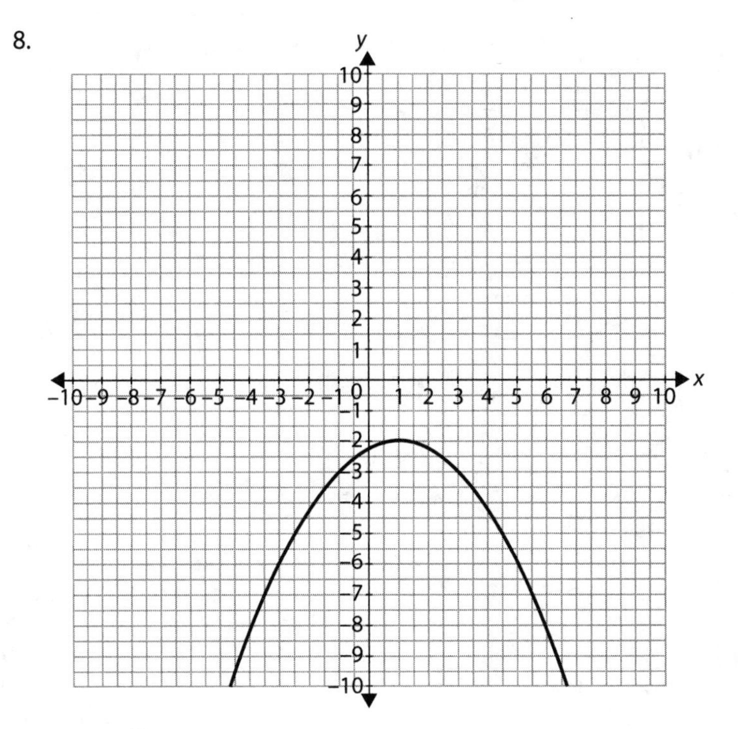

9. $x-4=\frac{1}{8}(y+3)^2$

10. $y-3=\frac{1}{8}(x+7)^2$

Exercise 6.2

1. $\frac{(x-2)^2}{25}+\frac{(y-1)^2}{16}=1$

2. $\frac{(x+3)^2}{25}+\frac{(y-2)^2}{169}=1$

3. $\frac{(x-4)^2}{4}+\frac{(y+7)^2}{1}=1$

4. Center: $(2,-1)$; vertices: $(-2,-1)$, $(6,-1)$; co-vertices: $(2,-4)$, $(2,2)$; foci: $(2\pm\sqrt{7},-1)$

5. Center: $(0,1)$; vertices: $(\pm5,1)$; co-vertices: $(0,-3)$, $(0,5)$; foci: $(\pm3,1)$

6. Center: $(0,0)$; vertices: $(\pm12,0)$; co-vertices: $(0,\pm5)$; foci: $(\pm\sqrt{119},0)$

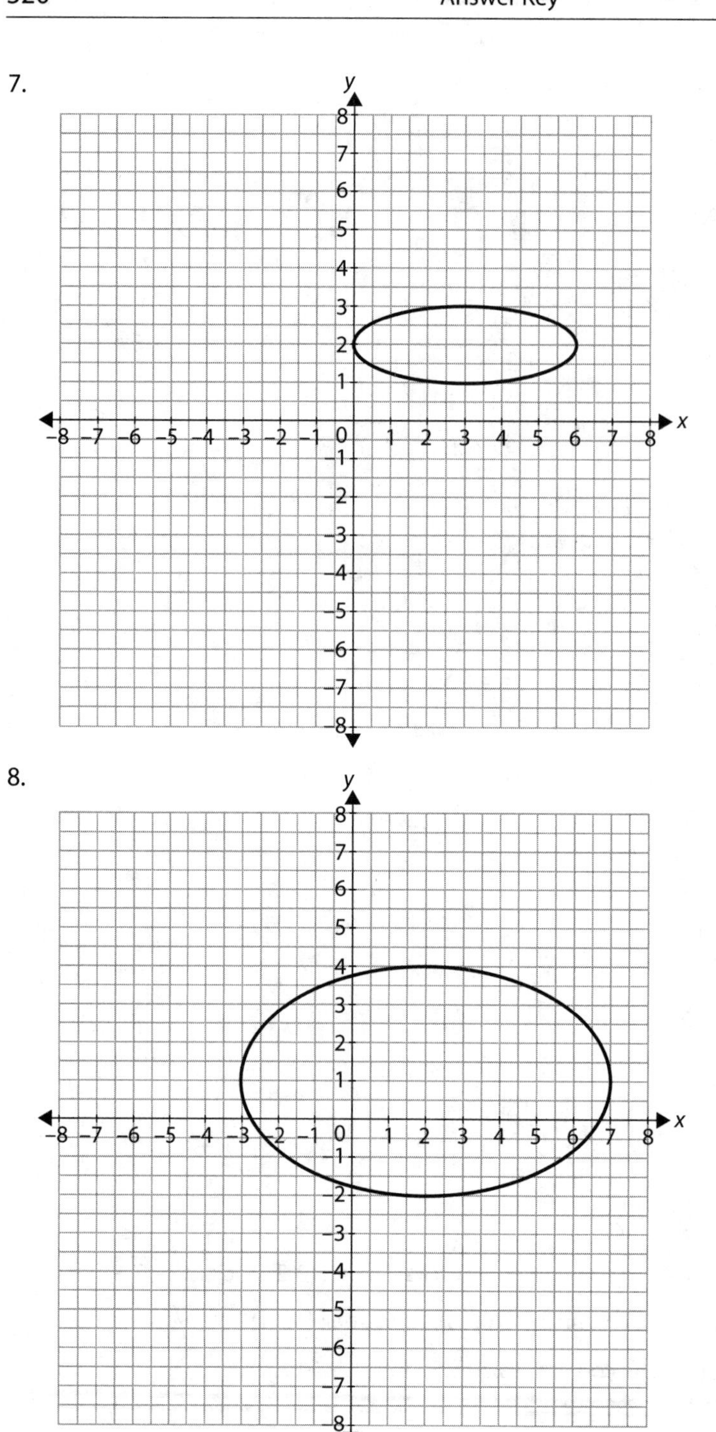
7.
y
8
7
6
5
4
3
2
1
−8 −7 −6 −5 −4 −3 −2 −1 0 1 2 3 4 5 6 7 8
x
−1
−2
−3
−4
−5
−6
−7
−8
8.
y
8
7
6
5
4
3
2
1
−8 −7 −6 −5 −4 −3 −2 −1 0 1 2 3 4 5 6 7 8
x
−1
−2
−3
−4
−5
−6
−7
−8

9.

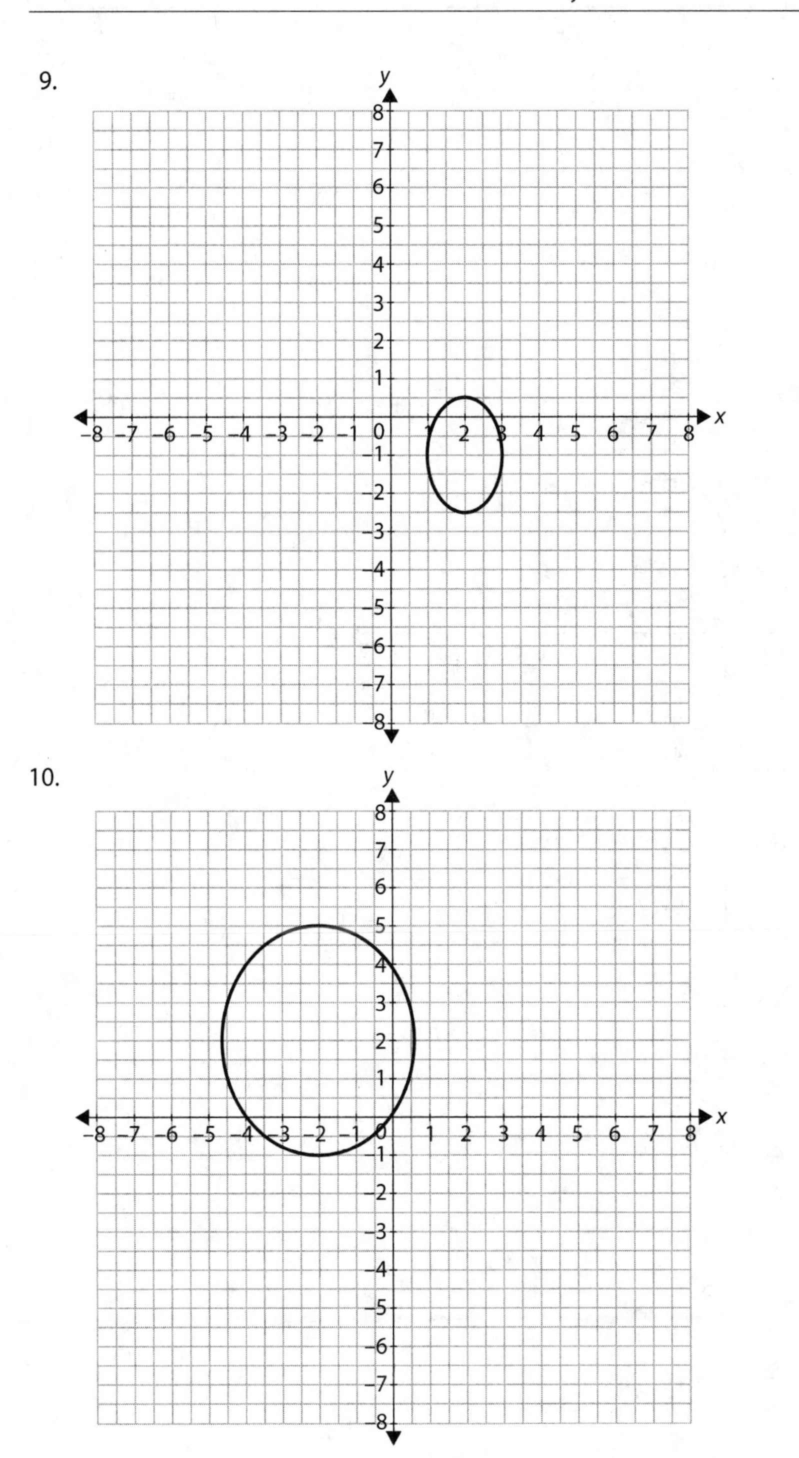
y
8
7
6
5
4
3
2
1
x
−8 −7 −6 −5 −4 −3 −2 −1 0 1 2 3 4 5 6 7 8
−1
−2
−3
−4
−5
−6
−7
−8
y
8
7
6
5
4
3
2
1
x
−8 −7 −6 −5 −4 −3 −2 −1 0 1 2 3 4 5 6 7 8
−1
−2
−3
−4
−5
−6
−7
−8

10.

Exercise 6.3

1. $(x-2)^2+(y+1)^2=25$

2. $(x+4)^2+(y-4)^2=144$

3. $(x-1)^2+\left(y-\frac{1}{4}\right)^2=2$

4.

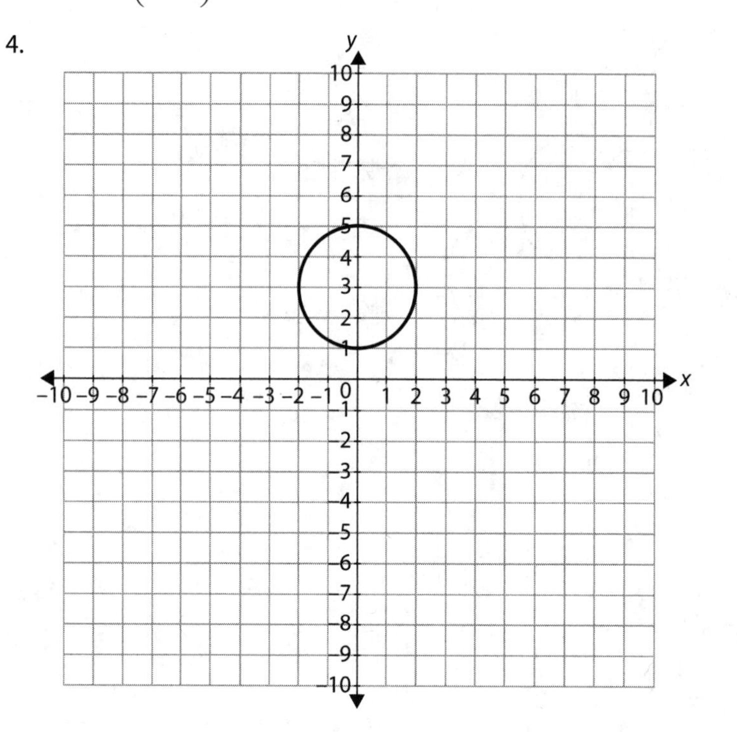

5.

6.

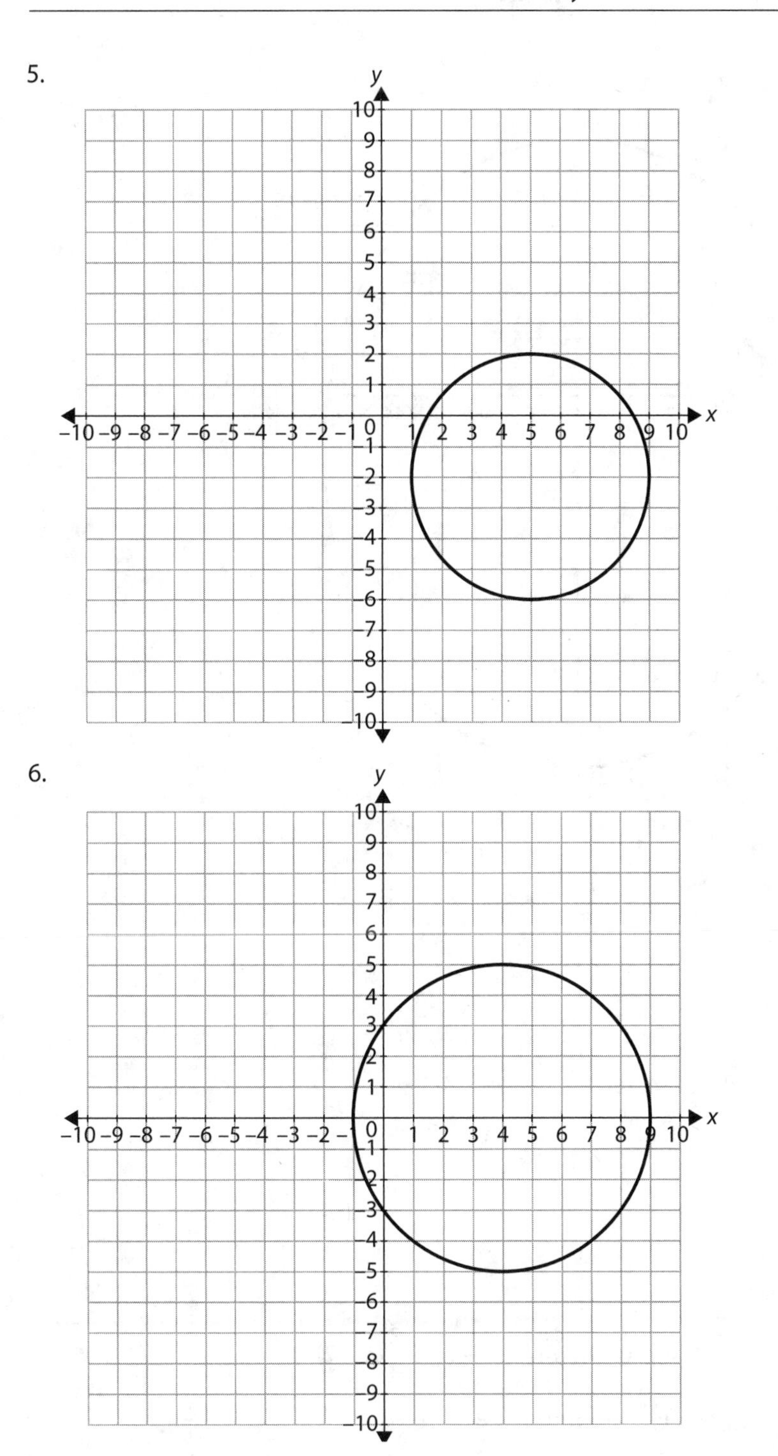

7.

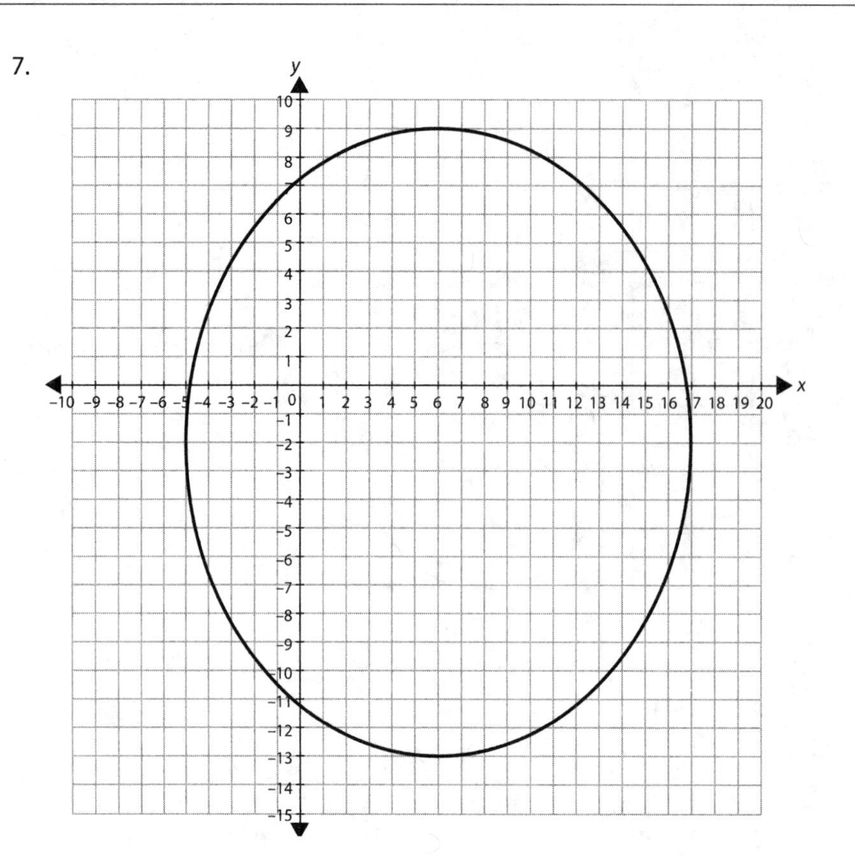

8. $(x-3)^2+(y-5)^2=49$

9. $(x+3)^2+(y-1)^2=169$

10. $x^2+(y+4)^2=64$

Exercise 6.4

1. $\dfrac{(x-5)^2}{4}-\dfrac{(y-1)^2}{5}=1$

2. $\dfrac{y^2}{4}-\dfrac{(x+3)^2}{21}=1$

3. $\dfrac{(y+7)^2}{1}-\dfrac{(x-2)^2}{48}=1$

4. Center: (2,–1); foci: (2,4), (2,–6); asymptotes: $y+1=\pm\dfrac{3}{4}(x-2)$

5. Center: (0,–1); foci: $(\pm\sqrt{41},-1)$; asymptotes: $y+1=\pm\dfrac{4}{5}x$

6. Center: (0,0); foci: (±13,0); asymptotes: $y=\pm\dfrac{12}{5}x$

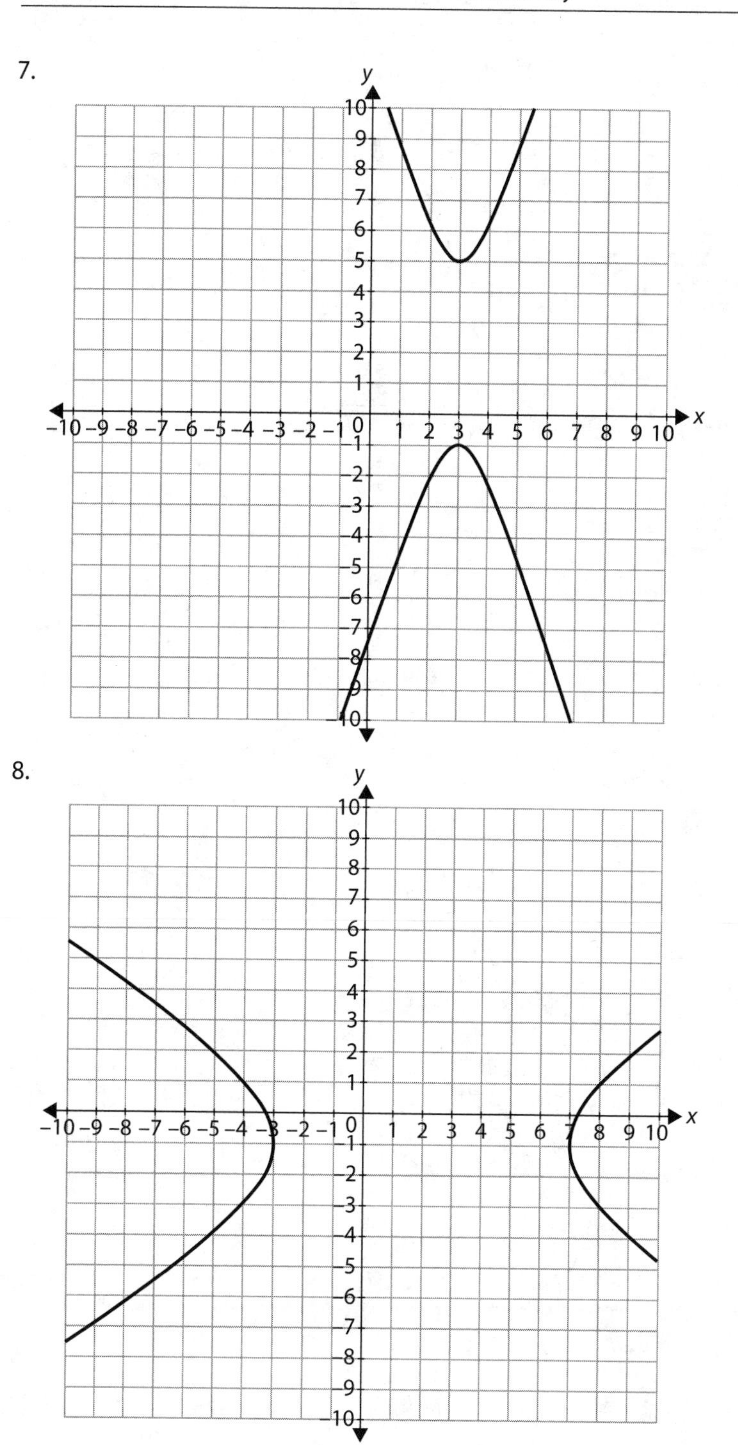
7.
y
x
8.
y
x

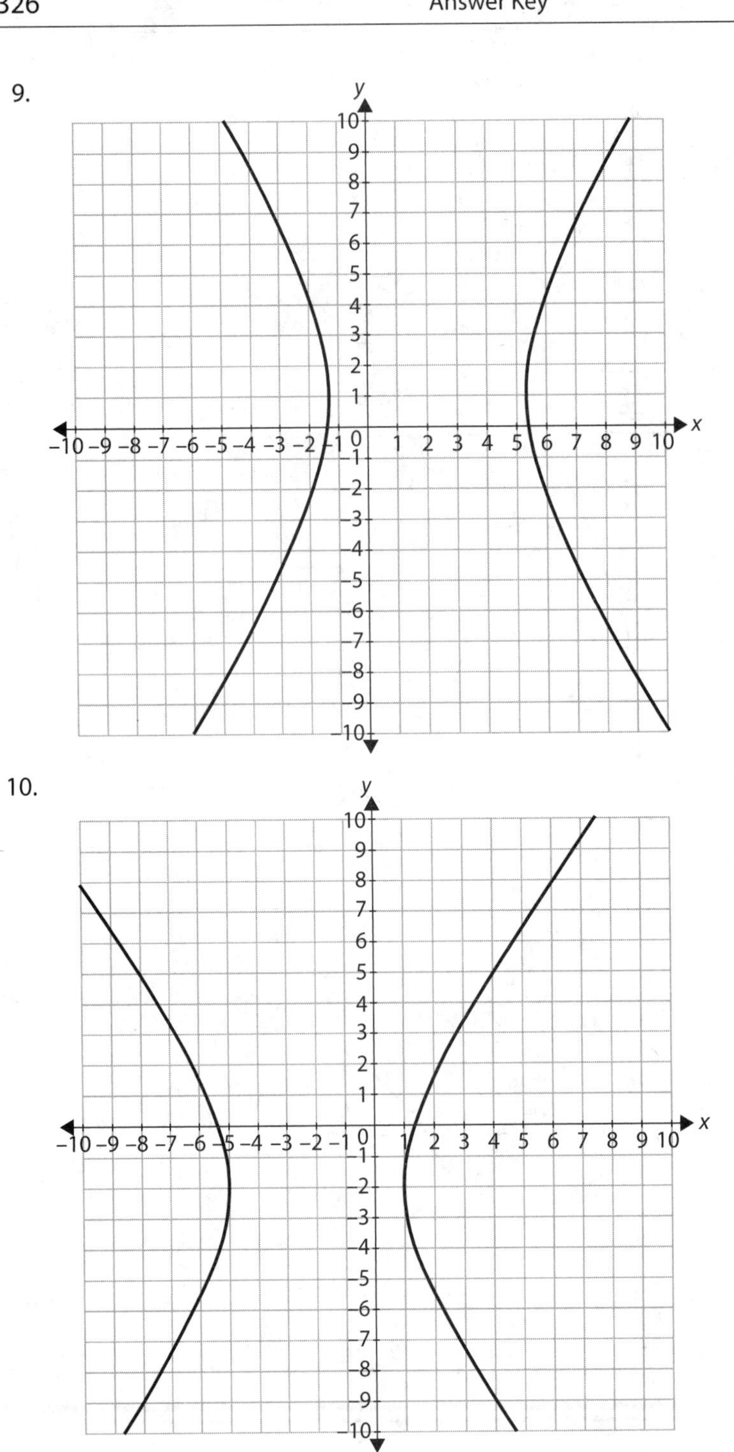
9.
y
x
10.
y
x

Exercise 6.5

1. 2 solutions

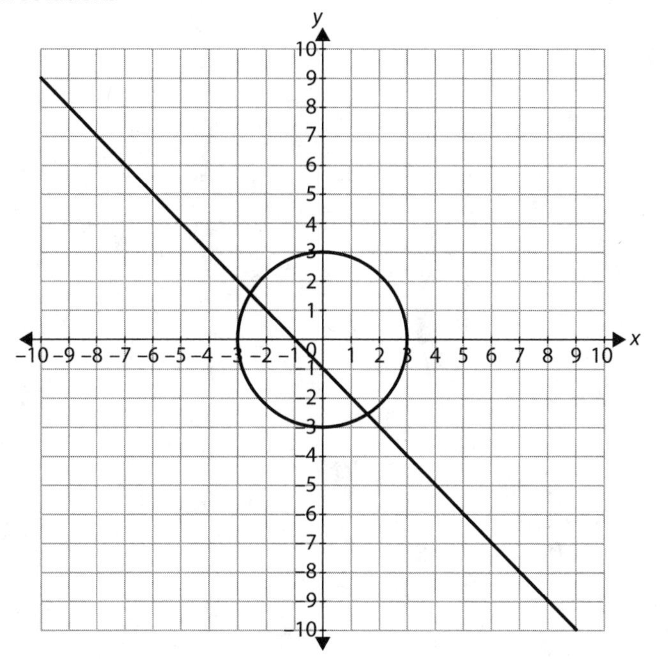

2. 2 solutions

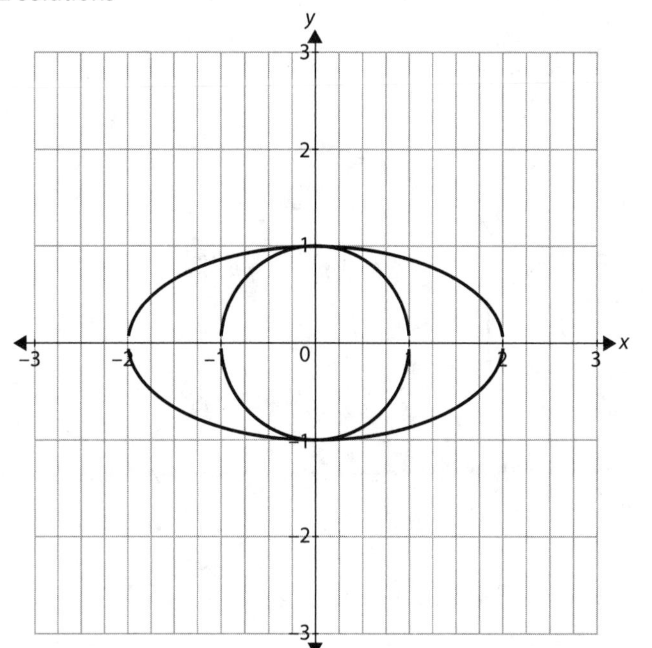

3. 2 solutions

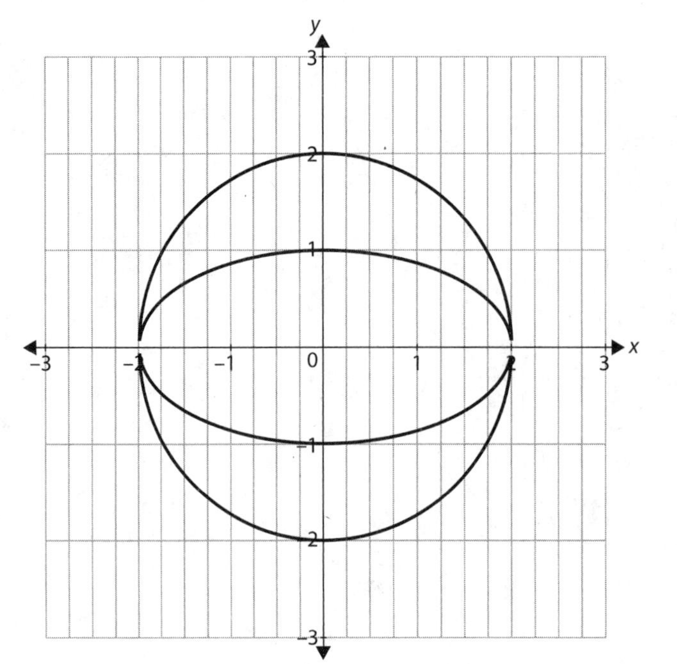

4. 4 solutions

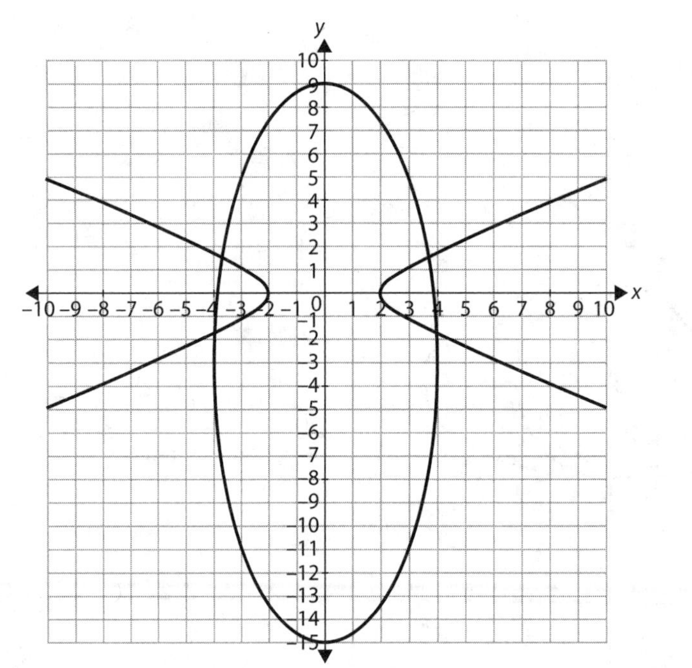

5. 4 solutions

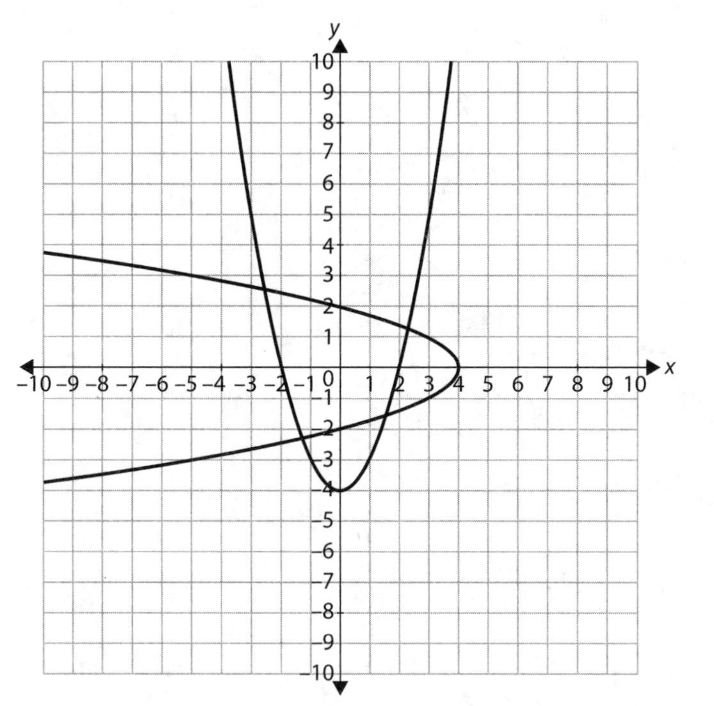

6. $x = 3, y = 4$
   $x = -4, y = -3$

7. $x = \frac{6\sqrt{5}}{5}, y = \frac{4\sqrt{5}}{5}$
   $x = -\frac{6\sqrt{5}}{5}, y = -\frac{4\sqrt{5}}{5}$

8. $x \approx 5.179, y \approx 2.276$
   $x \approx 5.179, y \approx -2.276$

9. $x = 0, y = 1$
   $x = 0, y = -1$

10. $x = \frac{3\sqrt{10}}{10}, y = \frac{3\sqrt{10}}{10}$
    $x = -\frac{3\sqrt{10}}{10}, y = \frac{3\sqrt{10}}{10}$
    $x = -\frac{3\sqrt{10}}{10}, y = -\frac{3\sqrt{10}}{10}$
    $x = \frac{3\sqrt{10}}{10}, y = -\frac{3\sqrt{10}}{10}$

Chapter 7 Exponential and Logarithmic Functions

Exercise 7.1

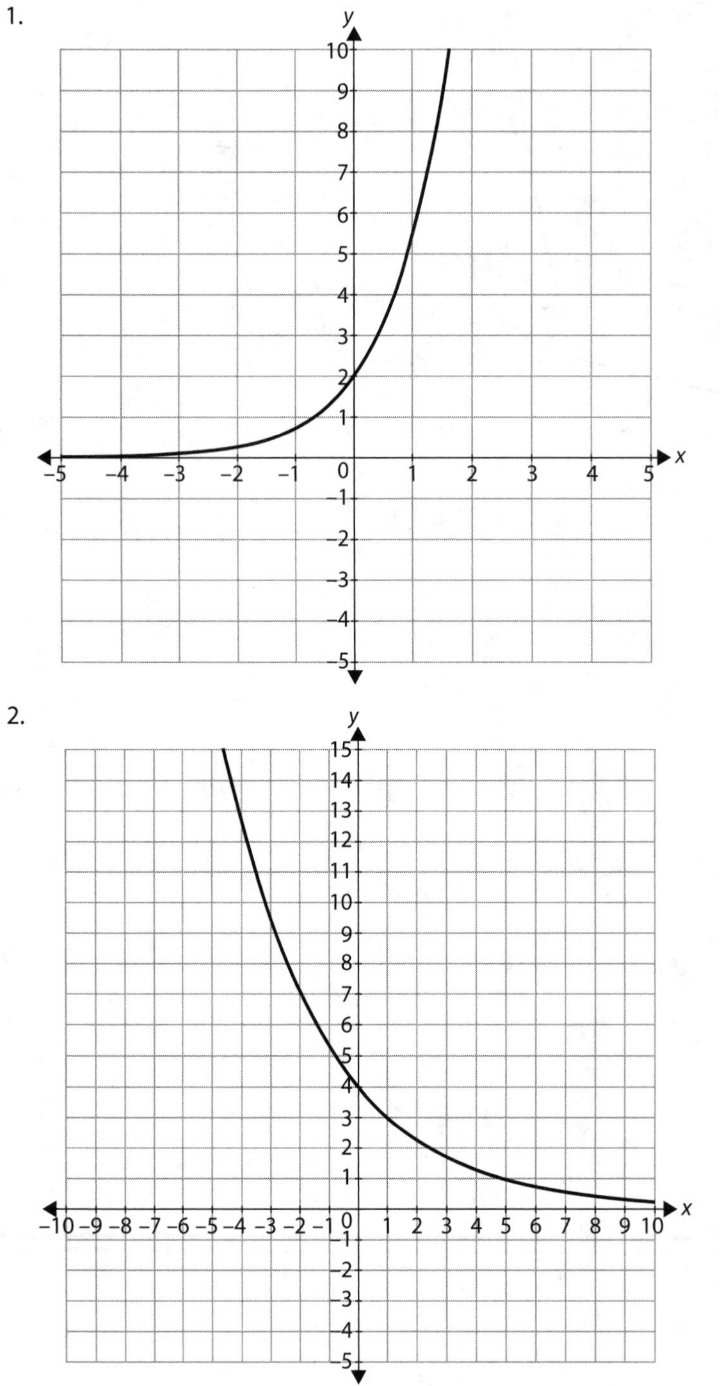

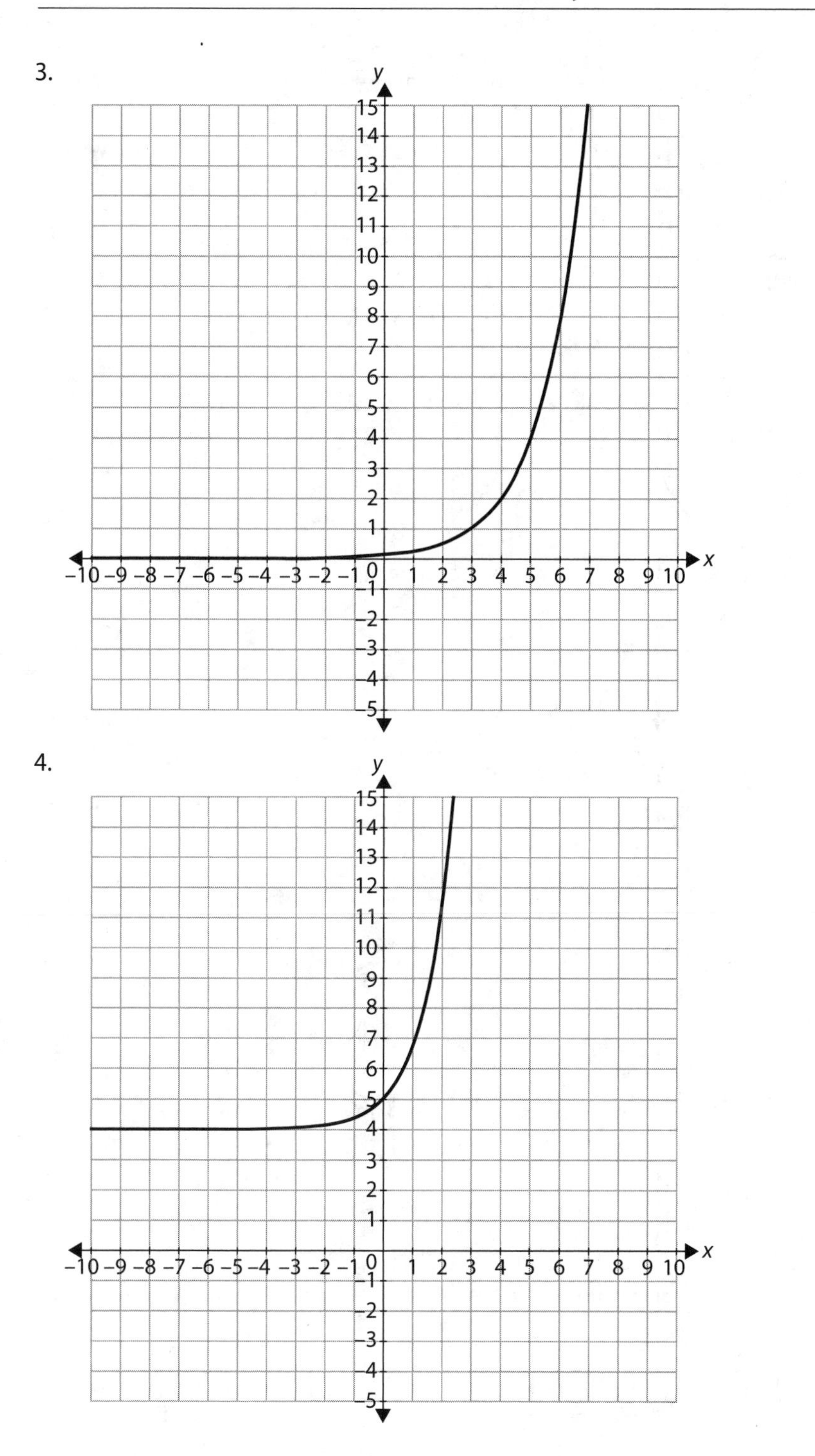
3.
4.
y
x
15
14
13
12
11
10
9
8
7
6
5
4
3
2
1
0
−1
−2
−3
−4
−5
−10 −9 −8 −7 −6 −5 −4 −3 −2 −1
1 2 3 4 5 6 7 8 9 10

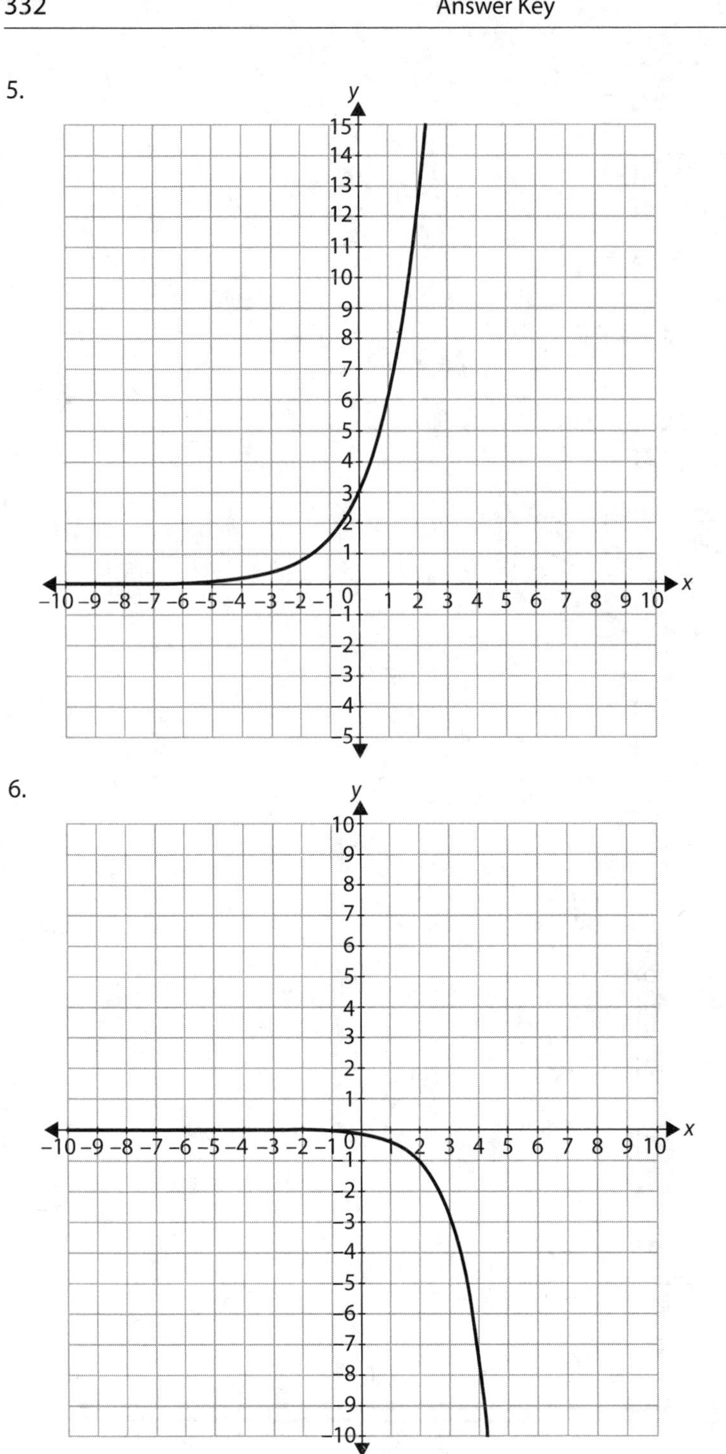
5.
y
x
15
14
13
12
11
10
9
8
7
6
5
4
3
2
1
0
−1
−2
−3
−4
−5
−10 −9 −8 −7 −6 −5 −4 −3 −2 −1
1 2 3 4 5 6 7 8 9 10
6.
y
x
10
9
8
7
6
5
4
3
2
1
0
−1
−2
−3
−4
−5
−6
−7
−8
−9
−10
−10 −9 −8 −7 −6 −5 −4 −3 −2 −1
1 2 3 4 5 6 7 8 9 10

7.

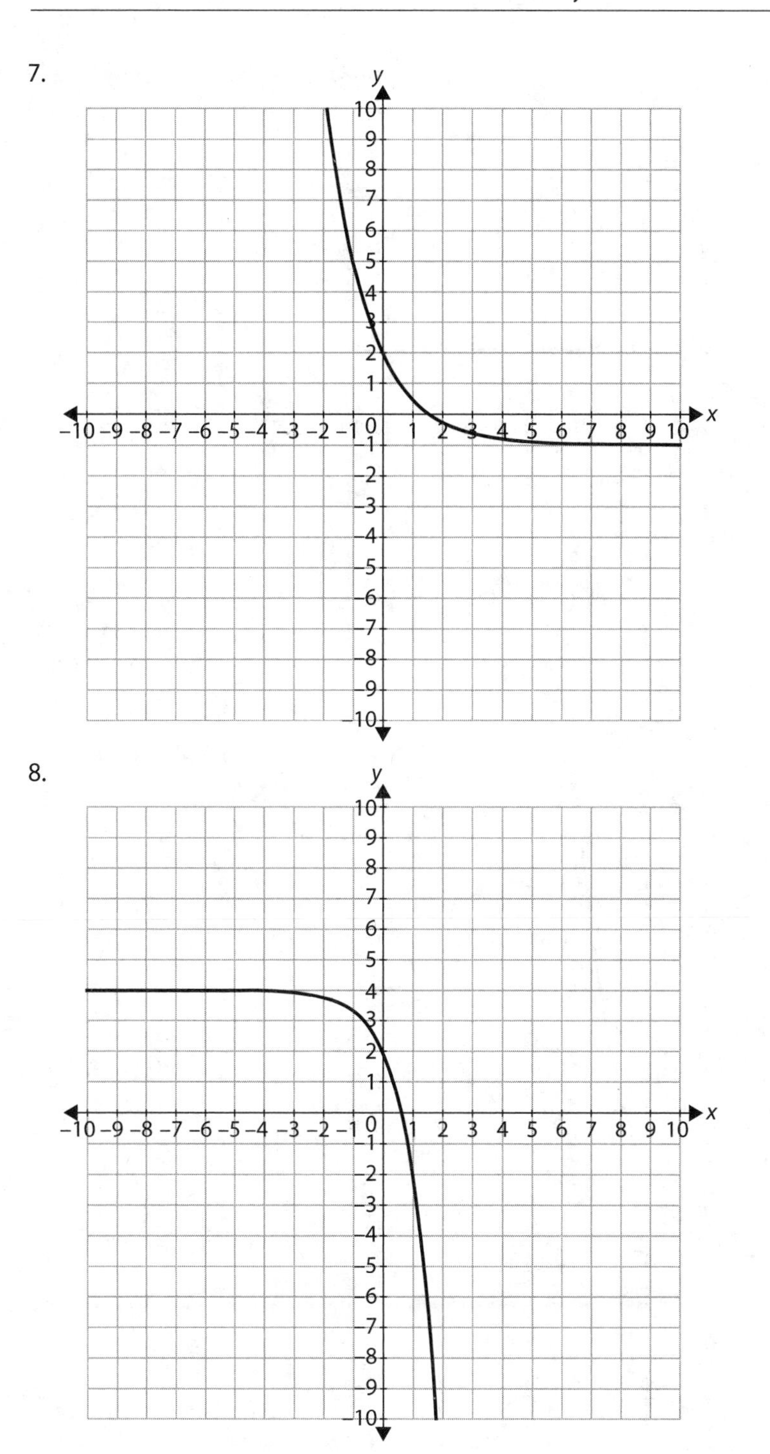

8.

9.

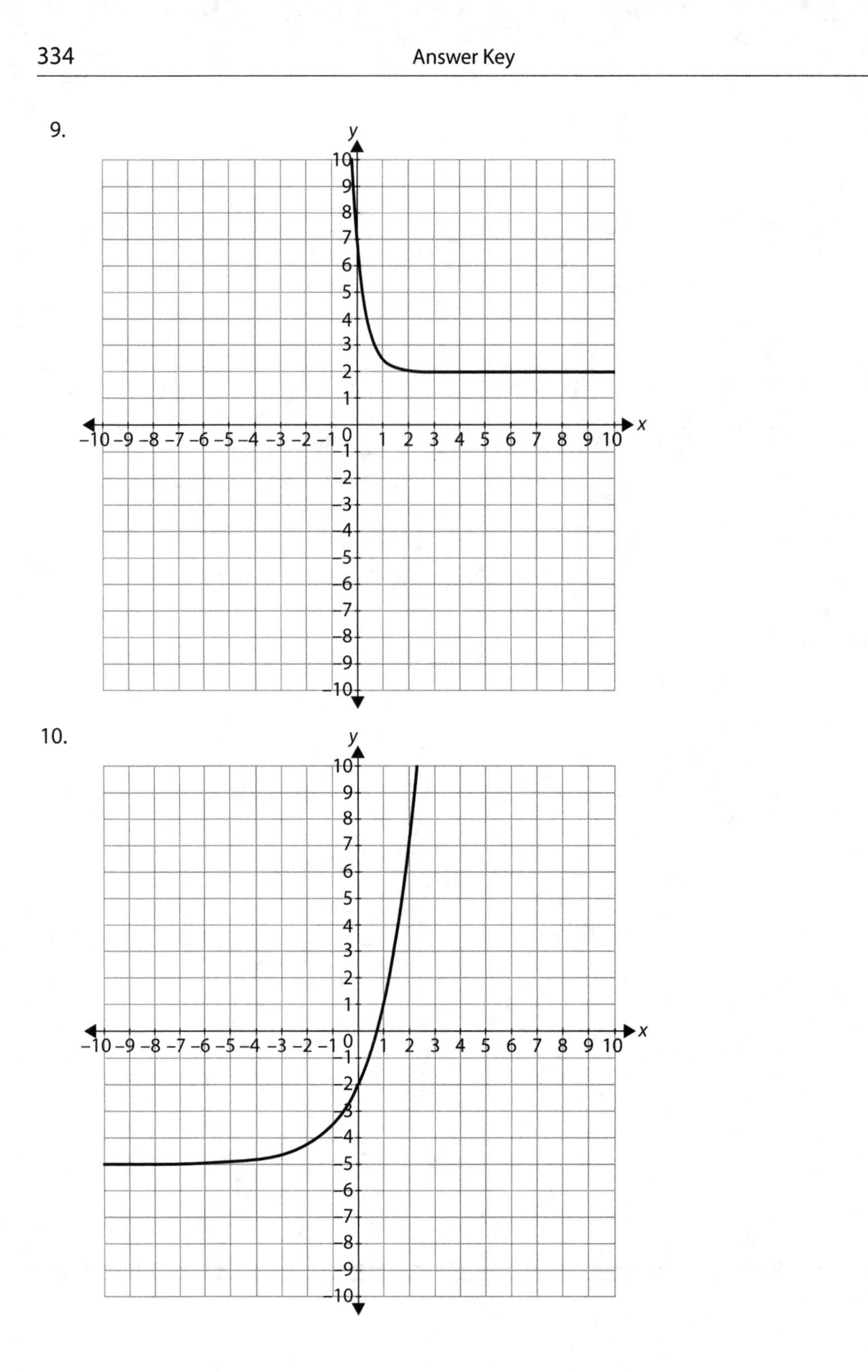
y
x
10
9
8
7
6
5
4
3
2
1
0
-1
-2
-3
-4
-5
-6
-7
-8
-9
-10
-10 -9 -8 -7 -6 -5 -4 -3 -2 -1 0 1 2 3 4 5 6 7 8 9 10

10.

y
x
10
9
8
7
6
5
4
3
2
1
0
-1
-2
-3
-4
-5
-6
-7
-8
-9
-10
-10 -9 -8 -7 -6 -5 -4 -3 -2 -1 0 1 2 3 4 5 6 7 8 9 10

Exercise 7.2

1. $f^{-1}(x)=\log(x)-1$; domain: $(0,\infty)$; range: $(-\infty,\infty)$
2. $f^{-1}(x)=\ln(x-5)$; domain: $(5,\infty)$; range: $(-\infty,\infty)$
3. $f^{-1}(x)=\log_2\left(\frac{x}{5}\right)+3$ or $f^{-1}(x)=\log_2 x-\log_2 5+3$; domain: $(0,\infty)$; range: $(-\infty,\infty)$
4. $f^{-1}(x)=4\log_{\frac{1}{2}}\left(\frac{x}{100}\right)=4\log_{\frac{1}{2}} x-4\log_{\frac{1}{2}} 100$ or

   $f^{-1}(x)=-4\log_2\left(\frac{x}{100}\right)=4\log_2 100-4\log_2 x$; domain: $(0,\infty)$; range: $(-\infty,\infty)$
5. $f^{-1}(x)=\frac{1}{2}\ln\left(\frac{5-x}{4}\right)-\frac{1}{2}=\frac{1}{2}\ln(5-x)-\frac{1}{2}\ln 4-\frac{1}{2}$; domain: $(-\infty,5)$; range: $(-\infty,\infty)$
6. $\log_3 27=3$
7. $\log_b\sqrt{b}=\frac{1}{2}$
8. $\log_8\frac{1}{64}=-2$
9. $\ln e^{4x}=4x$
10. $e^{(\ln e^x)}=e^x$

Exercise 7.3

1. $\ln[x(x+1)]=\ln x+\ln(x+1)$
2. $\ln\left(\frac{x}{x-3}\right)=\ln x-\ln(x-3)$
3. $$\begin{aligned}\log_b\left(\frac{xy}{z}\right)^2&=2\log_b\left(\frac{xy}{z}\right)\\&=2[\log_b x+\log_b y-\log_b z]\\&=2\log_b x+2\log_b y-\log_b z\end{aligned}$$
4. $$\begin{aligned}\ln\left(\frac{1}{x^2}\sqrt{x-2}\right)&=\ln\frac{1}{x^2}+\ln\sqrt{x-2}\\&=-2\ln x+\frac{1}{2}\ln(x-2)\end{aligned}$$
5. $4\ln x+3\ln y=\ln x^4+\ln y^3=\ln(x^4y^3)$
6. $\frac{1}{2}\log_b(x+5)-\log_b x=\log_b\frac{\sqrt{x+5}}{x}$

7. $\frac{1}{5}[\ln x + 3\ln y - 4\ln z] = \sqrt[5]{\frac{xy^3}{z^4}}$

8. $\log_3 5 = \frac{\ln 5}{\ln 3} = \frac{1.6094}{1.0986} \approx 1.4650$

9. $\log_3 \frac{4}{5} = \frac{\ln\frac{4}{5}}{\ln 3} = \frac{\ln 4 - \ln 5}{\ln 3} = \frac{2\ln 2 - \ln 5}{\ln 3} = \frac{2(0.6931) - 1.6094}{1.0986} \approx -0.2032$

10. $\log_7 1.5 = \log_7 \frac{3}{2} = \log_7 3 - \log_7 2 = \frac{\ln 3}{\ln 7} - \frac{\ln 2}{\ln 7} = \frac{1.0986}{1.9459} - \frac{0.6931}{1.9459} \approx 0.2084$

Exercise 7.4

1. $x = 3.6$
2. $x \approx -3.889$
3. $x \approx 6.945$
4. $r \approx 2.688$
5. $x \approx 2.079$
6. $x = 23$
7. $x = 5$
8. $x = 3$
9. no solution
10. no solution

Exercise 7.5

1. Decay: 42 mg
2. Growth: 308,025
3. Decay: 36,653 acres
4. Growth: 26.6 hours
5. Decay: 34 years
6. Growth: early in 2010
7. Decay: September 22, 2012 (265 days)
8. $8,498.61
9. $30,491.91
10. $61,490.08

Chapter 8 Radical Functions

Exercise 8.1

1. $\frac{2^{11}}{x^6}$ or $\frac{2048}{x^6}$
2. $\frac{1}{x^9y^3}$
3. $\frac{b^{18}}{a^{10}c^{10}}$
4. $\frac{3y^2}{x^2z^4}$
5. $\frac{27}{2x^4y^5}$
6. $4x^5y^2$

7. $\dfrac{243x^{25}}{y^{25}}$

8. $\dfrac{3y^2}{x^2}$

9. $\dfrac{(x+y)^2}{xy}$

10. $(x+y)^3(x-y)^2 = (x+y)(x^2-y^2)^2$

Exercise 8.2

1. $2a^3b^2 \cdot \sqrt[4]{8c}$

2. $\dfrac{3y^2z}{x}\sqrt[3]{\dfrac{3y^2z}{x^2}}$

3. $2x-1$

4. $(x+3)\sqrt{(x^2-4)}$

5. $t^2\sqrt{t^2-1}$

6. $\sqrt{x-2}$

7. $2(x-3)\sqrt[4]{(x-3)^2} = 2\sqrt{(x-3)}$

8. $\dfrac{x(\sqrt[3]{x^3+8})^2}{x^3+8}$

9. $\dfrac{x+\sqrt{30-x}}{x-5}$

10. $-2x(\sqrt{3x}+\sqrt{5x})$

Exercise 8.3

1.

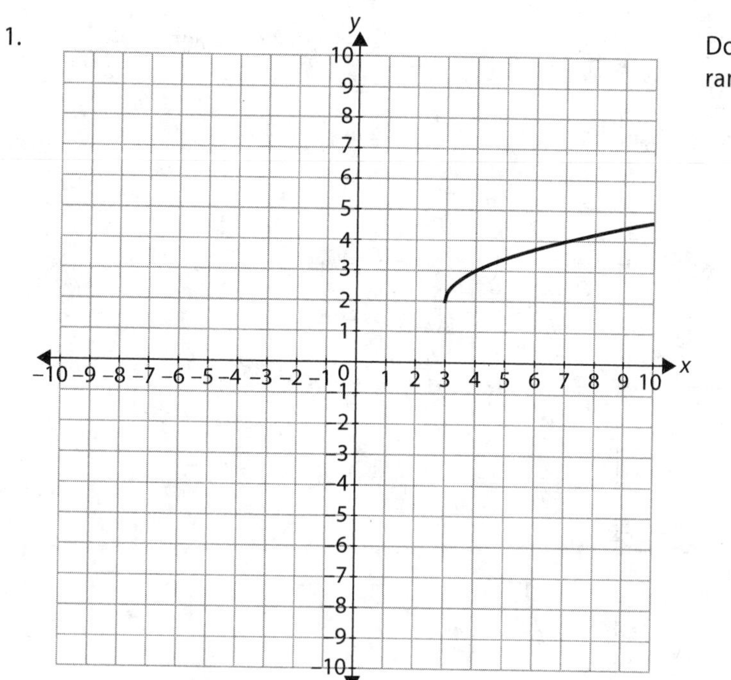

Domain: $[3,\infty)$, range: $[2,\infty)$

2.

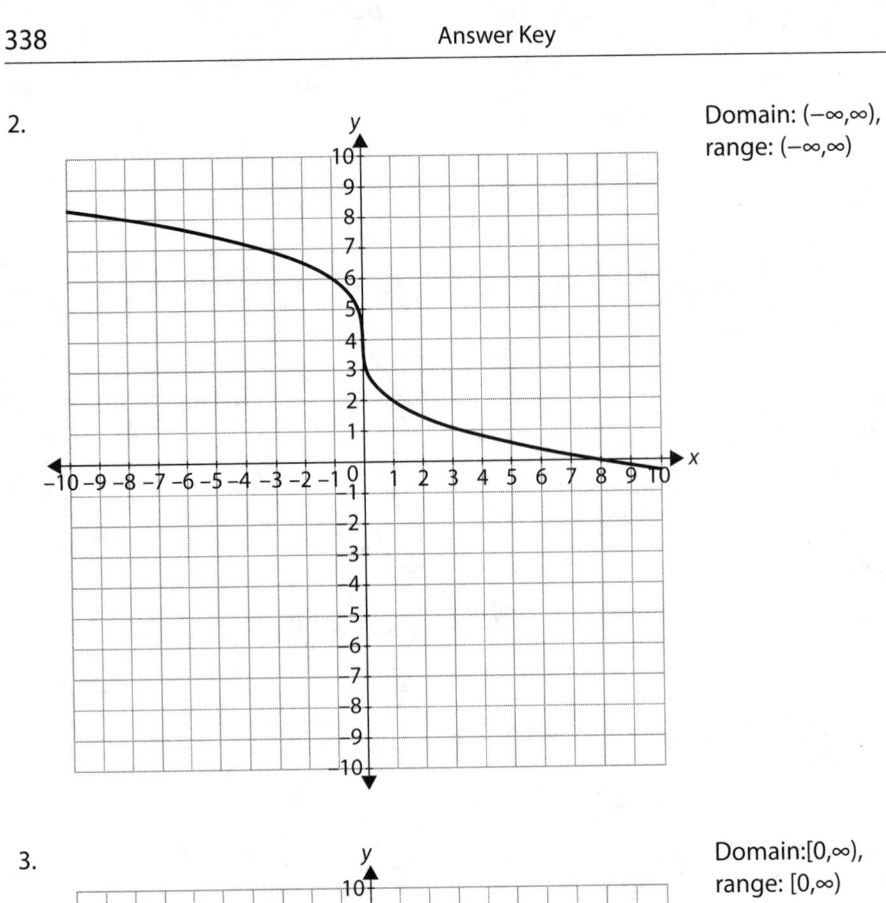

Domain: $(-\infty,\infty)$, range: $(-\infty,\infty)$

3.

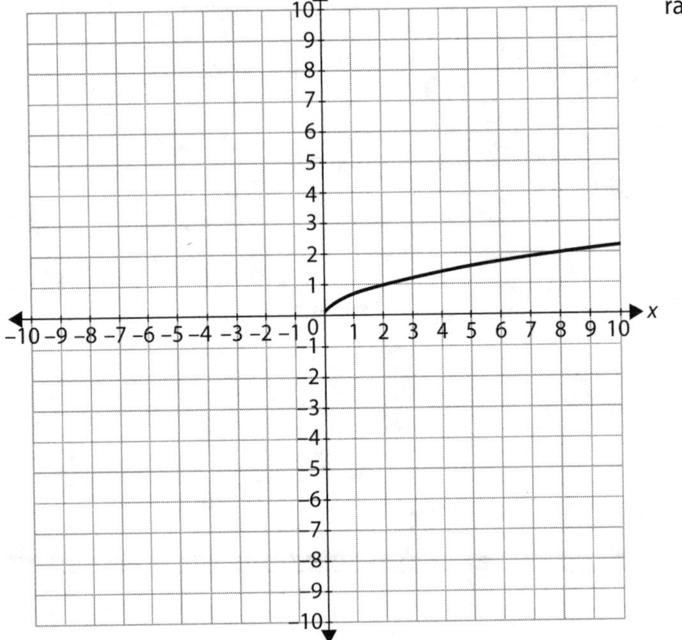

Domain: $[0,\infty)$, range: $[0,\infty)$

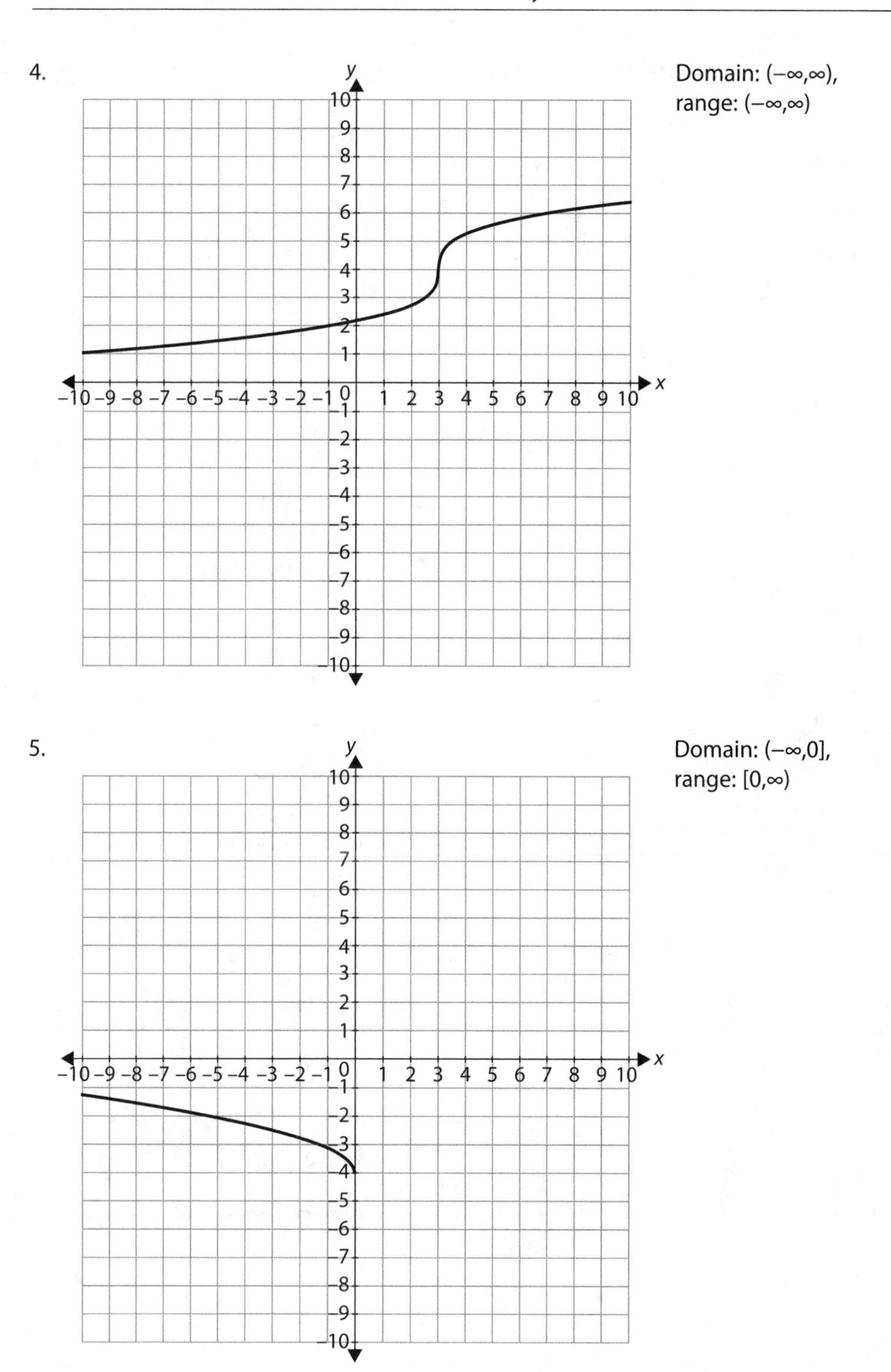

4. Domain: $(-\infty,\infty)$, range: $(-\infty,\infty)$

5. Domain: $(-\infty,0]$, range: $[0,\infty)$

Chapter 9 Systems of Equations

Exercise 9.1

1. $a=1, b=6$
2. $x=1, y=10$
3. $x=2.5, y=-2$
4. $x=7, y=3$
5. $x=5, y=2$
6. $x=0.2, y=1.2$
7. $x=3, y=-3$
8. $x=-2, y=-5$
9. $x=-\frac{1}{6}, y=\frac{4}{3}$
10. $x=0.6, y=6.6$

Exercise 9.2

1. $x=4, y=6, z=-8$
2. $x=3.5, y=-1, z=1.5$
3. $x=5, y=9, z=-3$
4. $x=4, y=-7, z=3$
5. $x=-2, y=3, z=-1$
6. $x=-3.6, y=-8.4, z=9.6$
7. $x=5, y=-1, z=0$
8. $x=\frac{58}{47}, y=\frac{9}{47}, z=-\frac{43}{47}$
9. $x=4, y=-2, z=3$
10. $x=-3, y=8, z=-4$

Exercise 9.3

1. $x=3,\ y=4$
   $x\approx 4.68,\ y\approx 1.76$
2. $x=3, y=0$
   $x=-3, y=0$
   $x=\frac{9}{5}, y=-\frac{8}{5}$
   $x=-\frac{9}{5}, y=-\frac{8}{5}$
3. $x=-\frac{1}{2}, y=\frac{\sqrt{14}}{2}$
   $x=-\frac{1}{2}, y=-\frac{\sqrt{14}}{2}$
4. $x=2\sqrt{2}, y=2\sqrt{2}$
   $x=-2\sqrt{2}, y=-2\sqrt{2}$
5. $x=1, y=\sqrt{2}$
   $x=-1, y=\sqrt{2}$
   $x=-1, y=-\sqrt{2}$
   $x=1, y=-\sqrt{2}$
6. $x=2, y=0$
   $x=-2, y=0$
7. $x=\frac{1}{2}, y=\ln\left(\frac{7}{2}\right)\approx 1.253$
8. $x=2, y=\sqrt{3}$
9. $x=4, y=\ln 2\approx 0.693$
10. $x\approx 2.710, y\approx 6.541$

Chapter 10 Matrices and Determinants

Exercise 10.1

1. $\begin{bmatrix} 8 & -12 \\ 4 & 16 \end{bmatrix}$

2. [12]

3. $\begin{bmatrix} 2 & 4 & 6 \\ -1 & -2 & -3 \\ 4 & 8 & 12 \end{bmatrix}$

4. $\begin{bmatrix} 4 & 5 & 8 \\ 2 & 4 & 10 \end{bmatrix}$

5. Not possible. Dimensions are different.

6. $\begin{bmatrix} 0 & -10 & 4 \\ 10 & 6 & 4 \end{bmatrix}$

7. $\begin{bmatrix} 1 & 7 \\ -5 & 9 \end{bmatrix}$

8. Not possible. Dimensions are different.

9. $\begin{bmatrix} 30 & -8 \\ -57 & 31 \end{bmatrix}$

10. $\begin{bmatrix} 45 & -6 & -6 \\ -36 & 14 & 8 \\ -3 & 5 & 2 \end{bmatrix}$

Exercise 10.2

1. $\begin{vmatrix} 4 & 2 \\ 9 & 6 \end{vmatrix} = 6$

2. $\begin{vmatrix} -5 & 6 \\ 5 & 4 \end{vmatrix} = -50$

3. $\begin{vmatrix} -3 & 1 \\ -5 & 8 \end{vmatrix} = -19$

4. $\begin{vmatrix} 12 & 0 \\ -3 & \frac{1}{2} \end{vmatrix} = 6$

5. $\begin{vmatrix} 5 & 3 \\ 2 & -9 \end{vmatrix} = -51$

6. $\begin{vmatrix} 11 & 10 \\ 2 & 1 \end{vmatrix} = -9$

7. $\begin{vmatrix} 3 & 2 \\ 1 & 0 \end{vmatrix} = -2$

8. $\begin{vmatrix} 4 & 1 & -2 \\ 3 & 0 & 1 \\ -1 & -2 & 5 \end{vmatrix} = 4$

9. $\begin{vmatrix} -3 & 7 & 2 \\ 0 & 4 & -1 \\ 3 & 2 & 0 \end{vmatrix} = -51$

10. $\begin{vmatrix} -1 & 2 & 1 \\ 2 & 0 & 0 \\ 3 & -4 & 2 \end{vmatrix} = -16$

Exercise 10.3

1. $x = 2.5, y = -1$
2. $x = 0.2, y = 1.2$
3. $x = -7, y = 2$
4. $x = 3, y = -3$
5. $x = 2, y = -1$
6. $x = 7, y = 3$
7. $x = -2, y = -3, z = 4$
8. $x = 0, y = 1, z = -1$
9. $x = -1, y = 3, z = 4$
10. $x = 4, y = 5, z = 10$

Exercise 10.4

1. Inverses
2. Not inverses
3. Inverses
4. Inverses
5. Not inverses
6. $\begin{bmatrix} 2 & 1 \\ 5 & 3 \end{bmatrix}$
7. 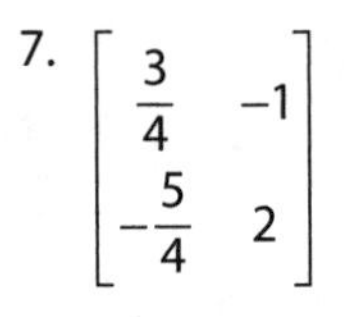
$\begin{bmatrix} \frac{3}{4} & -1 \\ -\frac{5}{4} & 2 \end{bmatrix}$
8. No inverse is possible because the matrix is not square.
9. $\begin{bmatrix} \frac{1}{3} & -\frac{1}{3} & \frac{1}{3} \\ \frac{1}{3} & \frac{2}{3} & -\frac{2}{3} \\ -\frac{1}{6} & \frac{1}{6} & \frac{1}{3} \end{bmatrix}$
10. No inverse is possible because the determinant is 0.

Exercise 10.5

1. $x = 5, y = 2$
2. $x = 0.2, y = 1.2$
3. $x = 3, y = -3$
4. $x = -2, y = -5$
5. $x = -\frac{1}{6}, y = \frac{4}{3}$
6. $x = 0.6, y = 6.6$
7. $x = 8, y = -3.5, z = 2.5$
8. $x = \frac{5}{94}, y = -\frac{483}{94}, z = \frac{365}{94}$
9. $x = 4, y = -2, z = 3$
10. $x = -3, y = 8, z = -4$

Chapter 11 Triangle Trigonometry

Exercise 11.1

1. $6\sqrt{2}$ ft and $6\sqrt{2}$ ft
2. $5\sqrt{6}$ m and $10\sqrt{3}$ m
3. $28\sqrt{6}$ ft and $42\sqrt{2}$ ft
4. $2\sqrt{21}$ cm and $6\sqrt{7}$ cm
5. 92.8 m and 109.5 m
6. 7.25 ft
7. Down the slope: 4727.3 ft
8. 204.4 m
9. 40°
10. 26°, 64°

Exercise 11.2

1. 32.1
2. 37.3
3. 18.8
4. 20.6
5. 3.8
6. 7.9
7. 33.8
8. 15.2
9. 17.4
10. 11.9

Exercise 11.3

1. 5.70 cm
2. 169.13 ft
3. 23.02 m
4. 280.67 in
5. 87.08 ft
6. 53.62°
7. 55.13°
8. 27.13°
9. 74.52°
10. 62.15°

Exercise 11.4

1. 39.58 in
2. 17.14 ft
3. 69.97 cm
4. 33.59°
5. 6.87°
6. 66.93° or 113.07°
7. 12.43°
8. 35.11 cm and 73.64 cm
9. 116.84 ft
10. 131.78 m

Chapter 12 Trigonometric Functions

Exercise 12.1

1. $\frac{2\pi}{9}$
2. $\frac{11\pi}{6}$
3. $\frac{2\pi}{3}$
4. $\frac{5\pi}{4}$
5. 150°
6. 40°
7. 15°

Questions 8–10 have multiple correct answers. These are sample answers:

8. $\frac{5\pi}{2}, \frac{-3\pi}{2}$
9. $\frac{23\pi}{6}, \frac{-\pi}{6}$
10. $\frac{\pi}{3}, -\frac{11\pi}{3}$

Exercise 12.2

1. $\left(-\frac{\sqrt{3}}{2}, -\frac{1}{2}\right)$
2. $\left(-\frac{\sqrt{2}}{2}, \frac{\sqrt{2}}{2}\right)$
3. Quadrant III
4. Quadrant II
5. $\sin\theta = \frac{5}{13}, \cos\theta = \frac{12}{13}, \tan\theta = \frac{5}{12}, \csc\theta = \frac{13}{5}, \sec\theta = \frac{13}{12}, \cot\theta = \frac{12}{5}$
6. $\sin\theta = -\frac{4}{5}, \cos\theta = \frac{3}{5}, \tan\theta = -\frac{4}{3}, \csc\theta = -\frac{5}{4}, \sec\theta = \frac{5}{3}, \cot\theta = -\frac{3}{4}$
7. sine: $-\frac{\sqrt{2}}{2}$, cosine: $\frac{\sqrt{2}}{2}$, tangent: -1
8. sine: 1, cosine: 0, tangent: undefined
9. sine: $-\frac{2\sqrt{7}}{7}$, cosine: $\frac{\sqrt{21}}{7}$, tangent: $-\frac{2\sqrt{3}}{3}$
10. sine: $\frac{5\sqrt{29}}{29}$, cosine: $-\frac{2\sqrt{29}}{29}$, tangent: $-\frac{5}{2}$

Exercise 12.3

1.

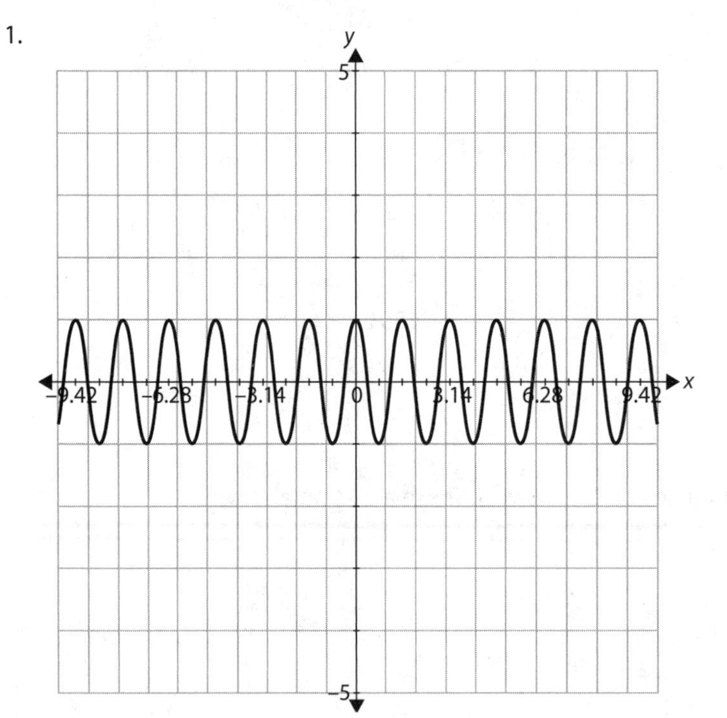

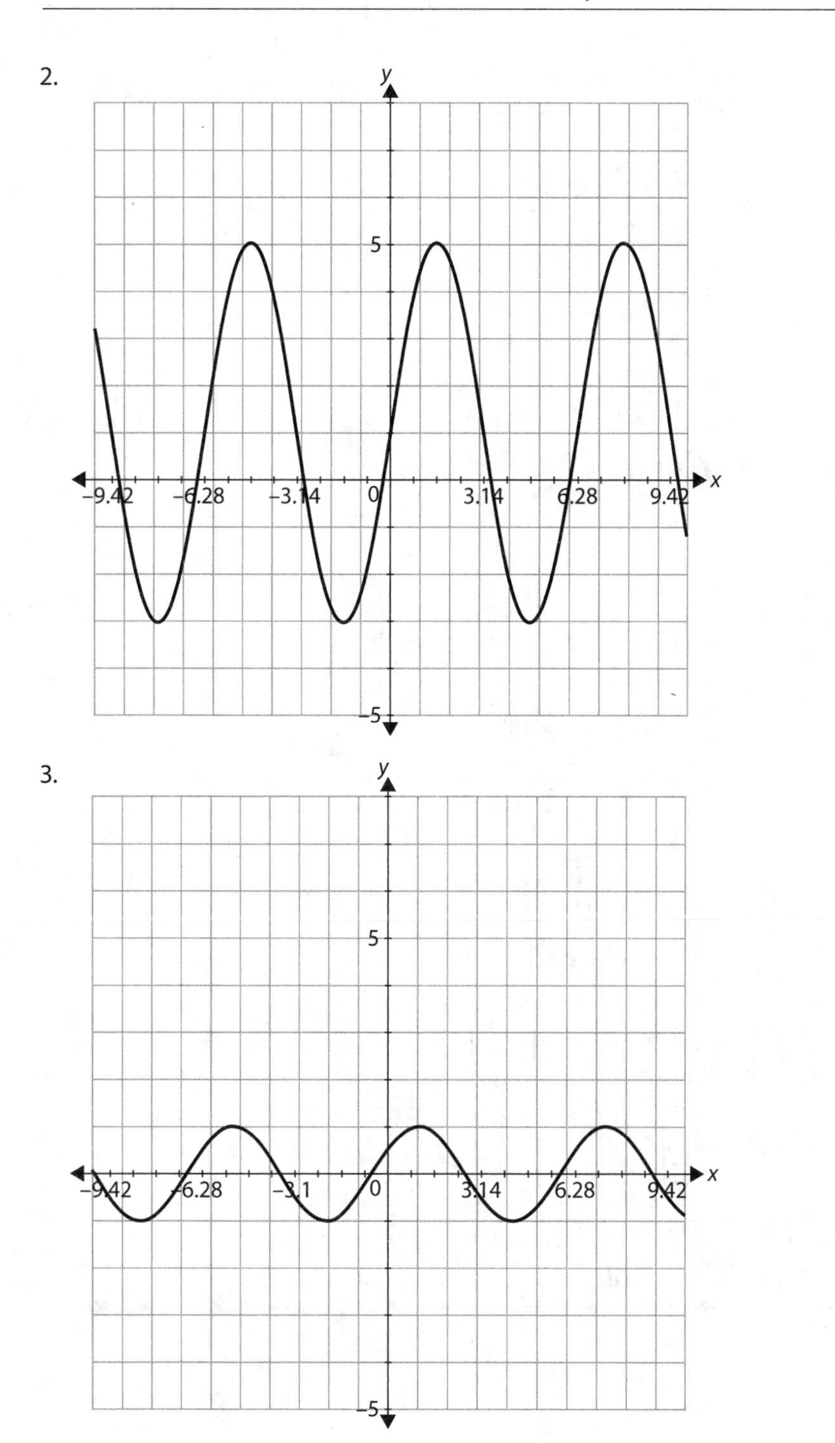
2.
y
5
−9.42
−6.28
−3.14
0
3.14
6.28
9.42
x
−5
3.
y
5
−9.42
−6.28
−3.1
0
3.14
6.28
9.42
x
−5

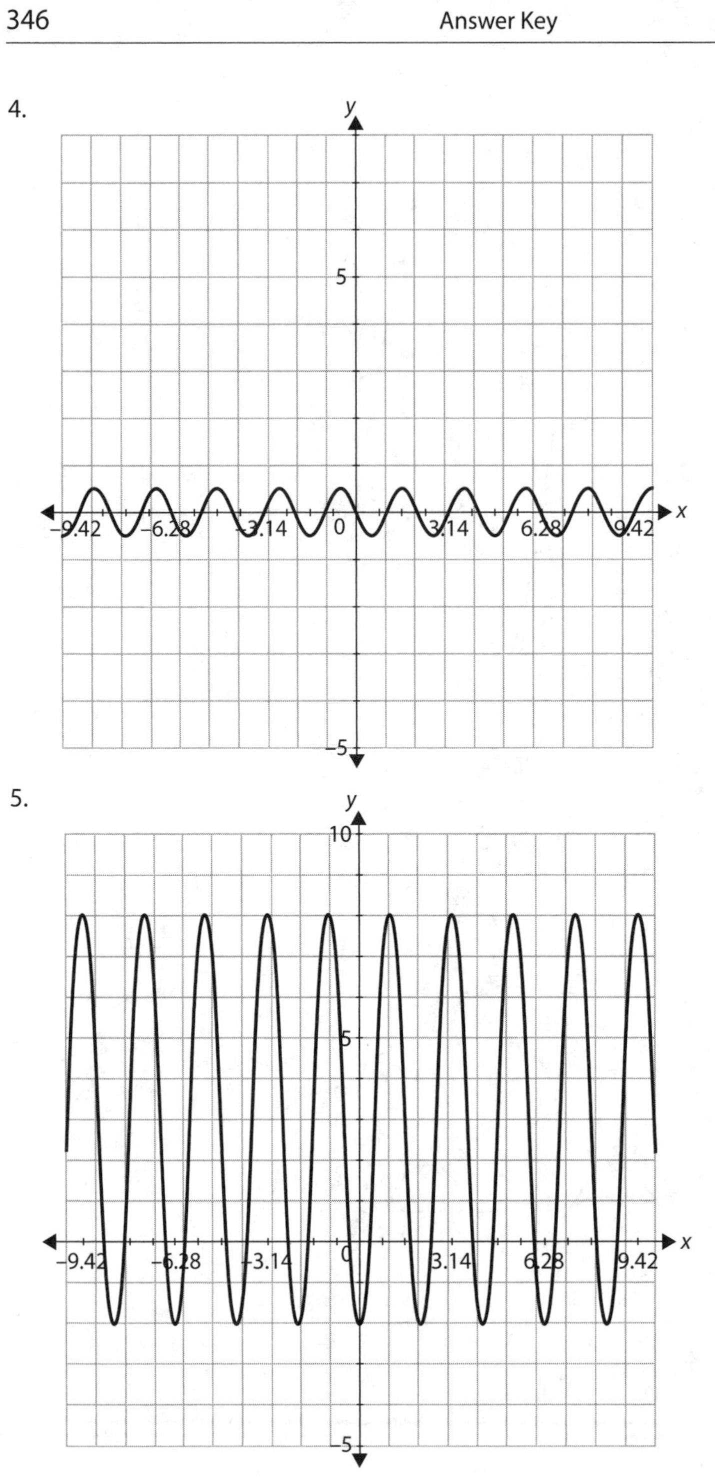
4.
y
5
−5
x
−9.42
−6.28
−3.14
0
3.14
6.28
9.42
5.
y
10
5
−5
x
−9.42
−6.28
−3.14
0
3.14
6.28
9.42

6.

7.

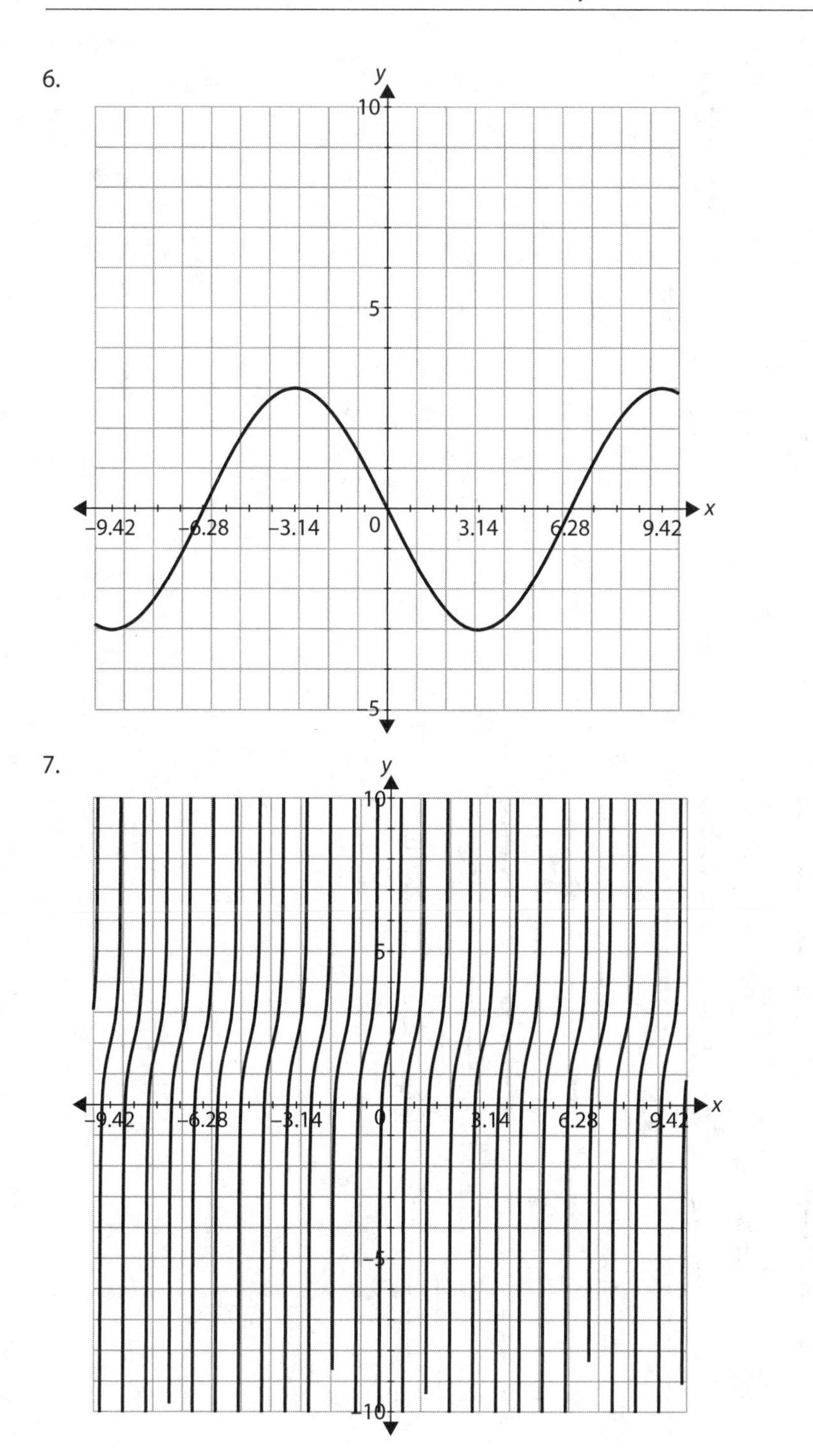

8.

9.

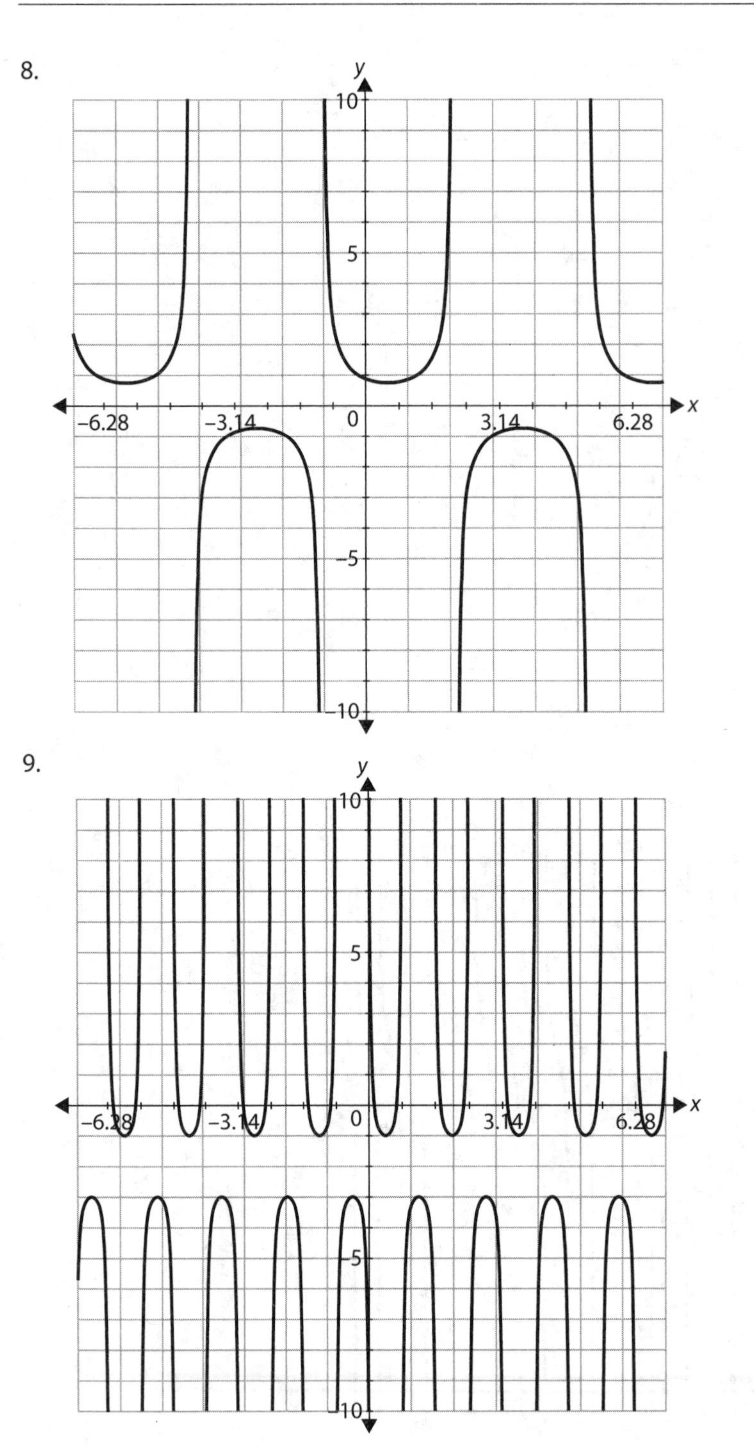

10.

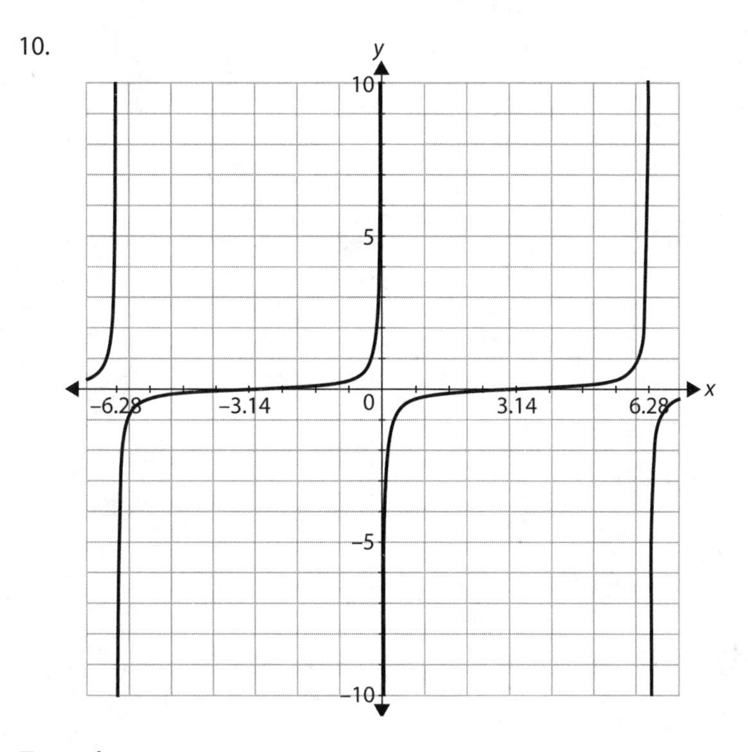

Exercise 12.4

1. $\frac{-\sqrt{3}}{2}$
2. $\sqrt{3}$
3. $\frac{2\pi}{3}$
4. 1
5. $-\frac{\pi}{4}$
6. 0
7. $-\sqrt{3}$
8. $\frac{\pi}{3}$
9. $\frac{\pi}{2}$
10. $\frac{\pi}{2}$

Exercise 12.5

1. $$\csc x + \cot x \cdot \sec x = 2\csc x$$
$$\frac{1}{\sin x} + \frac{\cancel{\cos x}}{\sin x} \cdot \frac{1}{\cancel{\cos x}} = 2\csc x$$
$$\frac{1}{\sin x} + \frac{1}{\sin x} = 2\csc x$$
$$\frac{2}{\sin x} = 2\csc x$$
$$2\csc x = 2\csc x$$

2. $$\frac{\cos x \csc x}{\cot^2 x} = \tan x$$
$$\frac{\cos x \cdot \dfrac{1}{\sin x}}{\cot^2 x} = \tan x$$
$$\frac{\cot x}{\cot^2 x} = \tan x$$
$$\frac{1}{\cot x} = \tan x$$
$$\tan x = \tan x$$

3. $$\tan\left(\frac{\pi}{2} - \theta\right) \cdot (1 - \cos^2\theta) \cdot \sec\theta = \sin\theta$$
$$\cot\theta \cdot \sin^2\theta \cdot \sec\theta = \sin\theta$$
$$\frac{\cos\theta}{\sin\theta} \cdot \frac{\sin^2\theta}{1} \cdot \frac{1}{\cos\theta} = \sin\theta$$
$$\sin\theta = \sin\theta$$

4. $$\cos\left(\frac{\pi}{2} - \theta\right) \cdot (\sin\theta + \cot\theta\cos\theta) = 1$$
$$\sin\theta \cdot \left(\sin\theta + \frac{\cos\theta}{\sin\theta}\cos\theta\right) = 1$$
$$\sin\theta \cdot \left(\sin\theta + \frac{\cos^2\theta}{\sin\theta}\right) = 1$$
$$\sin^2\theta + \cos^2\theta = 1$$

5. $$\frac{\sin x - 1}{\csc x} \cdot (\csc x + 1) = -\cos^2 x$$
$$\frac{1 - \csc x + \sin x - 1}{\csc x} = -\cos^2 x$$
$$\frac{\sin x - \csc x}{\csc x} = -\cos^2 x$$
$$\frac{\sin x - \dfrac{1}{\sin x}}{\dfrac{1}{\sin x}} = -\cos^2 x$$
$$\sin^2 x - 1 = -\cos^2 x$$
$$-(1 - \sin^2 x) = -\cos^2 x$$
$$-\cos^2 x = -\cos^2 x$$

6. $\csc\theta\tan\theta - \sin\theta = \dfrac{\tan\theta - \sin^2\theta}{\sin\theta}$

$$\csc\theta\tan\theta - \sin\theta = \frac{\frac{\sin\theta}{\cos\theta} - \sin^2\theta}{\sin\theta}$$

$$\csc\theta\tan\theta - \sin\theta = \left(\frac{\sin\theta}{\cos\theta} - \sin^2\theta\right)\cdot\frac{1}{\sin\theta}$$

$$\csc\theta\tan\theta - \sin\theta = \frac{1}{\cos\theta} - \sin\theta$$

$$\csc\theta\tan\theta - \sin\theta = \frac{1}{\cos\theta}\cdot\frac{\sin\theta}{\sin\theta} - \sin\theta$$

$$\csc\theta\tan\theta - \sin\theta = \frac{1}{\sin\theta}\cdot\frac{\sin\theta}{\cos\theta} - \sin\theta$$

$$\csc\theta\tan\theta - \sin\theta = \csc\theta\tan\theta - \sin\theta$$

7. $\sec^2\theta(\cos^2\theta - 1) = -\tan^2\theta$

$$-\sec^2\theta(1-\cos^2\theta) = -\tan^2\theta$$

$$-\sec^2\theta\sin^2\theta = -\tan^2\theta$$

$$-\frac{1}{\cos^2\theta}\cdot\sin^2\theta = -\tan^2\theta$$

$$-\frac{\sin^2\theta}{\cos^2\theta} = -\tan^2\theta$$

$$-\tan^2\theta = -\tan^2\theta$$

8. $\sec\theta\tan\theta - \csc^2\theta = \dfrac{\sin^3\theta - \cos^2\theta}{(\sin\theta\cos\theta)^2}$

$$\sec\theta\tan\theta - \csc^2\theta = \frac{\sin^3\theta}{\sin^2\theta\cos^2\theta} - \frac{\cos^2\theta}{\sin^2\theta\cos^2\theta}$$

$$\sec\theta\tan\theta - \csc^2\theta = \frac{\sin\theta}{\cos^2\theta} - \frac{1}{\sin^2\theta}$$

$$\sec\theta\tan\theta - \csc^2\theta = \frac{1}{\cos\theta}\cdot\frac{\sin\theta}{\cos\theta} - \frac{1}{\sin^2\theta}$$

$$\sec\theta\tan\theta - \csc^2\theta = \sec\theta\tan\theta - \csc^2\theta$$

9.

$$\sin(\alpha+\beta)\cdot\sin(\alpha-\beta) = \sin^2\alpha - \sin^2\beta$$

$$(\sin\alpha\cos\beta + \cos\alpha\sin\beta)(\sin\alpha\cos\beta - \cos\alpha\sin\beta) = \sin^2\alpha - \sin^2\beta$$

$$\sin^2\alpha\cos^2\beta + \cos\alpha\sin\beta\sin\alpha\cos\beta - \cos\alpha\sin\beta\sin\alpha\cos\beta - \cos^2\alpha\sin^2\beta = \sin^2\alpha - \sin^2\beta$$

$$\sin^2\alpha\cos^2\beta - \cos^2\alpha\sin^2\beta = \sin^2\alpha - \sin^2\beta$$

$$\sin^2\alpha(1-\sin^2\beta) - (1-\sin^2\alpha)\sin^2\beta = \sin^2\alpha - \sin^2\beta$$

$$\sin^2\alpha - \sin^2\alpha\sin^2\beta - \sin^2\beta + \sin^2\alpha\sin^2\beta = \sin^2\alpha - \sin^2\beta$$

$$\sin^2\alpha - \sin^2\beta = \sin^2\alpha - \sin^2\beta$$

10.

$$\csc(2\theta) = \frac{\sec\theta\csc\theta}{2}$$

$$\frac{1}{\sin(2\theta)} = \frac{\sec\theta\csc\theta}{2}$$

$$\frac{1}{2\sin\alpha\cos\alpha} = \frac{\sec\theta\csc\theta}{2}$$

$$\frac{1}{2}\cdot\frac{1}{\sin\alpha}\cdot\frac{1}{\cos\alpha} = \frac{\sec\theta\csc\theta}{2}$$

$$\frac{\sec\theta\csc\theta}{2} = \frac{\sec\theta\csc\theta}{2}$$

Exercise 12.6

1. $\frac{\sqrt{2}-\sqrt{6}}{4}$
2. $2+\sqrt{3}$
3. $\frac{-\sqrt{2}-\sqrt{6}}{4}$
4. $\frac{\sqrt{2+\sqrt{3}}}{2}$
5. $\frac{\sqrt{2+\sqrt{3}}}{2}$

Exercise 12.7

1. $x = 0, x = \pi, x = 2\pi, x = \frac{\pi}{3}, x = \frac{5\pi}{3}$
2. $x = \frac{2\pi}{3}, x = \frac{4\pi}{3}, x = \pi$
3. $\theta = \frac{\pi}{6}, \theta = \frac{5\pi}{6}, \theta = \frac{7\pi}{6}, \theta = \frac{11\pi}{6}$
4. $\theta \approx 2.21, \theta \approx 4.07, \theta = \frac{\pi}{2}, \theta = \frac{3\pi}{2}$
5. $\theta = \frac{\pi}{3}, \theta = \frac{5\pi}{3}$
6. $\theta = \pm 2n\pi, \theta = \frac{\pi}{3} \pm n\pi, \theta = \frac{2\pi}{3} \pm n\pi$
7. no solution
8. $\theta \approx 0.322 \pm n\pi, \theta \approx 0.464 \pm n\pi$
9. $\theta = \frac{\pi}{8} \pm n\pi, \theta = \frac{3\pi}{8} \pm n\pi$
10. $\theta = \frac{\pi}{3} \pm 2n\pi, \theta = \frac{5\pi}{3} \pm 2n\pi$

Chapter 13 Polar and Parametric Equations

Exercise 13.1

1.

2.

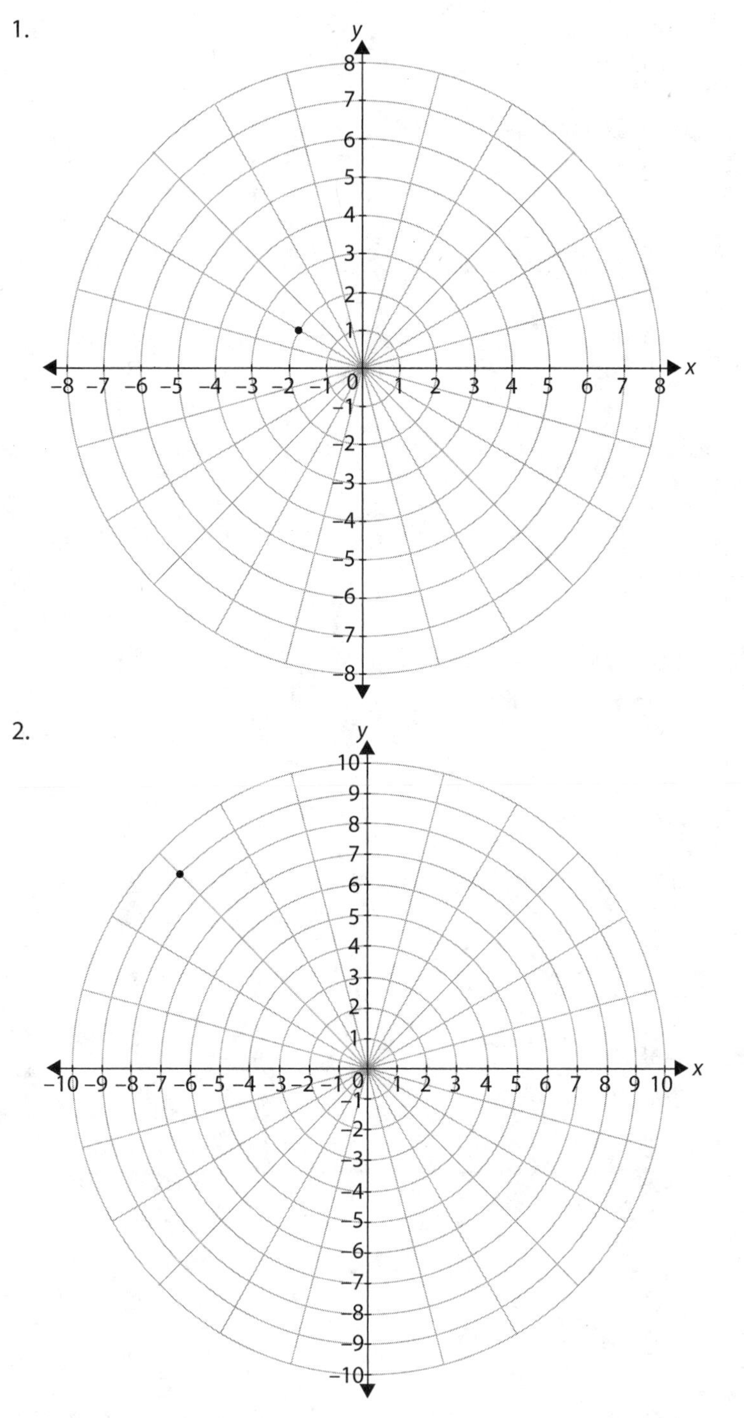

3.

4.

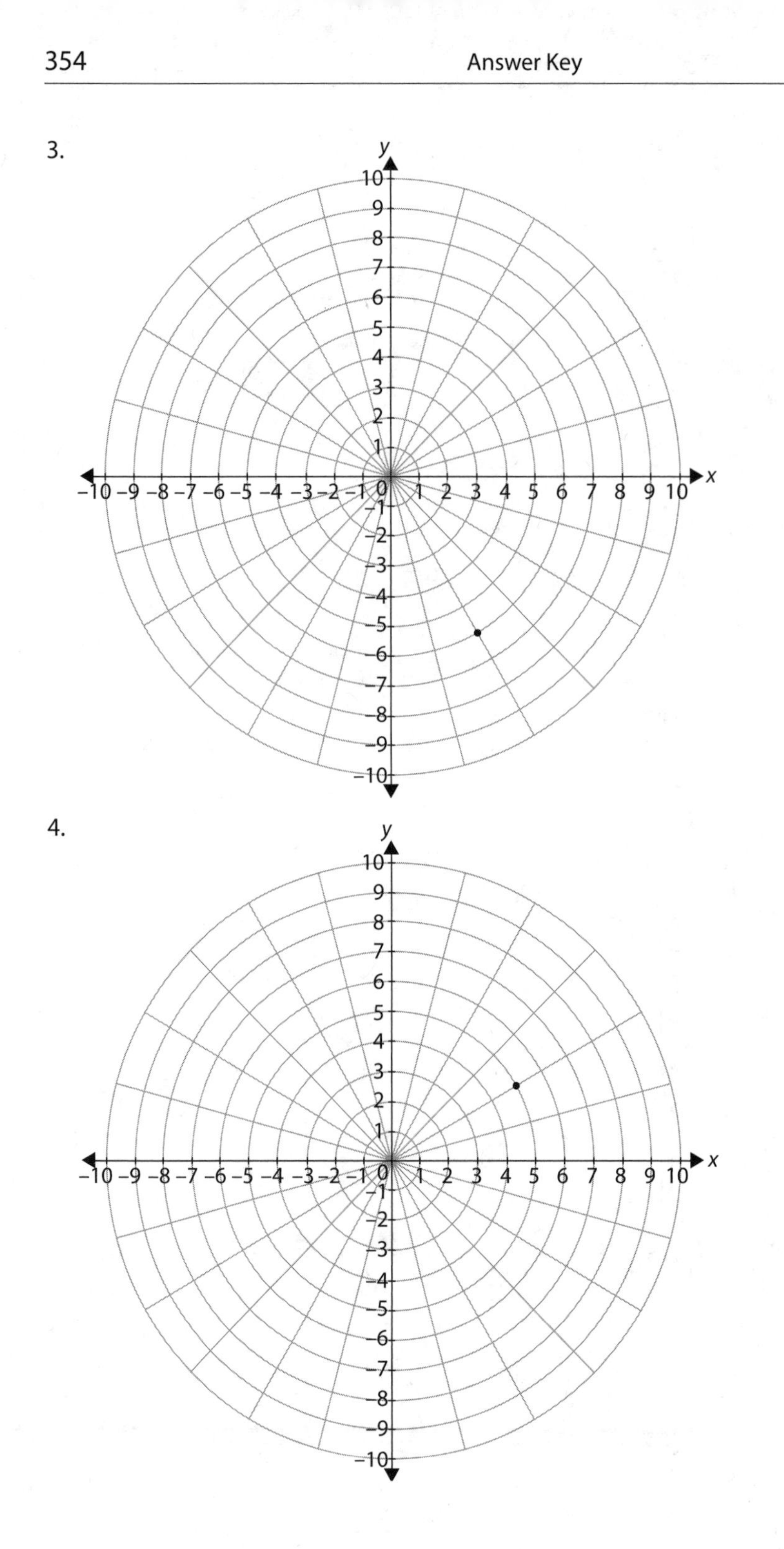

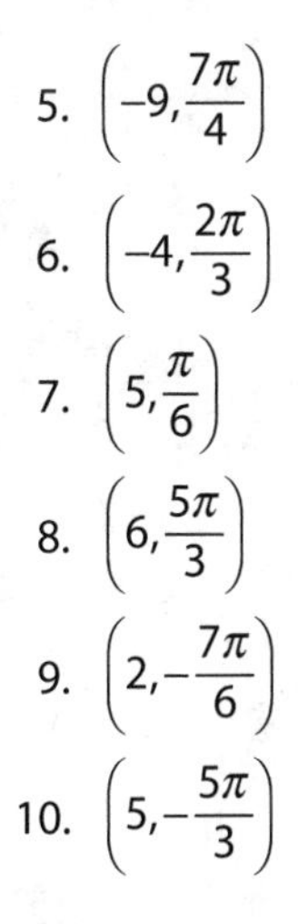

5. $\left(-9, \frac{7\pi}{4}\right)$
6. $\left(-4, \frac{2\pi}{3}\right)$
7. $\left(5, \frac{\pi}{6}\right)$
8. $\left(6, \frac{5\pi}{3}\right)$
9. $\left(2, -\frac{7\pi}{6}\right)$
10. $\left(5, -\frac{5\pi}{3}\right)$

Exercise 13.2

1. $\left(8, \frac{4\pi}{3}\right)$
2. $\left(1, \frac{3\pi}{2}\right)$
3. $\left(4\sqrt{2}, \frac{7\pi}{4}\right)$
4. $\left(12, \frac{11\pi}{6}\right)$
5. $\left(18\sqrt{2}, \frac{11\pi}{6}\right)$
6. $\left(\frac{9\sqrt{2}}{2}, -\frac{9\sqrt{2}}{2}\right)$
7. $(0, -2)$
8. $(-4, 4\sqrt{3})$
9. $(-6\sqrt{3}, 6)$
10. $(8\sqrt{2}, -8\sqrt{2})$

Exercise 13.3

1.

2.

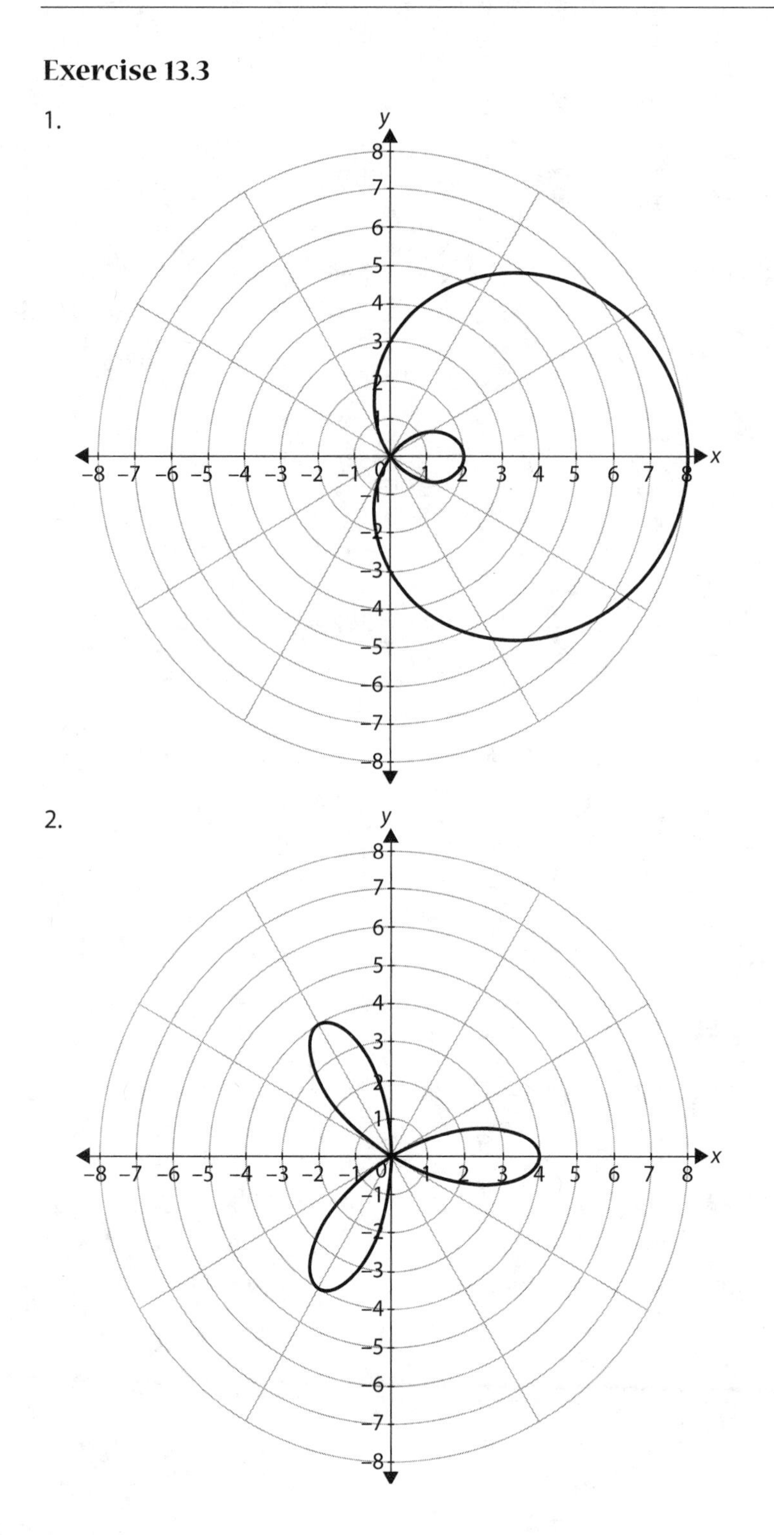

3.

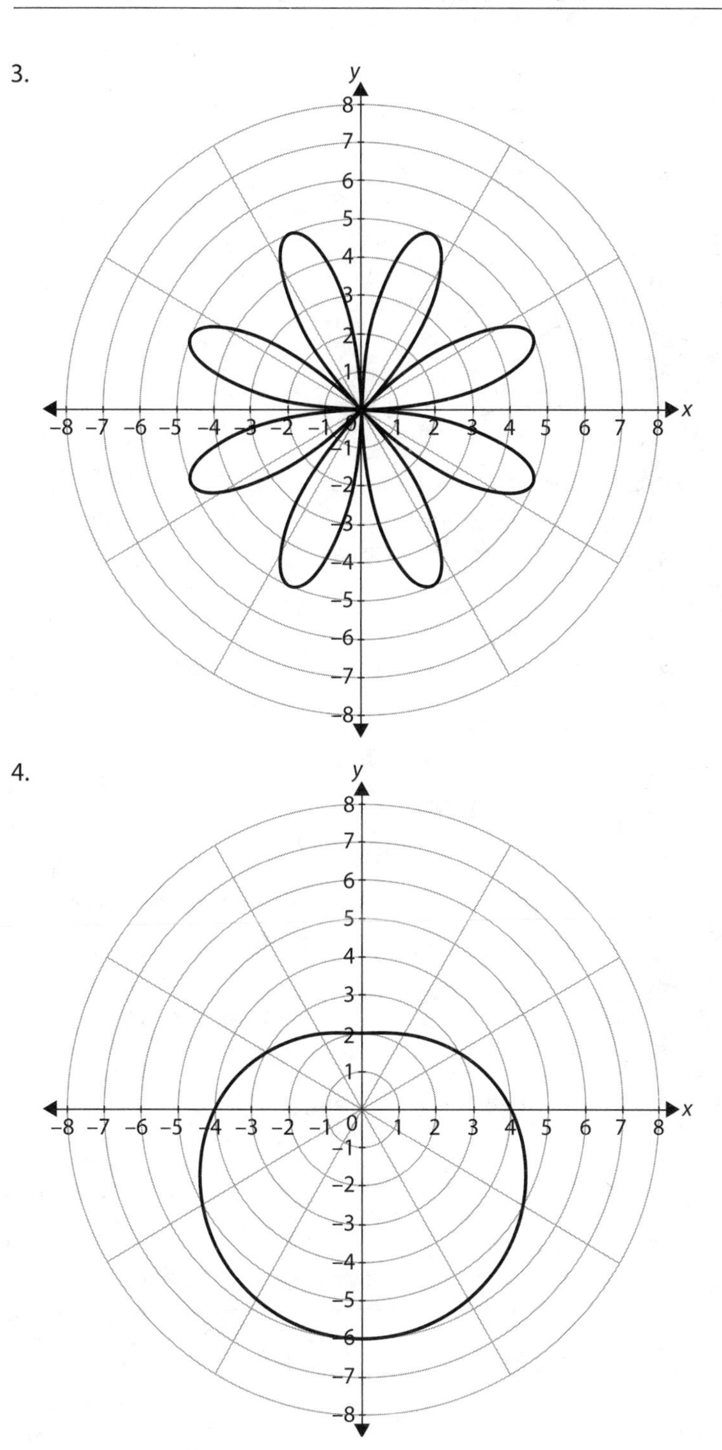

4.

5.

6.

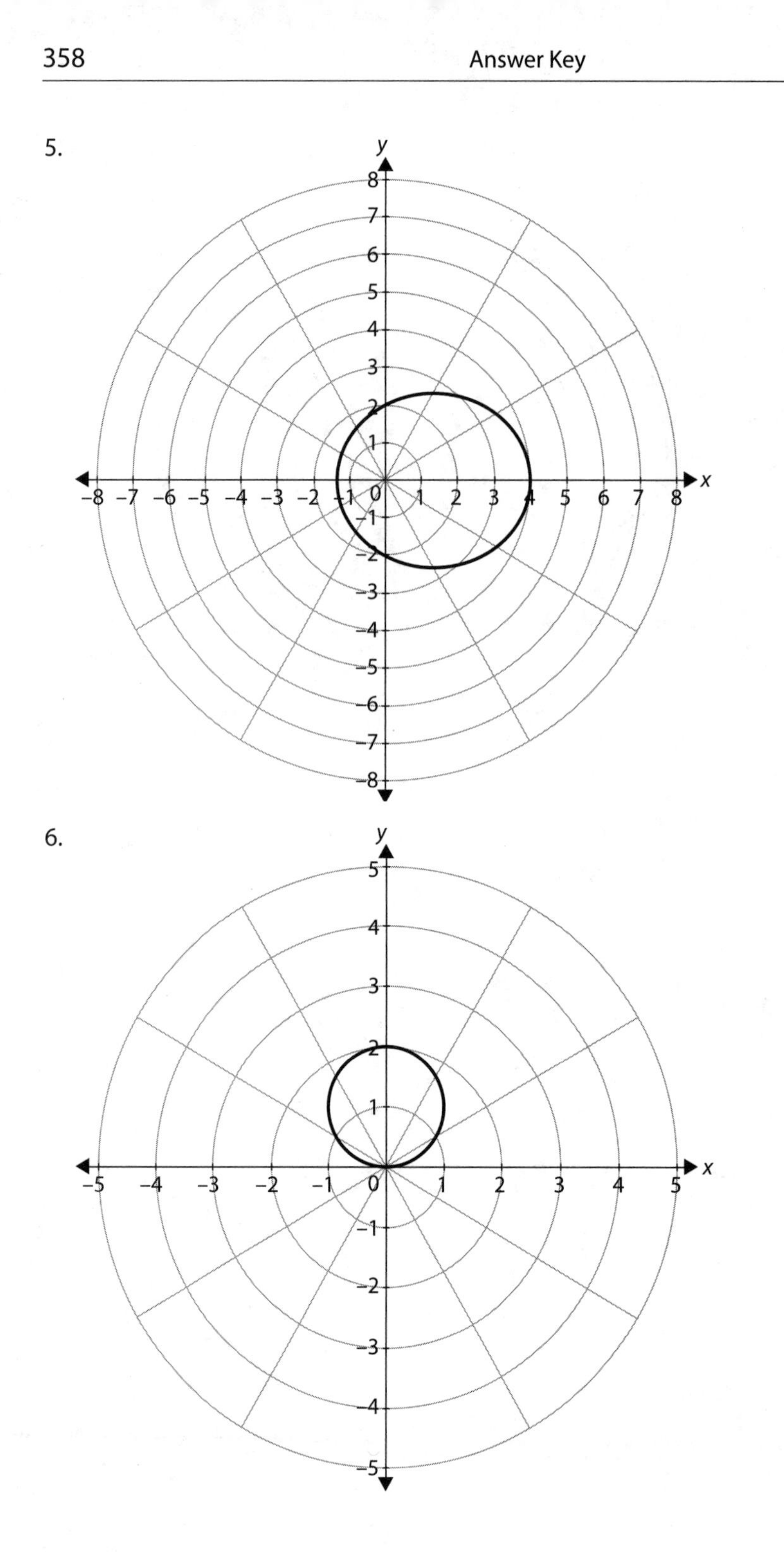
y
x
8
7
6
5
4
3
2
1
0
−1
−2
−3
−4
−5
−6
−7
−8
−8 −7 −6 −5 −4 −3 −2 −1 0 1 2 3 4 5 6 7 8
y
x
5
4
3
2
1
0
−1
−2
−3
−4
−5
−5 −4 −3 −2 −1 0 1 2 3 4 5

7.

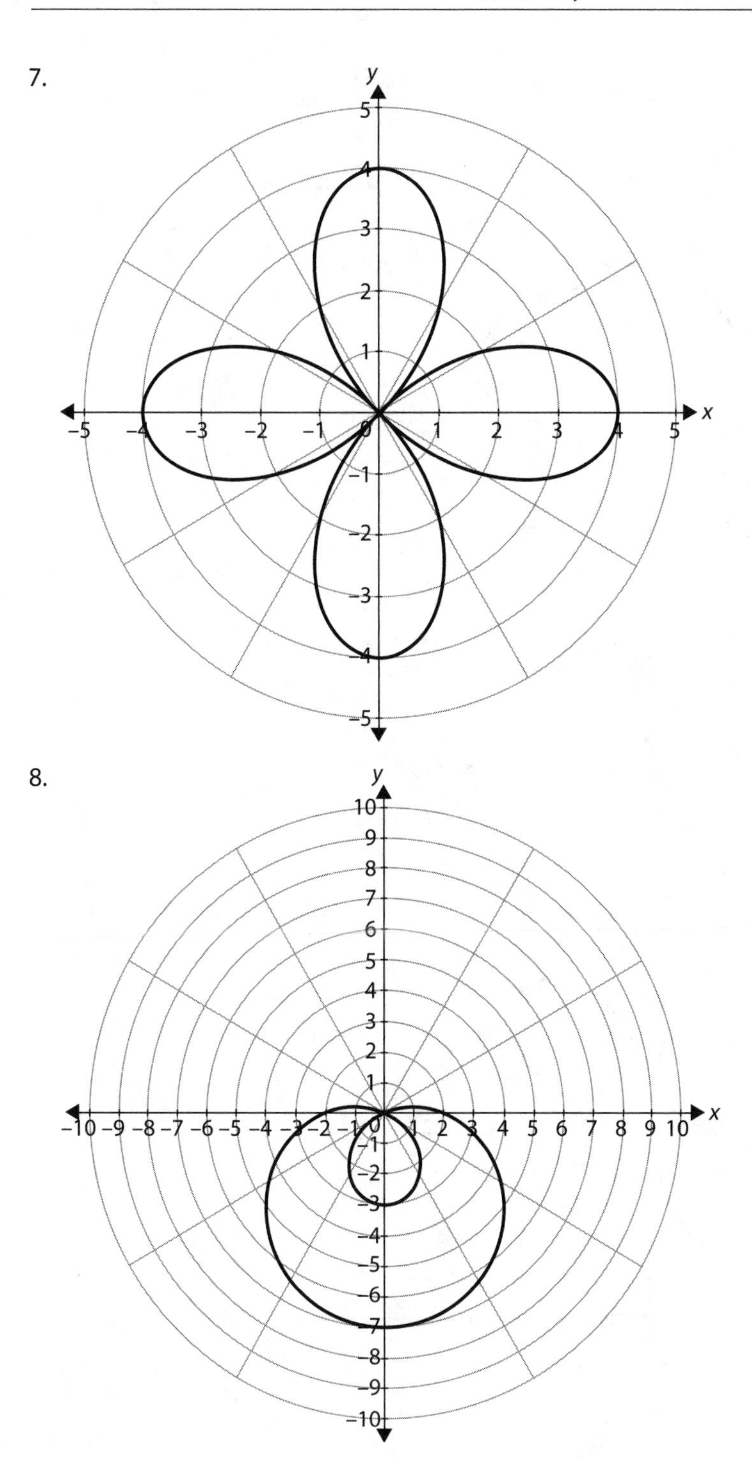

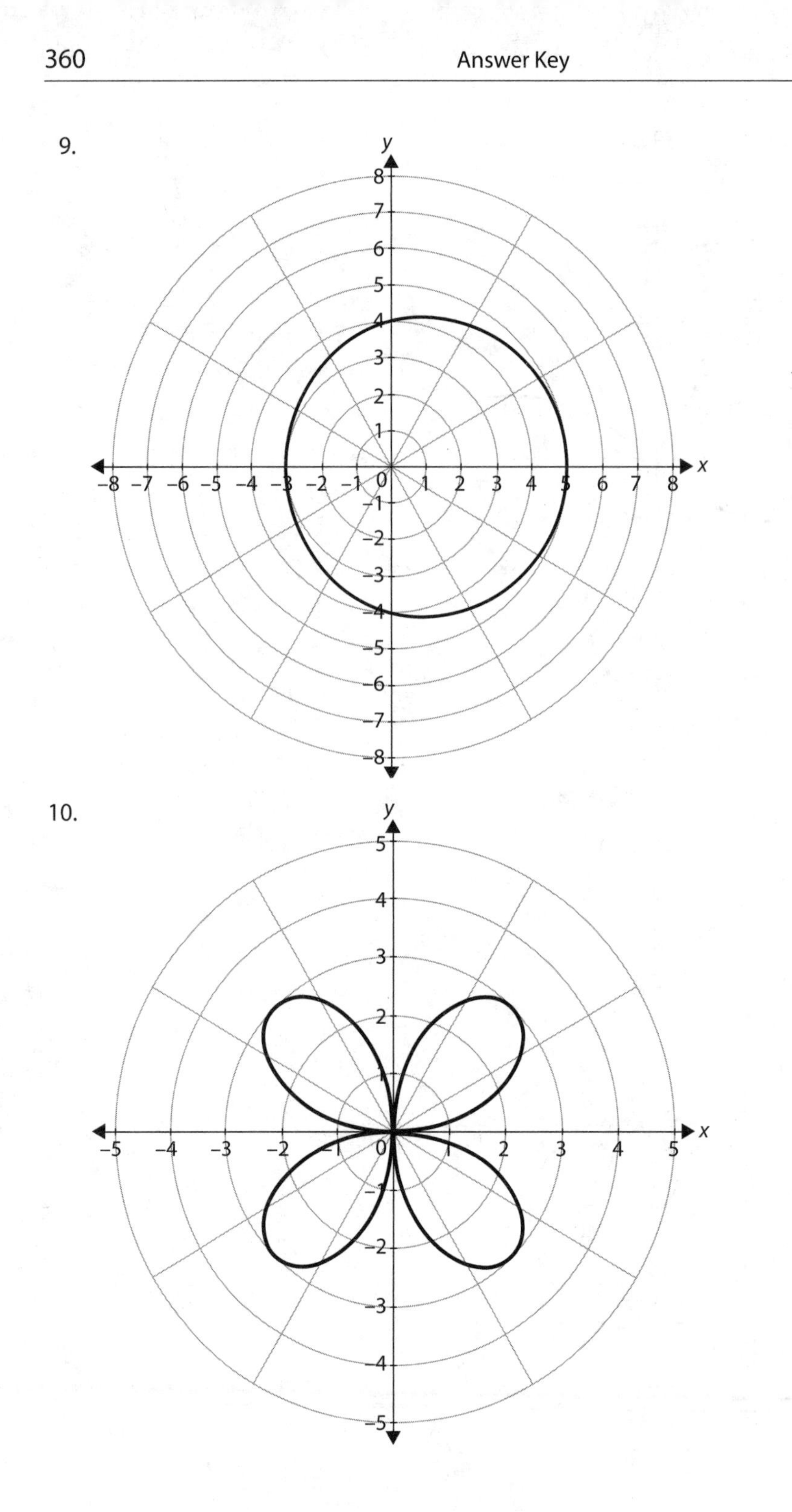
9.
y
x
8
7
6
5
4
3
2
1
0
−1
−2
−3
−4
−5
−6
−7
−8
10.
y
x
5
4
3
2
1
0
−1
−2
−3
−4
−5

Exercise 13.4

1. $r = \dfrac{7}{\cos\theta}$
2. $r^2 - 2r\cos\theta - 24 = 0$
3. $r^2 - r^2\sin^2\theta - r\sin\theta - 4 = 0$
4. $r^2 = \dfrac{36}{4 + 5\sin^2\theta}$
5. $r^2 = \dfrac{16}{5\sin^2\theta - 1}$
6. $y = 3$
7. $y = \sqrt{3}x$
8. $(x^2 + 3x + y^2)^2 = 4(x^2 + y^2)$
9. $(x + y)^2 = x^2(x^2 + y^2)$
10. $9(x^2 + y^2) = (5 - 2y)^2$

Exercise 13.5

1. $y = x^2 - 2x + 1$
2. $y = x - 4$
3. $y = x^2 + 4$
4. $y = \dfrac{1 - x}{x}$
5. $y = \cos\left(\dfrac{x}{3}\right)$
6. $y = \sin\left(\dfrac{x}{\pi}\right)$
7. $y = 3\cos\left(x + \dfrac{\pi}{4}\right)$
8. $y = \sqrt{1 - x^2}$
9. $y = \dfrac{1}{x}$
10. $y = (x + 4)^2$

Exercise 13.6

1.

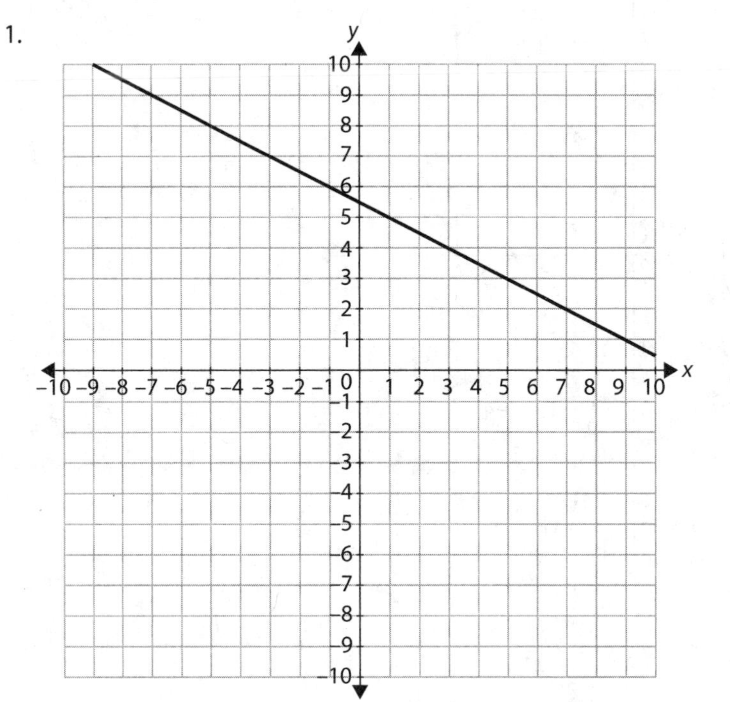

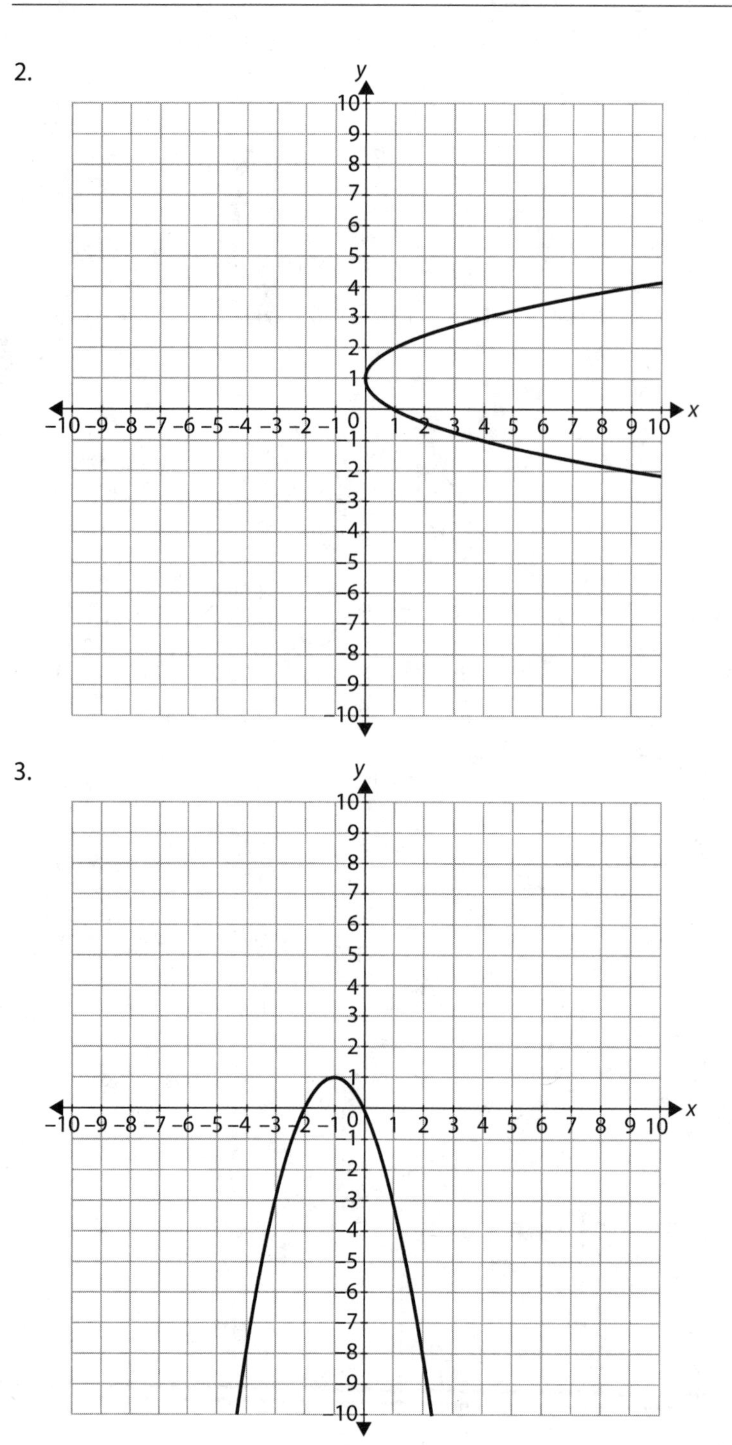
2.
y
x
−10 −9 −8 −7 −6 −5 −4 −3 −2 −1 0 1 2 3 4 5 6 7 8 9 10
3.
y
x
−10 −9 −8 −7 −6 −5 −4 −3 −2 −1 0 1 2 3 4 5 6 7 8 9 10

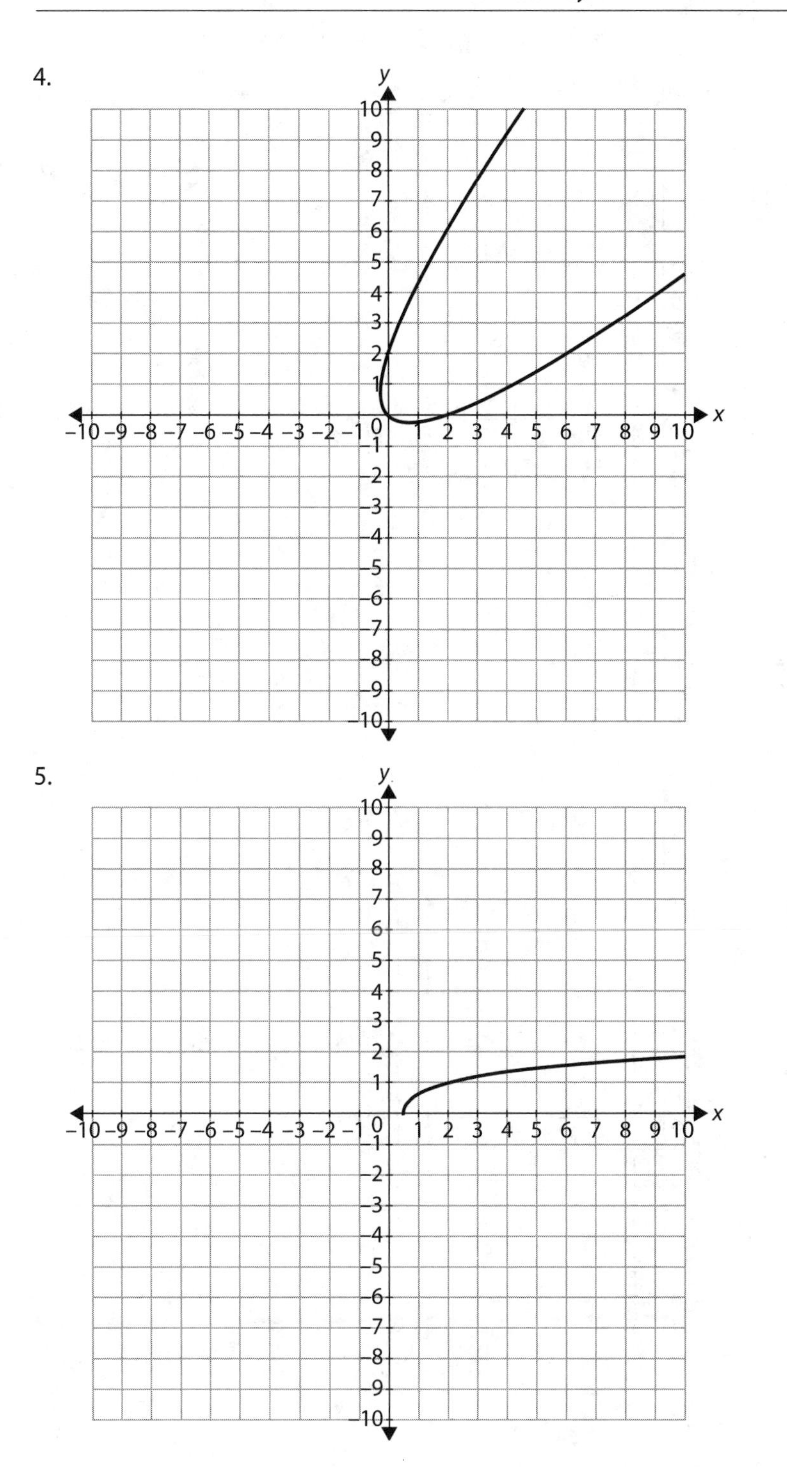
4.
y
x
5.
y
x

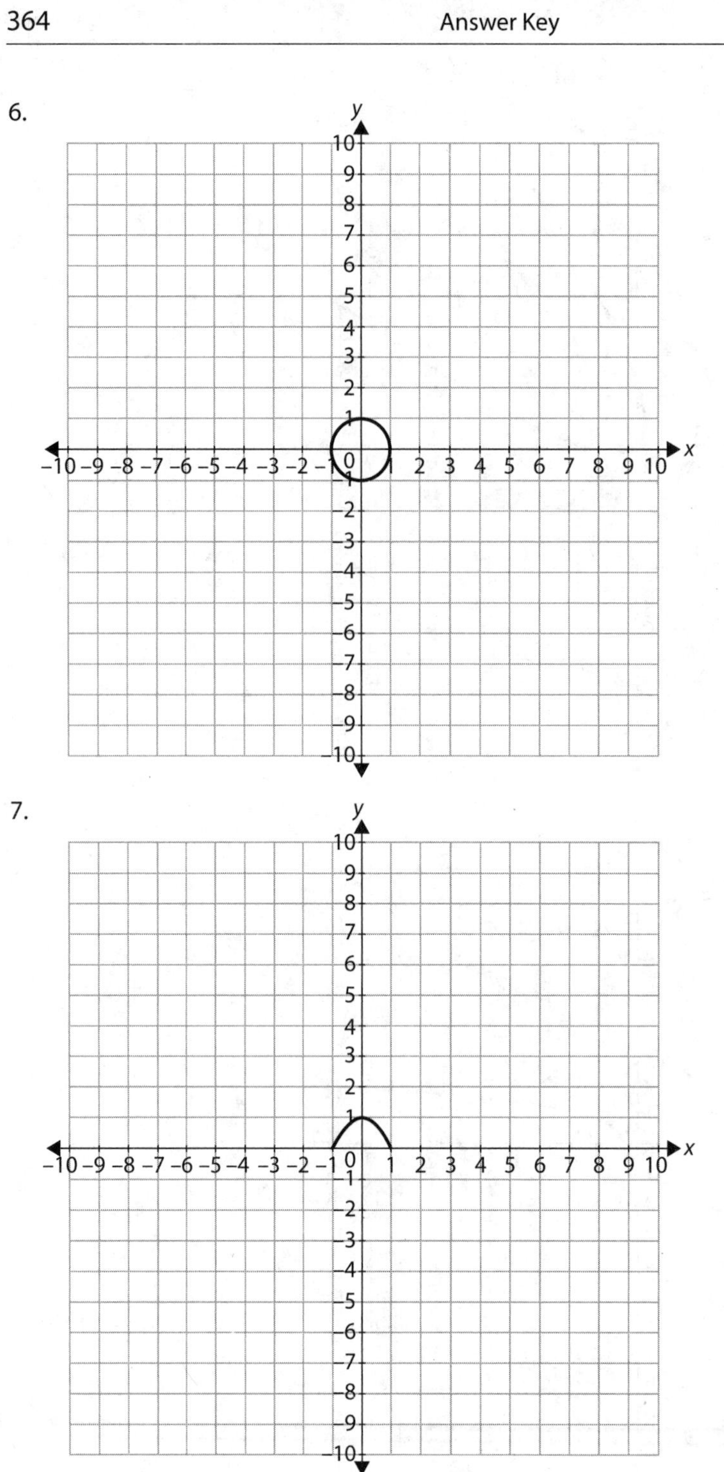
6.
y
x
−10 −9 −8 −7 −6 −5 −4 −3 −2 −1 0 1 2 3 4 5 6 7 8 9 10
10 9 8 7 6 5 4 3 2 1 −1 −2 −3 −4 −5 −6 −7 −8 −9 −10
7.
y
x
−10 −9 −8 −7 −6 −5 −4 −3 −2 −1 0 1 2 3 4 5 6 7 8 9 10
10 9 8 7 6 5 4 3 2 1 −1 −2 −3 −4 −5 −6 −7 −8 −9 −10

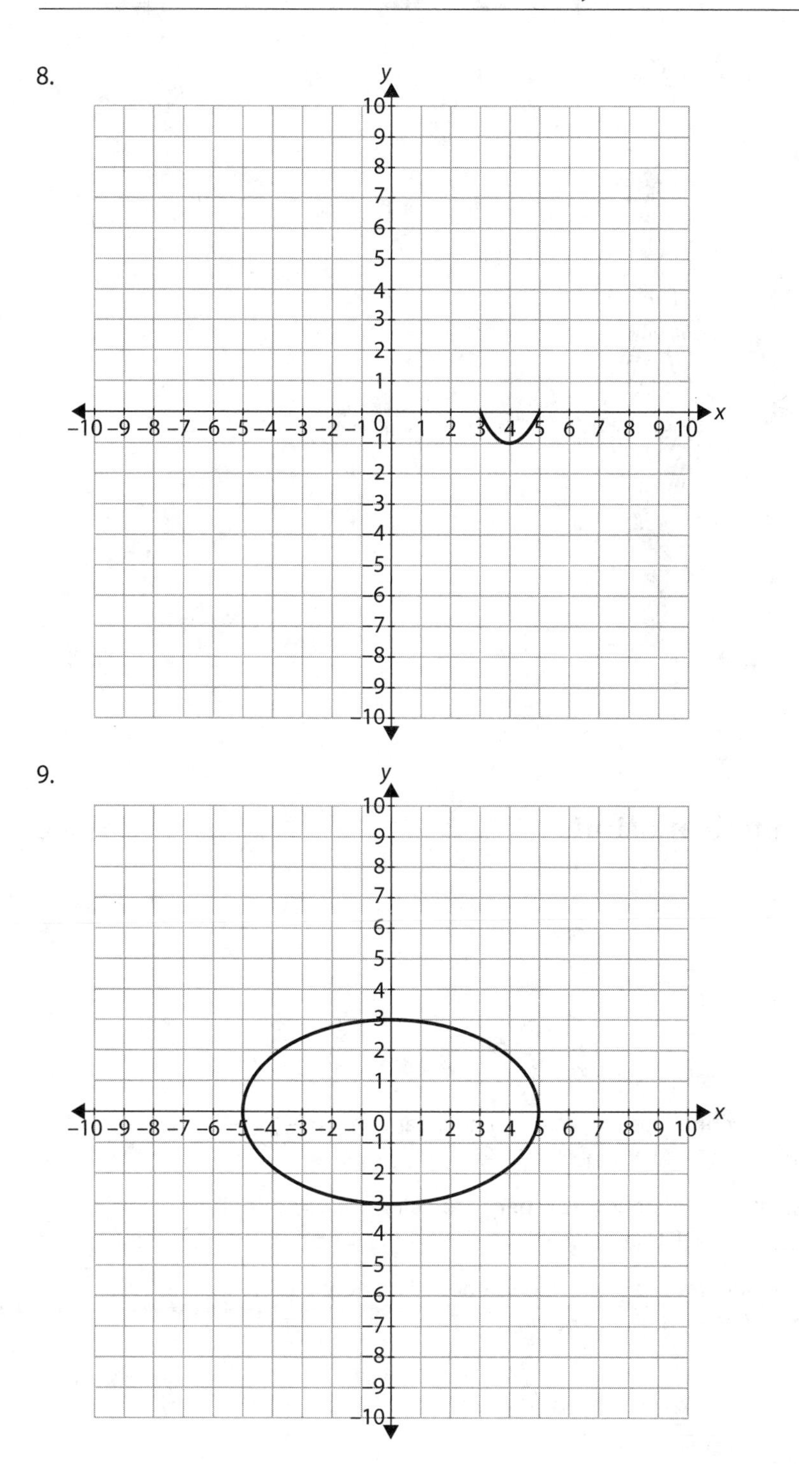
8.
y
x
9.
y
x

10.

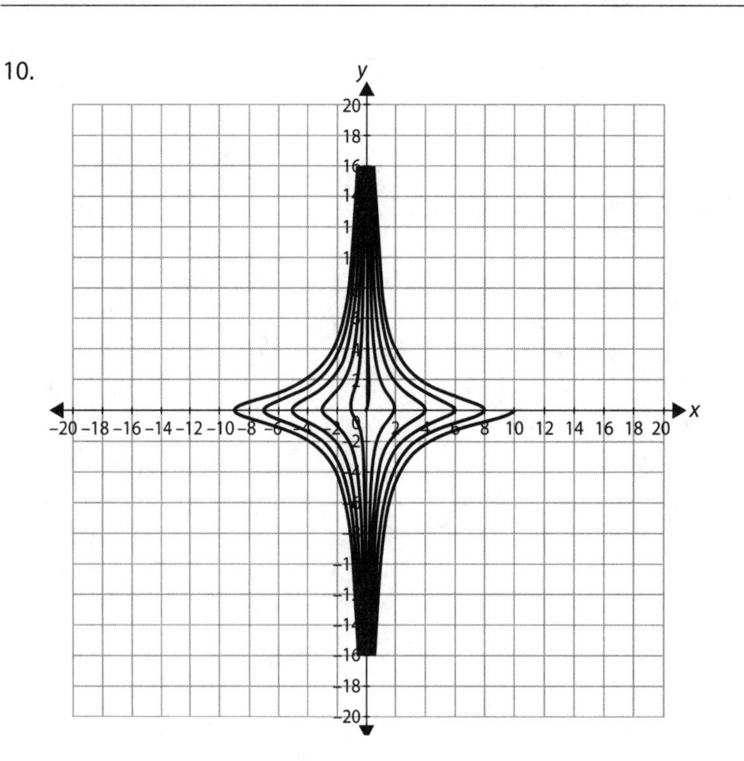

Chapter 14 Transformations

Exercise 14.1

1. (–3, –7)
2. (–2, 5)
3. (–5, 6)
4. A′(1, 4), B′(–6, 1), C′(–1, –3)
5. Pre-image (2, –2), reflecting line $y = 4x$, perpendicular line $y = -\frac{1}{4}(x-2)-2$. Point of intersection $\left(-\frac{6}{17}, -\frac{24}{17}\right)$. Image $\left(-\frac{46}{7}, -\frac{34}{7}\right)$
6. Pre-image (0, –4), reflecting line $y = -2x$, perpendicular line $y = \frac{1}{2}x - 4$. Point of intersection $\left(\frac{8}{5}, -\frac{14}{5}\right)$. Image $\left(\frac{16}{5}, -\frac{12}{5}\right)$
7. Pre-image (–5, 3), reflecting line $y = \frac{1}{3}x$, perpendicular line $y = -3(x+5)+3$. Point of intersection (–3.6, –1.2). Image (–2.2, –5.4)
8. $y = (x+4)^2$
9. $x^2 + (y-6)^2 = 4$
10. $\frac{(x-6)^2}{4} + \frac{(y-4)^2}{9} = 1$

Exercise 14.2

1. 12 units right
2. 6 units up
3. R′(−3, 1), S′(0, −3), T′(−1, −8)
4. R′(4, 8), S′(7, 4), T′(6, −1)
5. R′(−2, 0), S′(1, −4), T′(0, −9). Straight-line distance $5\sqrt{2}$ units (≈ 7.07 units)
6. Based on special right triangle relationships, translate 5 units right and $5\sqrt{3}$ units up.
7. 9 units left
8. 7 units down
9. $y = x^2 - 6$
10. $y = (x - 6)^2 - 6$

Exercise 14.3

1. 90° clockwise
2. 90° counterclockwise
3. 90° clockwise
4. X′(4, −2), Y′(3, −7), Z′(−1, −5)
5. X′(−2, −4), Y′(−7, −3), Z′(−5, 1)
6. P(4, −2), Q(3, −7), R(−1, −5)
7. $y = 2x - 3$
8. $y = -|x + 3|$
9. $\frac{x^2}{9} + \frac{y^2}{16} = 1$
10. $x = y^2 - 4$

Exercise 14.4

1. A′(−6, 15), B′(0, −9)
2. C′(1, 2), D′(3, −2)
3. X′(1.5, 6), Y′(−18, 11)
4. A(−5, 9), B(4, 8), C(6, −10)

5. To reduce the area to $\frac{1}{4}$ of its original size, reduce base and height to half of their original size. The scale factor is $\frac{1}{2}$.

6. If the area is multiplied by 17.64, the side lengths are multiplied by $\sqrt{17.64}$ or 4.2. The scale factor is 4.2.

7. Translate the center and the segment 2 units left and 1 unit down to put the center at the origin. Then multiply by the scale factor. A(–4, 5) translates to (–6, 4) and that scales to (–9, 6). B(2, 1) translates to (0, 0) and that scales to (0, 0). Translate 2 units right and 1 unit up. The image has vertices A′(–7, 7) and B′(2, 1).

8. Translate 5 left and 5 down. R(–2, 6)→(–7, 1), S(3, 5)→(–2, 0), T(1, –1)→(–4, –6)
   Apply scale factor of 3. (–7, 1)→(–21, 3), (–2, 0)→(–6, 0), (–4, –6)→(–12, –18)
   Translate 5 right and 5 up. (–21, 3)→(–16, 8), (–6, 0)→(–1, 5), (–12, –18)→(–7, –13)

9. $x^2+(y+6)^2=9$

10. $\frac{x^2}{36}+\frac{(y-2)^2}{16}=1$

Exercise 14.5

1. $\begin{bmatrix}0 & -1\\ -1 & 0\end{bmatrix}\begin{bmatrix}2 & 4 & 8 & 8 & 4\\ 4 & 6 & 5 & 3 & 1\end{bmatrix}=\begin{bmatrix}-4 & -6 & -5 & -3 & -1\\ -2 & -4 & -8 & -8 & -4\end{bmatrix}$

2. $\begin{bmatrix}1 & 0\\ 0 & -1\end{bmatrix}\begin{bmatrix}2 & 4 & 8 & 8 & 4\\ 4 & 6 & 5 & 3 & 1\end{bmatrix}=\begin{bmatrix}2 & 4 & 8 & 8 & 4\\ -4 & -6 & -5 & -3 & -1\end{bmatrix}$

   $\begin{bmatrix}0 & 1\\ 1 & 0\end{bmatrix}\begin{bmatrix}2 & 4 & 8 & 8 & 4\\ -4 & -6 & -5 & -3 & -1\end{bmatrix}=\begin{bmatrix}-4 & -6 & -5 & -3 & -1\\ 2 & 4 & 8 & 8 & 4\end{bmatrix}$

3. $\begin{bmatrix}0 & 1\\ -1 & 0\end{bmatrix}\begin{bmatrix}2 & 4 & 8 & 8 & 4\\ 4 & 6 & 5 & 3 & 1\end{bmatrix}=\begin{bmatrix}4 & 6 & 5 & 3 & 1\\ -2 & -4 & -8 & -8 & -4\end{bmatrix}$

4. $\begin{bmatrix}1.5 & 0\\ 0 & 1.5\end{bmatrix}\begin{bmatrix}2 & 4 & 8 & 8 & 4\\ 4 & 6 & 5 & 3 & 1\end{bmatrix}=\begin{bmatrix}3 & 6 & 12 & 12 & 6\\ 6 & 9 & 7.5 & 4.5 & 1.5\end{bmatrix}$

5. $\begin{bmatrix}-1 & 0\\ 0 & -1\end{bmatrix}\begin{bmatrix}2 & 4 & 8 & 8 & 4\\ 4 & 6 & 5 & 3 & 1\end{bmatrix}=\begin{bmatrix}-2 & -4 & -8 & -8 & -4\\ -4 & -6 & -5 & -3 & -1\end{bmatrix}$

   $\begin{bmatrix}0 & -1\\ -1 & 0\end{bmatrix}\begin{bmatrix}-2 & -4 & -8 & -8 & -4\\ -4 & -6 & -5 & -3 & -1\end{bmatrix}=\begin{bmatrix}4 & 6 & 5 & 3 & 1\\ 2 & 4 & 8 & 8 & 4\end{bmatrix}$

   $\begin{bmatrix}0.2 & 0\\ 0 & 0.2\end{bmatrix}\begin{bmatrix}4 & 6 & 5 & 3 & 1\\ 2 & 4 & 8 & 8 & 4\end{bmatrix}=\begin{bmatrix}0.8 & 1.2 & 1 & 0.6 & 0.2\\ 0.4 & 0.8 & 1.6 & 1.6 & 0.8\end{bmatrix}$

6. Solve $\begin{cases}2a+4b=8\\ -a-b=11\end{cases}$ to get $a=-26$ and $b=15$.

7. The correct transformation maps (–2, 2) to (12, 2) and (–1, –1) to (11, –1). The matrix multiplication $\begin{bmatrix} -26 & 15 \\ 0 & 1 \end{bmatrix}\begin{bmatrix} -2 & -1 \\ 2 & -1 \end{bmatrix} = \begin{bmatrix} 82 & 11 \\ 2 & -1 \end{bmatrix}$ does not produce the correct coordinates.

8. To find the original, rotate the image in the clockwise direction:
$\begin{bmatrix} 0 & 1 \\ -1 & 0 \end{bmatrix}\begin{bmatrix} 3 & 0 & 2 \\ 5 & 2 & -3 \end{bmatrix} = \begin{bmatrix} 5 & 2 & -3 \\ -3 & 0 & -2 \end{bmatrix}$

9. Rotating 90° counterclockwise then reflecting over the *y*-axis:

$\begin{bmatrix} -1 & 0 \\ 0 & 1 \end{bmatrix} \times \left[\begin{bmatrix} 0 & -1 \\ 1 & 0 \end{bmatrix} \times \begin{bmatrix} 5 & 2 & -3 \\ -3 & 0 & -2 \end{bmatrix}\right] = \begin{bmatrix} 0 & 1 \\ 1 & 0 \end{bmatrix} \times \begin{bmatrix} 5 & 2 & -3 \\ -3 & 0 & -2 \end{bmatrix}$ is equivalent to reflecting over the line $y = x$.

10. If the order is reversed, $\begin{bmatrix} 0 & -1 \\ 1 & 0 \end{bmatrix}\begin{bmatrix} -1 & 0 \\ 0 & 1 \end{bmatrix} = \begin{bmatrix} 0 & -1 \\ -1 & 0 \end{bmatrix}$ is equivalent to reflecting over the line $y = -x$.

Chapter 15 Rotating Conics

Exercise 15.1

1. $\left(3\sqrt{2}, -\sqrt{2}\right)$
2. $\left(\dfrac{-6-\sqrt{3}}{2}, \dfrac{-1+6\sqrt{3}}{2}\right)$
3. $\left(-2+2\sqrt{3}, -2-2\sqrt{3}\right)$
4. (8, –4)
5. (–10, 6)
6. (4, –4)
7. $x^2 + 2\sqrt{3}xy + 3y^2 - 2\sqrt{3}x + 2y - 36 = 0$
8. $5x^2 + 6xy + 5y^2 = 8$
9. $-2x^2 + y^2 = 2$
10. $x^2 - 2\sqrt{3}xy + 3y^2 - 2\sqrt{3}x - 4 = 0$

Exercise 15.2

1. 13.3°
2. 26.6°
3. 45°

4. 22.5°
5. 30°
6. $y = 0.85x$ and $y = -1.18x$
7. $y = x$ and $y = -x$
8. $y = 0.41x$ and $y = -2.41x$
9. $y = 0.54x$ and $y = -1.87x$
10. $y = \frac{\sqrt{3}}{3}x$ and $y = -\sqrt{3}x$

Exercise 15.3

1. $x^2 - 3y^2 + 3 = 0$

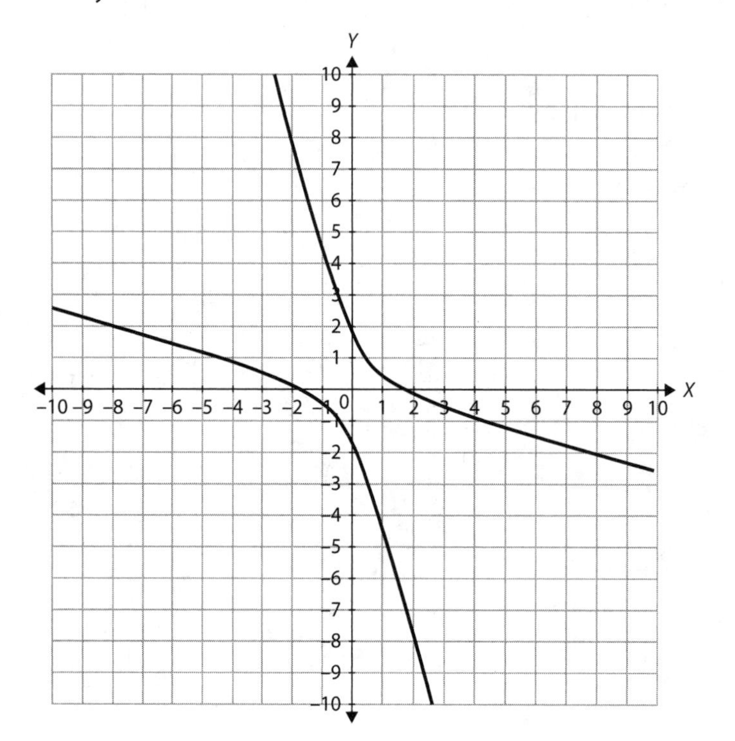

2. $3x^2 + y^2 = 4$

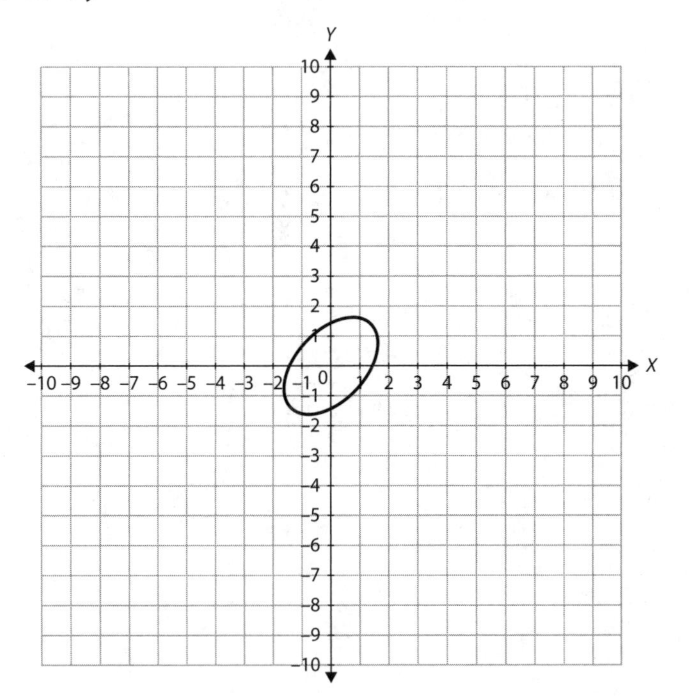

3. $4y^2 - 9x^2 = 36$

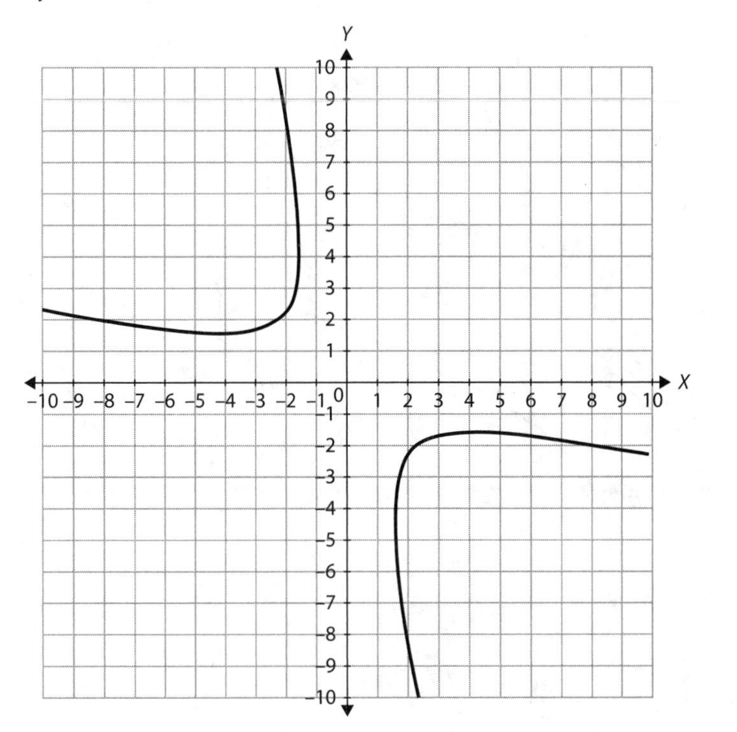

4. $x^2 + 16y^2 = 16$

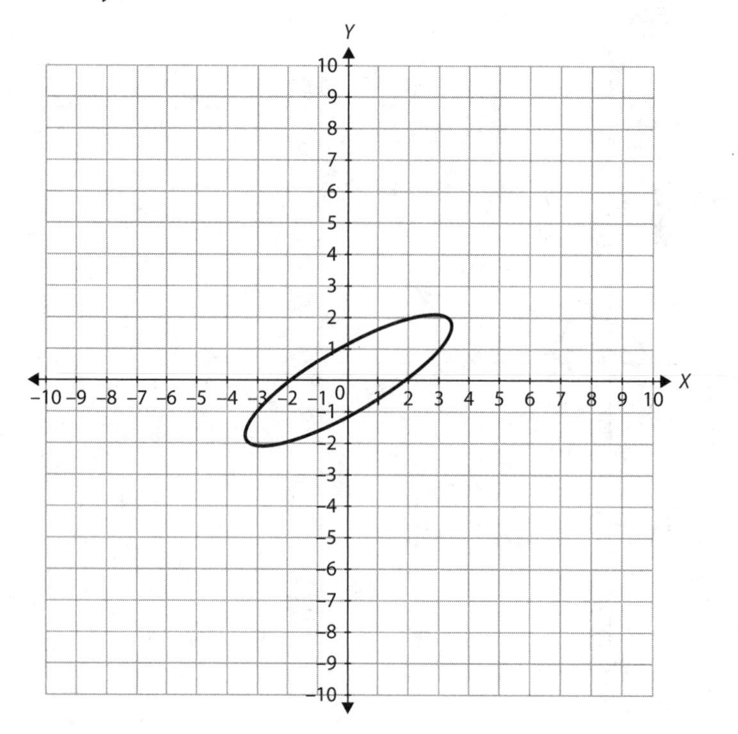

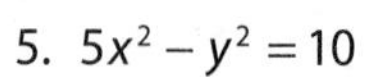
5. $5x^2 - y^2 = 10$

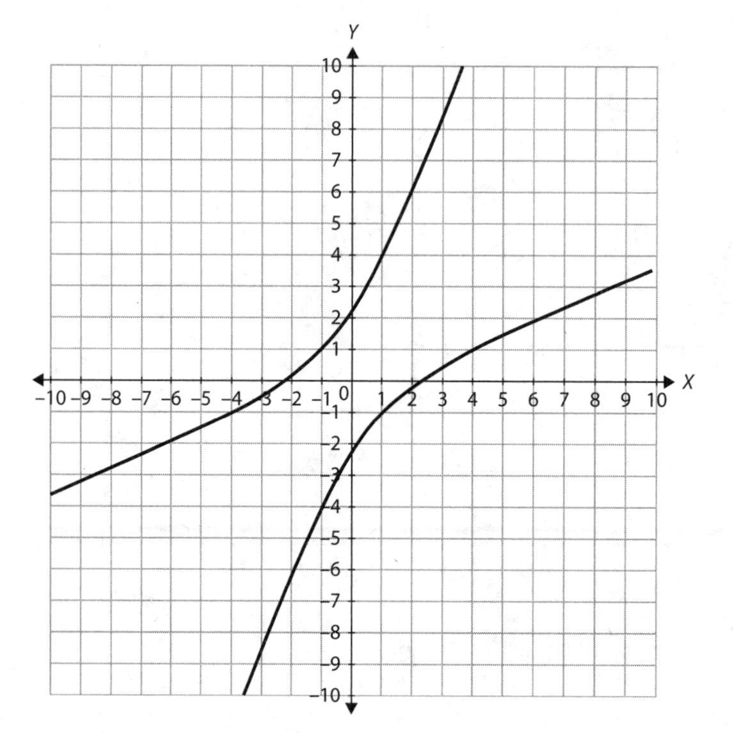

6. $2x^2 + 4y^2 = 5$

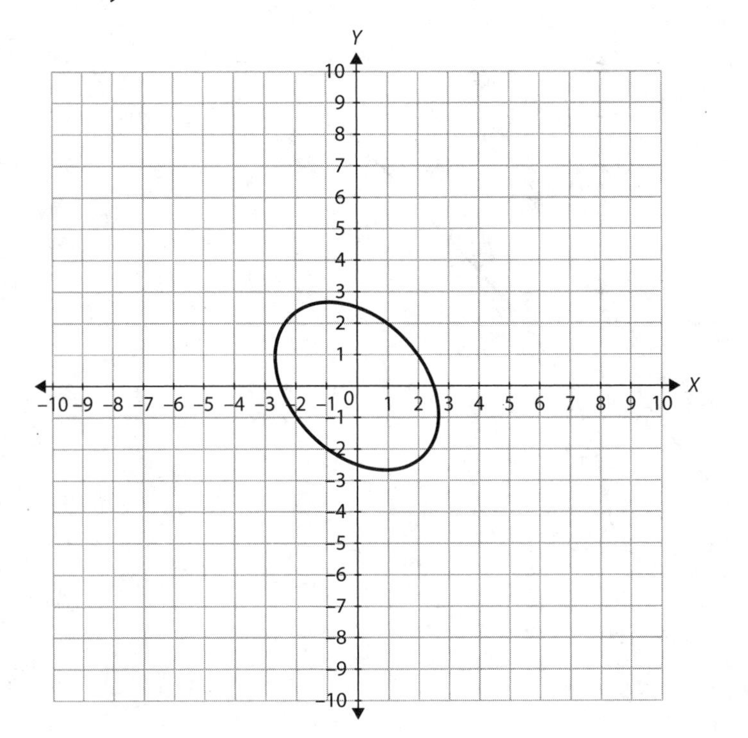

7. $y = x^2 - 4x + 4$

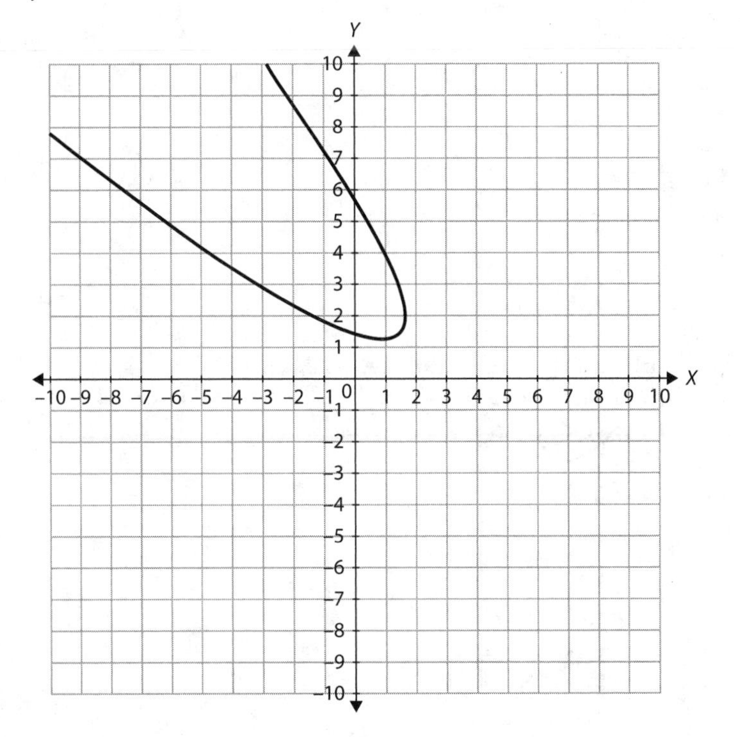

8. $-5x^2 + 11y^2 = 9$

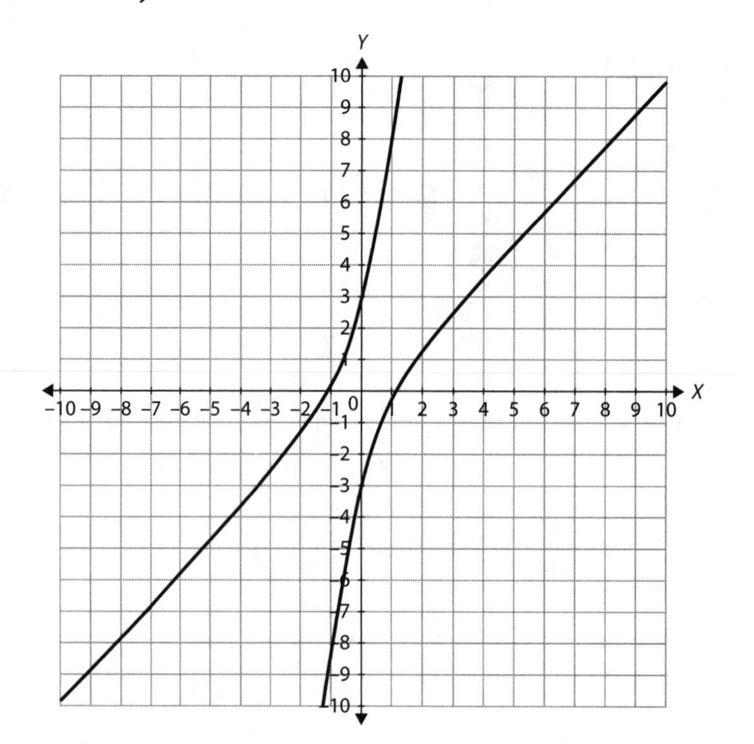

9. $10x^2 + 2y^2 = 5$

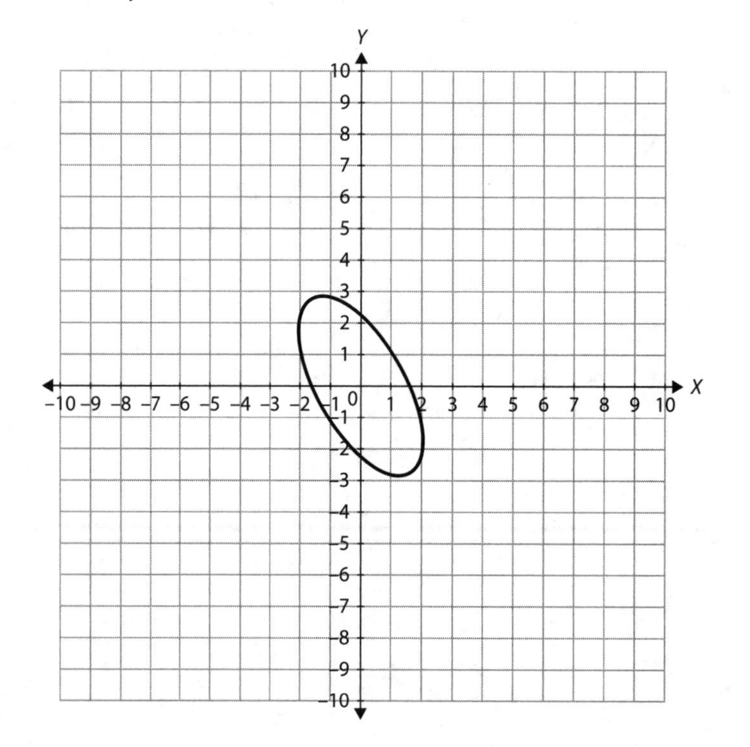

10. $x = y^2$

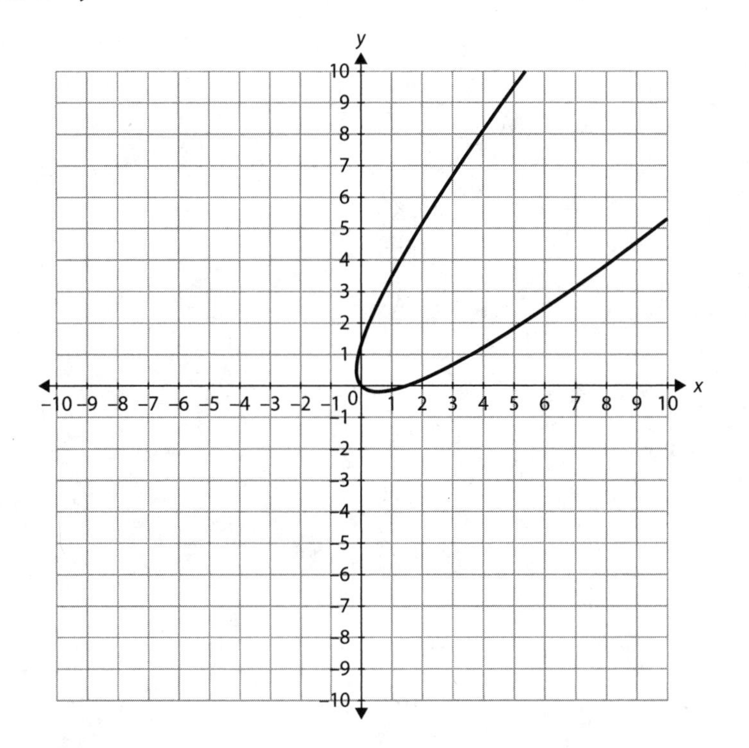

Exercise 15.4

1. $B^2 - 4AC = -3$, Ellipse
2. $B^2 - 4AC = 0$, Parabola
3. $B^2 - 4AC = 2$, Hyperbola
4. $B^2 - 4AC = -3$, Ellipse
5. $B^2 - 4AC = 648$, Hyperbola
6. $B^2 - 4AC = 3$, Hyperbola
7. $B^2 - 4AC = 1$, Hyperbola
8. $B^2 - 4AC = -31$, Ellipse
9. $B^2 - 4AC = 120$, Hyperbola
10. $B^2 - 4AC = 0$, Parabola

Chapter 16 Complex Numbers

Exercise 16.1

1. $4\sqrt{2}\left(\cos\left(-\frac{\pi}{4}\right)+i\sin\left(-\frac{\pi}{4}\right)\right)$
2. $2\left(\cos\left(\frac{\pi}{3}\right)+i\sin\left(\frac{\pi}{3}\right)\right)$
3. $5\left(\cos\left(\frac{3\pi}{2}\right)+i\sin\left(\frac{3\pi}{2}\right)\right)$
4. $10\left(\cos\left(\frac{\pi}{6}\right)+i\sin\left(\frac{\pi}{6}\right)\right)$
5. $5(\cos(-53.1°)+i\sin(-53.1°))$
6. $3+3\sqrt{3}i$
7. $4\sqrt{2}+4\sqrt{2}i$
8. $\frac{5\sqrt{3}}{2}+\frac{5}{2}i$
9. $2i$
10. $-6\sqrt{3}-6i$
11. $\frac{\sqrt{3}}{2}+\frac{3}{2}i$
12. $\frac{\sqrt{3}}{2}+\frac{3}{2}i$
13. $\frac{\sqrt{3}}{2}+\frac{3}{2}i$
14. $3\sqrt{2}+3\sqrt{2}i$
15. $3\sqrt{2}+3\sqrt{2}i$

Exercise 16.2

1. $7+2i$
2. -4
3. $5-8i$
4. $-3-12i$
5. $3+6i$
6. $16-i$
7. $6-9i$
8. $12i$
9. -3
10. $2-5i$

Exercise 16.3

1. $26+7i$
2. $8\left(\cos\frac{\pi}{2}+i\sin\frac{\pi}{2}\right)=8i$
3. $-27-15i$
4. $60(\cos 110°+i\sin 110°)$
5. 13
6. $3(\cos 45°+i\sin 45°)$
7. $\frac{8}{5}-\frac{1}{5}i$
8. $21e^{\frac{3\pi}{4}i}$
9. 6
10. $4e^{\frac{\pi}{12}i}$
11. $2-4i$
12. $63\left(\cos\frac{5\pi}{12}+i\sin\frac{5\pi}{12}\right)$
13. $3i$
14. $-\frac{1}{2}+\frac{7}{6}i$
15. $28(\cos 150°+i\sin 150°)$

Exercise 16.4

1. $-46-9i$
2. 256
3. $625(\cos 80° + i\sin 80°)$
4. $32(\cos 300° + i\sin 300°)$
5. $-1{,}000{,}000$
6. $256\left(\cos\frac{5\pi}{6} + i\sin\frac{5\pi}{6}\right)$
7. -512
8. $\frac{1}{64}(\cos(\pi) + i\sin(\pi)) = -\frac{1}{64}$
9. $81e^{\frac{4\pi}{3}i}$
10. $4{,}096e^{\frac{18\pi}{2}i} = 4{,}096e^{9\pi i} = 4{,}096e^{\pi i} = 4{,}096$

Exercise 16.5

1. $1(\cos(0°) + i\sin(0°)) = 1$

   $1(\cos(120°) + i\sin(120°)) = -\frac{1}{2} + \frac{\sqrt{3}}{2}i$

   $1(\cos(240°) + i\sin(240°)) = -\frac{1}{2} - \frac{\sqrt{3}}{2}i$
2. $2(\cos(54°) + i\sin(54°))$

   $2(\cos(126°) + i\sin(126°))$

   $2(\cos(198°) + i\sin(198°))$

   $2(\cos(270°) + i\sin(270°))$

   $2(\cos(342°) + i\sin(342°))$
3. $4(\cos(45°) + i\sin(45°)) = 2\sqrt{2} + 2\sqrt{2}i$

   $4(\cos(225°) + i\sin(225°)) = -2\sqrt{2} - 2\sqrt{2}i$
4. $3(\cos(45°) + i\sin(45°)) = \frac{3\sqrt{2}}{2} + \frac{3\sqrt{2}}{2}i$

   $3(\cos(135°) + i\sin(135°)) = -\frac{3\sqrt{2}}{2} + \frac{3\sqrt{2}}{2}i$

   $3(\cos(225°) + i\sin(225°)) = -\frac{3\sqrt{2}}{2} - \frac{3\sqrt{2}}{2}i$

   $3(\cos(315°) + i\sin(315°)) = \frac{3\sqrt{2}}{2} - \frac{3\sqrt{2}}{2}i$
5. $10(\cos(0°) + i\sin(0°)) = 10$

   $10(\cos(60°) + i\sin(60°)) = 5 + 5\sqrt{3}i$

   $10(\cos(120°) + i\sin(120°)) = -5 + 5\sqrt{3}i$

   $10(\cos(180°) + i\sin(180°)) = -10$

   $10(\cos(240°) + i\sin(240°)) = -5 - 5\sqrt{3}i$

   $10(\cos(300°) + i\sin(300°)) = 5 - 5\sqrt{3}i$

6. $7(\cos(45°)+i\sin(45°))=\frac{7\sqrt{2}}{2}+\frac{7\sqrt{2}}{2}i$

   $7(\cos(225°)+i\sin(225°))=-\frac{7\sqrt{2}}{2}-\frac{7\sqrt{2}}{2}i$

7. $4(\cos(60°)+i\sin(60°))=2+2\sqrt{3}$

   $4(\cos(180°)+i\sin(180°))=-4$

   $4(\cos(300°)+i\sin(300°))=2-2\sqrt{3}$

8. $3(\cos(0°)+i\sin(0°))$

   $3(\cos(72°)+i\sin(72°))$

   $3(\cos(144°)+i\sin(144°))$

   $3(\cos(208°)+i\sin(208°))$

   $3(\cos(280°)+i\sin(280°))$

9. $10(\cos(30°)+i\sin(30°))=5\sqrt{3}+5i$

   $10(\cos(150°)+i\sin(150°))=-5\sqrt{3}+5i$

   $10(\cos(270°)+i\sin(270°))=-10i$

10. $5(\cos(22.5°)+i\sin(22.5°))$

    $5(\cos(112.5°)+i\sin(112.5°))$

    $5(\cos(202.5°)+i\sin(202.5°))$

    $5(\cos(292.5°)+i\sin(292.5°))$

Chapter 17 Limits

Exercise 17.1

1. $\lim_{x\to 4} f(x)=2$
2. $\lim_{x\to -2} f(x)=1$
3. $\lim_{x\to 3} f(x)=0$
4. $\lim_{x\to 0} f(x)$ does not exist.
5. $\lim_{x\to 2}\frac{x-2}{\sqrt{x^2-4}}=0$
6. $\lim_{x\to 0}\left(\frac{3x-2}{x+4}\right)=-\frac{1}{2}$
7. $\lim_{x\to 4}\left(\frac{x^2-16}{x-4}\right)=8$
8. $\lim_{x\to -3}\left(\frac{x^2+5x+6}{x^2-x-12}\right)=\frac{1}{7}$
9. $\lim_{x\to 25}\left(\frac{x-25}{\sqrt{x}-5}\right)=10$
10. $\lim_{x\to -5}\left(\frac{\sqrt{x+6}-1}{x+5}\right)=\frac{1}{2}$

Exercise 17.2

1. $\lim_{x\to 3^+} \frac{1}{x-3} = \infty$

2. $\lim_{x\to 2^+} \frac{x^2-4}{x+2} = 0$

3. $\lim_{x\to 2^+} \frac{x}{x-2} = \infty$

4. $\lim_{x\to 2^-} \frac{x-2}{x} = 0$

5. $\lim_{x\to 3^+} \sqrt{x-3} = 0$

6. $\lim_{x\to 0}\left[\sin\left(\frac{1}{x}\right)\right]$ does not exist because the function oscillates.

7. $\lim_{x\to -4} \frac{|x+4|}{x+4}$ does not exist because the limit from the left and the limit from the right do not agree.

8. $\lim_{x\to 0} \frac{4}{4-x} = 0$

9. $\lim_{x\to 4} \frac{4}{4-x}$ does not exist because the limit from the left goes to $+\infty$ while the limit from the right goes to $-\infty$.

10. $\lim_{x\to 0}(\sin|x|) = 0$

Exercise 17.3

1. $\lim_{x\to 2}(x^2-2x) = 0$

2. $\lim_{x\to 1}(4x^2-3x+1) = 2$

3. $\lim_{x\to 0}[[(x-3)(2x+5)]] = -15$

4. $\lim_{x\to \frac{\pi}{2}}\left(\frac{\sin x}{x}\right) = \frac{2}{\pi}$

5. $\lim_{x\to 2}\left(\frac{x^2-5}{x+3}\right) = -\frac{1}{5}$

6. $\lim_{x\to 4}\left(\frac{x-3}{5-x}\right)^{10} = 1$

7. $\lim_{x\to 3}\sqrt{x^2-4x+7} = 2$

8. $\lim_{x\to 1}\frac{x^2-1}{x} = 0$

9. $\lim_{x\to 2}(6x^2-18) = 6$

10. $\lim_{x\to 6}\sqrt[5]{x^2-4} = 2$

Exercise 17.4

1. $\lim_{x\to\infty}\frac{5x-2}{x+3}=5$

2. $\lim_{x\to\infty}\frac{x^4-1}{2x^3+1}=\infty$

3. $\lim_{x\to\infty}\frac{4-3x^2}{2x+4x^2}=-\frac{3}{4}$

4. $\lim_{x\to\infty}\frac{5x^2-3x+1}{x^2+x+3}=5$

5. $\lim_{x\to\infty}2^{-x}=0$

6. $\lim_{x\to-\infty}(2+e^x)=2$

7. $\lim_{x\to 3}\frac{1}{(x-3)^2}=\infty$

8. $\lim_{x\to 0}(\ln x)=-\infty$

9. $\lim_{x\to 0}\left(\frac{8}{x^2}\right)=\infty$

10. $\lim_{x\to\frac{\pi}{2}}(\tan x)$ does not exist because $\lim_{x\to\frac{\pi^+}{2}}(\tan x)\neq\lim_{x\to\frac{\pi^-}{2}}(\tan x)$.

Chapter 18 Sequences and Series

Exercise 18.1

1. Neither
2. Arithmetic
3. Geometric
4. $f(10)=\frac{11}{100}$
5. $f(50)=\frac{1}{7}$
6. $f(20)=380$
7. $f(10)=\frac{45}{512}$
8. $3, 3\frac{1}{2}, 3\frac{2}{3}, 3\frac{3}{4}, 3\frac{4}{5}$

9. $1, \frac{3}{2}, \frac{9}{4}, \frac{27}{8}, \frac{81}{16}$

10. $-2, 2, -\frac{4}{3}, \frac{2}{3}, -\frac{4}{15}$

Exercise 18.2

1. Converges to 0
2. Does not converge
3. Converges to 0
4. Does not converge
5. Converges to 0
6. Converges to $\frac{1}{2}$
7. Converges to 0
8. Does not converge
9. Converges to 0
10. Converges to 0

Exercise 18.3

1. $\sum_{i=1}^{10} [8+12(n-1)] = 620$

2. $\sum_{i=1}^{10} \frac{1}{2^n} = 0.999023$

3. $\sum_{i=1}^{20} 3\left(\frac{1}{2}\right)^{n-1} = 5.99999$

4. $\sum_{i=1}^{10} (-2)^n = 682$

5. $\sum_{i=1}^{20} (1.55 - .05n) = 20.5$

6. $\sum_{i-1}^{10} 1024\left(\frac{1}{4}\right)^{n-1} = 1365.33$

7. $\sum_{i=1}^{12} \left(\frac{2}{3}\right)^n = \frac{1{,}054{,}690}{531{,}441} \approx 1.98459$

8. $\sum_{i=1}^{100} n = 5050$

9. $\sum_{i-1}^{15}(100-2n)=1260$

10. $\sum_{i=1}^{20}\left(10-\frac{2}{n}\right)=\frac{1,496,115,265}{7,759,752}\approx 192.805$

Exercise 18.4

1. $\sum_{n=1}^{\infty}\frac{1}{2^{n-1}}=2$

2. $\sum_{n=1}^{\infty}\frac{1}{n}$ diverges

3. $\sum_{n=1}^{\infty}6\left(\frac{1}{2}\right)^{n}=6$

4. $\sum_{n=1}^{\infty}\frac{3^{n-1}}{243}$ diverges

5. $\sum_{n=1}^{\infty}8\left(\frac{3}{4}\right)^{n-1}=32$

6. $\sum_{n=1}^{\infty}n^2$ diverges

7. $\sum_{n=1}^{\infty}3^n$ diverges

8. $\sum_{n=1}^{\infty}\frac{2}{4n^2-1}=1$

9. $\sum_{n=1}^{\infty}\left(\frac{1}{4}\right)^{n}=\frac{1}{3}$

10. $\sum_{n=1}^{\infty}(10-n)$ diverges

Index